Praise for *Fixing Niagara Falls:*

"How do you write an original book about Niagara Falls, when so many excellent books about the Falls have already been written? Macfarlane shows it's possible. In this fascinating and well-crafted study, Macfarlane weaves together energy histories, toxic histories, and cultural readings in his analysis, while foregrounding the waterfall itself. He shows that Niagara Falls today is a mesmerizing mixture of nature and culture, radically re-made in service of industrial capitalism. This is truly a transboundary analysis, paying close attention to evolving ideas about the public good and the role of nature in industrial North America."

– Nancy Langston, professor of environmental history, Department of Social Sciences, Michigan Technological University

"*Fixing Niagara Falls* is unlike any other book that I know of, framing the Niagara landscape as an example of the 'technological sublime' devoted both to beauty and power."

– Kurk Dorsey, professor, Department of History, University of New Hampshire

"Daniel Macfarlane has surely read most – if not all – relevant books on Niagara Falls in his research for this truly cross-border history of one of the most important natural sites in North America."

– James Murton, associate professor, Department of History, Nipissing University

The Nature | History | Society series is devoted to the publication of high-quality scholarship in environmental history and allied fields. Its broad compass is signalled by its title: *nature* because it takes the natural world seriously; *history* because it aims to foster work that has temporal depth; and *society* because its essential concern is with the interface between nature and society, broadly conceived. The series is avowedly interdisciplinary and is open to the work of anthropologists, ecologists, historians, geographers, literary scholars, political scientists, sociologists, and others whose interests resonate with its mandate. It offers a timely outlet for lively, innovative, and well-written work on the interaction of people and nature through time in North America.

General Editor: Graeme Wynn, University of British Columbia

A list of titles in the series appears at the end of the book.

Fixing Niagara Falls

Environment, Energy, and Engineers at the World's Most Famous Waterfall

DANIEL MACFARLANE

FOREWORD BY GRAEME WYNN

UBC Press • Vancouver • Toronto

29 28 27 26 25 24 23 22 21 20   5 4 3 2 1

Printed in Canada on FSC-certified ancient-forest-free paper (100% post-consumer recycled) that is processed chlorine- and acid-free.

Library and Archives Canada Cataloguing in Publication

Title: Fixing Niagara Falls : environment, energy, and engineers at the world's most famous waterfall / Daniel Macfarlane.
Names: Macfarlane, Daniel, author.
Series: Nature, history, society.
Description: Series statement: Nature, history, society | Includes bibliographical references.
Identifiers: Canadiana (print) 20200193244 | Canadiana (ebook) 20200193287 | ISBN 9780774864220 (hardcover) | ISBN 9780774864244 (PDF) | ISBN 9780774864251 (EPUB) | ISBN 9780774864268 (Kindle)
Subjects: LCSH: Niagara Falls (N.Y. and Ont.) – History. | LCSH: Niagara Falls (N.Y. and Ont.) – Environmental conditions. | LCSH: Water-power – Niagara Falls (N.Y. and Ont.)
Classification: LCC FC3095.N5 M25 2020 | DDC 971.3/39—dc23

Canada

UBC Press gratefully acknowledges the financial support for our publishing program of the Government of Canada (through the Canada Book Fund), the Canada Council for the Arts, and the British Columbia Arts Council.

This book has been published with the help of a grant from the Canadian Federation for the Humanities and Social Sciences, through the Awards to Scholarly Publications Program, using funds provided by the Social Sciences and Humanities Research Council of Canada.

Printed and bound in Canada by Friesens
Set in Garamond by Artegraphica Design Co.
Copy editor: Frank Chow
Proofreader: Judith Earnshaw
Cover designer: Will Brown

UBC Press
The University of British Columbia
2029 West Mall
Vancouver, BC V6T 1Z2
www.ubcpress.ca

This book is dedicated to the memory of Bill Macfarlane.

Contents

Illustrations

Figures

Table

FIGURE 0.1 Niagara, 1857 (oil on canvas), by Frederic Edwin Church.
National Gallery of Art/NGA Images, Corcoran Collection (Museum Purchase, Gallery Fund), Washington, DC

FOREWORD

Iconic Falls, Contrived Landscapes, and Tantalizing Opportunities

Graeme Wynn

NIAGARA FALLS IS ONE of North America's iconic landscapes. The first description of this "vast and prodigious Cadence of Water" began to circulate in Europe in the 1690s. According to Father Louis Hennepin, a Recollect priest who visited the Falls in the winter of 1678–79, and whose account of them was translated into English in 1698, the waters that tumbled over "this horrible Precipice, do foam and boyl after the most hideous manner imaginable, making an outrageous Noise, more terrible than that of Thunder."[1] Little wonder that the Falls drew early travellers through the interior. The Jesuit Pierre Francois Xavier de Charlevoix came this way in 1721, the Swedish botanist Pehr Kalm in 1750, and the French American writer J. Hector St. John de Crèvecoeur in 1785, three years after the publication of his *Letters from an American Farmer*.[2] Yet few newcomers settled in the Niagara vicinity before the American War of Independence. One retrospective account, looking back three decades from the 1830s, averred that Niagara – difficult to access and rarely visited – was then "still the Cataract of the wilderness."[3] Indeed, the Falls remained remote from America's major cities until the completion of the Erie Canal in 1825.

Initial accounts – Hennepin had the river plummeting near 600 feet (180 metres) – were tempered with time, but the reputation of the Falls seemed beyond restraint. English author and traveller Isabella Lucy Bird marked her excitement at the prospect of seeing the Falls in 1854, declaring that since her "earliest infancy," she had become so "familiar with the name of Niagara" and seen so many pictures of it that she believed she could sketch "a very accurate likeness."[4] One twenty-first-century art historian figured that Niagara was "the most popular, the most often treated, and the tritest single item of subject matter to appear in eighteenth- and nineteenth-century European and American landscape painting."[5] Charles Dickens wrote about his visit to the Falls in 1842, and his words have been quoted in guidebooks and pamphlets ever since.[6] Interpretations of the cataract varied widely. Many early nineteenth-century accounts likened the Falls to Eden; others discovered chaos there. Some, influenced by prevailing conceptions of the sublime, found rapture in the overwhelming, even terrifying, spectacle of Niagara.[7] The nineteenth-century's romantic fascination with death and the afterlife led some to associate the Falls with the moment of decease; at least one poet found "that mysterious line/That separates eternity from time" in the lip of the Falls.[8]

When Frederic Edwin Church, a central figure in the Hudson River School of American landscape painters known for the grand scale of his works, came to depict the Falls in the middle decades of the nineteenth century, he produced several sketches and three major canvases, each larger than its predecessor.[9] The first, encompassing both the Horseshoe and the American Falls from a vantage point downstream, was completed in 1844 and filled a 111 cm by 119 cm canvas. The second, completed in 1857, offered a panorama of the Horseshoe Falls. The large and unusual dimensions of this depiction of "Niagara" (twice as wide as it was high at 106 cm by 230 cm) emphasized the horizontal grandeur of the scene. A decade later, a view of "Niagara Falls from the American Side" emphasized the verticality of the Falls. Measuring 255 cm by 226 cm, it was Church's largest work. At once both sublime and picturesque, all three of these paintings won acclaim. Creating the sense that the viewer stood at the edge of the precipice, "Niagara" was the focus of a single-painting exhibition held in Manhattan in May 1857. People in the tens of thousands paid 25 cents to view it in a darkened gallery in which it was the only object illuminated. For many, the painting was a fair substitute for a visit to the Falls themselves. Strikingly realistic and "without 'manner,'" Church's images of Niagara seem to encapsulate the mighty scale, pristine grandeur, and unharnessed potential of the continent (see Figure 0.1).

In retrospect, these magnificent paintings caught Niagara on the cusp; they mark a decisive moment in the history of the Falls. For Church, Niagara was a spectacle of nature. A handful of Native Americans perch on a rocky outcrop in his 1844 work, but there are neither human figures nor (save an indistinct cluster of distant dwellings) recognizable human works in the later paintings. But these depictions were made as Niagara –

and much of North America – stood on the brink of a new world.[10] Change had been in the air for several decades on both sides of the Atlantic, as railroads broke the bounds of time and space, and new methods of production allowed ever-larger factories to expand their output. Writing of his own country, the English historian Asa Briggs called the eighty-five years before 1867 the "age of improvement," and recognized how strongly the idea of progress shaped the minds and actions of influential Britons in the middle of the nineteenth century. Industrialization came later to North America, but, as in Britain, the idea of progress was in the air, and coal's capacity to produce thermal and mechanical power was integral to its early advance; industry concentrated on the coalfields or developed where coal and other resources could be brought together. Despite their spellbinding power, the Falls of Niagara were of little energetic value to early nineteenth-century settlers and industrialists. Small enterprises built waterwheels along the shoreline above the Falls to harness the power of the rapids, but the main cataracts were simply too formidable to tame in this way. A mid-century effort to divert water from above the Falls, to produce traditional hydraulic power below, faltered for three decades.

By the time the challenges of a 200-foot declivity were overcome, a new technology was revolutionizing the "energetic landscape." The English scientist Michael Faraday had discovered how to generate electricity in the 1820s and 1830s, but this new form of power was not easily transmitted beyond the generator. Its use was necessarily small in scale. Thomas Edison's incandescent light bulbs (patented in 1880) typically ran on the output of a small dynamo in each house they lit, until he opened the Pearl Street [Generating] Station in Manhattan in 1882. That station housed a cluster of steam-powered dynamos that eventually served about 500 customers and lit 10,000 lamps with direct current. At much the same time, small, hydro-electric power stations, newly built in England and the American Midwest, provided enough energy to run a single-arc lamp or to light a couple of mills and a house. Alternating current provided the key that would unlock electricity's potential. Favoured by inventor Nikola Tesla and industrialist George Westinghouse, it offered enormous advantages for long-distance transmission once transformers were developed to step-up the voltage at source and then reduce it at the point of use. This conferred the benefits of scale, as larger generators were able to serve more extensive areas.

Suddenly "White Coal" – electricity – was poised to light the world and challenge its carboniferous namesake as a driving force. By May 1893, when thousands of lightbulbs illuminated the Chicago World's Fair,

AC had triumphed over DC in "the battle of the currents," and electricity's transformative potential was plain to see. That very year, a small hydroelectric generating plant was built on the Niagara River, above the Horseshoe Falls, to supply an electric railway that ran seventeen kilometres between Queenston and Chippawa. Other technological refinements – from Tesla's invention of the induction motor and the use of polyphase current, to improvements in both the size and efficiency of hydraulic turbines and the capacity of transformers – increased the distribution and utility of electric power.[11] In the same year as it lit the Chicago World's Fair, the Westinghouse Electric Company signed a contract to install AC generators at Niagara Falls. Three years later, an 11,000-volt line carried electricity from the Adams Generating Plant below Niagara Falls to Buffalo, thirty-two kilometres away. Less than a decade later, the expanded Adams station produced 10 percent of all the electricity generated in the United States. The sublime force, "the resistless tide," of Niagara's plunging waters, so magnificently captured in Church's paintings, appeared set to usher in "a new, totally human order" – or so it seemed to the poet Benjamin Copeland, whose 1904 paean to the Falls included the lines: "With power unrivaled thy proud flood shall speed/The New World's progress toward Time's perfect day."[12]

The subsequent invention of electric-powered washing machines (1907), vacuum cleaners (1908), and household refrigerators (1912) might be taken, by some, to confirm the coming of this new day. There can be no doubt that these and other inventions, and new industrial uses for electricity (coupled with shrewd marketing campaigns) massively increased the demand for hydroelectric power (HEP). Through the 1920s engineers, abetted by politicians, wrung more and more power from the Niagara River. In 1961, the new Niagara Falls generating plant (in New York) was the largest hydro-power facility in the Western world. By 1970, nigh on 700 chemical plants, steel and aluminum mills, oil refineries, and other industries in its vicinity placed the area among the leading electrochemical and electrometallurgical centres in the United States. On the Canadian side, a similar mix of industries had turned this part of Ontario into an important manufacturing centre (though global competition and an economic recession drove operations on both sides of the border into difficult times in the 1980s). Today, generating stations on the American and Canadian banks of the river produce over 4 million kilowatts of electricity, a quarter of all the power used in New York State and Ontario.

Yet Niagara Falls remains, as it was in the middle of the nineteenth century, a major tourist attraction. The first hotels had been built near the

Falls in the 1820s, and the number of tourist visits increased as access to the area improved; by the late 1840s, more than 40,000 people visited each year. Most of them were relatively well-to-do. Until the Civil War, wrote Karen Dubinsky in her history of tourism at the Falls, "everyone who could afford it went to Niagara, often as part of the Northern Tour," following an itinerary that also included Boston and Quebec City.[13] Several grand hotels with views of the Falls were built on both sides of the border, and as tourist traffic increased so did their number. In 1892, Niagara Falls, NY, alone counted forty-two hotels. On the Canadian side, the grand and venerable Clifton Hotel was joined in 1894 by the Hotel Lafayette, which included a six-storey circular turret with a magnificent view of the American and Horseshoe Falls.[14] As the tide of tourists rose, however, so "crowds of sharks, hucksters, and pedlars" congregated to prey on them, and threatened – in the minds of many – to spoil the effect of the Falls and reduce the lucrative flood of visitors.[15]

So, the story goes, Lord Dufferin, the governor general of Canada, publicly deplored the desecration of Niagara and called for the establishment of a park, on both sides of the border, to restore the area to its original glory. If the intricate politics of the discussion and the precise details behind the establishment of reserved lands along the river were more complicated than this, the general outlines of an ongoing struggle were laid down by these developments in the last quarter of the nineteenth century. Rhetoric ran loose and sometimes furious. In simple terms there were, on the one hand, those who wished to preserve "the beauties of the frontier" in order that visitors might enjoy this "land of enchantment" and, on the other, those who wished to foster development: "Factories, smoke, and a real estate boom."[16] The problem thus framed deepened as the twentieth century dawned. As the enormous, transformative capacity of hydroelectric power generation at the Falls became evident, the eponymous cities near the Falls drew increasing numbers of visitors and became ever-more-popular honeymoon destinations. And here, in the resolution of this dilemma, in the choice between enchantment and development, Daniel Macfarlane finds the subject of his book.

Placing this work in the context of others in North American water history, I think immediately of Matthew Evenden's study, *Fish vs Power*.[17] Each book deals with a river. Evenden is concerned with British Columbia's Fraser, which runs 1,375 kilometres from the Rocky Mountains to the sea; Macfarlane with the Niagara, which divides Canada and the United States along its 58-kilometre length. Surprisingly, the latter has the greater average discharge – 5,796 m^3/s (204,700 cu ft/s) compared to 3,475 m^3/s

(122,700 cu ft/s) for the Fraser – but the flow of the British Columbia river has much greater seasonal variability. Buffered by the upstream Great Lakes, the highest mean monthly flow on the Niagara River in the twentieth century was 7,740m^3/sec in June, 1986; the lowest 4,070 m^3/sec in January, 1964. On the Fraser River, running free from its source, peak mean monthly flows (between 1913 and 1990), twice reached approximately 11,000 m^3/sec (in 1948 and 1972) and in most years of this span mean monthly low flows were in the 500 to 750 m^3/sec range. In Evenden's telling, the Fraser remained undammed to preserve the annual runs of salmon upstream. Unlike the Columbia River to the south, it would not be turned – in Richard White's oft-repeated phrase – into an "Organic Machine" with dams to generate electricity, and various forms of artifice to sustain fish (organic) life. In the often-heated debate about the future of the Fraser, the fate of the fish trumped the potential of power. At Niagara, both the challenge and its resolution were different. Although I sometimes imagine that Macfarlane might have called his book (in refracted echo of Evenden's) "Spectacle vs Power," the conflict between tourism and electricity generation along the US-Canadian border was not an either or proposition. There it seemed entirely possible both to have your power and look at the view. The question was how?

The answer is prefigured in Macfarlane's title: *Fixing Niagara Falls.* Although we might think of "fixing" first in the sense of repairing something broken, the word carries many meanings. Any reasonable thesaurus gives us the following list of synonyms in addition to "mend": *adapt, adjust, arrange, manipulate, order,* and *prepare*. To these we might add *attach, secure,* and *connect; improve, decide on,* and *correct; remedy, renovate,* and *reshape*. There are many ways "to fix," but all imply action of one sort or another. It should be no surprise, then, that the chapter titles of this book form a hymn to change. Here we find Niagara serially harnessed, saved, negotiated, empowered, disguised, preserved, and fabricated. Through these pages we come to understand the Falls as an extremely complex, ever-changing entity. Far from being a timeless feature, the Falls are an evolving part of a complicated landscape. Left alone, they had changed in shape and configuration through the centuries, as the erosive force of running, tumbling water worked on a complicated geological stratigraphy to undermine the lip of the Falls, move the precipice upstream, and leave a narrow gorge below the cataract. Manipulated, adapted, re-arranged, and reshaped by humans for their own purposes in the twentieth century, the Falls were by no means a natural spectacle by the 1960s. Nature and technology were here so intertwined and fused that they could not be disentangled.

Most commentators missed the implications of these circumstances. For Wilber Brucker, US secretary of the army speaking at the opening of the International Control Structure in 1957, the increasingly contrived river was simply "one more impressive example of the wonders man has wrought through the application of his engineering skill and genius in altering the physical world to fit the pattern of his expanding requirements" (see p. 154) To some, less boosterish, locals it seemed that the Falls were well on their way to becoming a "completely man-made and artificial cataract."[18] But perhaps the Canadian minister of natural resources, Alvin Hamilton, best encapsulated the effects of human interference with the river and Falls when he reflected, in 1957, without the slightest sense of incongruity, that

> the American side has been enhanced with a greater flow of water, the flanks of the Horseshoe Falls have been re-clothed and the ugly scars removed, an unbroken crestline has been achieved, and the mist cloud, which so often hid much of the beauty of the Horseshoe Falls, has been lessened, new vantage points for sightseers have been provided and landscaped and the central concentration of flow which formed the threat to the future of the Falls has been tapped and dispersed. In the face of increased diversions of water for power there has resulted for posterity a scenic spectacle more captivating than that which we have known in the past. Surely this is conservation at its best. (see p. 154)

To explain how this came about – to show how the Falls of today are no more than a simulacrum of those portrayed by Frederic Church – Macfarlane focuses his discussion on what he describes as "the interface of the fluid, terrestrial, and infrastructural elements that together constitute the defined space, or micro geography, which we call 'Niagara Falls'" (see p. 5). In the process, he ranges widely, across land and water, upstream and downstream of the Falls, through the sometimes arcane debates of engineers, and into transborder diplomacy, while keeping a steady eye on the roles played by "the state" and industrial capitalism in turning the twin cascades of the Horseshoe and the American Falls into "a tangled blend of nature and culture." *Fixing Niagara Falls* finds its centre in the physical manipulation of these Falls – in the diminution of the flows over their crests, in the engineering works above the Falls that diverted river water into massive tunnels to drive the turbines of hydroelectric power stations below them. It is also centred in the structural reinforcements and other measures applied to the cataracts to prevent rock slides and maintain the familiar appearance of the Falls. In the end, this book surprises with its argument that the messy, receding Horseshoe Falls have, to all intents and purposes, been stabilized, and to some extent fossilized, in an idealized and artificial form to serve the interests of both tourism and HEP-generation.

By stressing the engineered, and thus the fabricated or artificial nature of Niagara Falls as we now encounter them, Macfarlane exposes an intriguing paradox. Concerted efforts to preserve the Falls during the twentieth century changed them significantly. In the epic contest of Spectacle vs Power neither interest could be allowed to triumph. As the number of people visiting Niagara rose, and the "attractions" built to serve them became more numerous, elaborate, and out of the ordinary, local and corporate commercial interests staunchly resisted any developments likely to slay the tourism goose that produced their golden egg. At the same time, industrial and political (state) interests were unwilling to forego the enormous development potential promised by the flow and fall of the river. In the capitalist, market-orientated, development ethos of the times, compromise was necessary, and geographical (physiographic) realities meant that it had to be worked out along a small section of the Niagara River above and below the cataracts. Deploying both the authority and increasingly sophisticated knowledge that was accruing to their professions, experts (engineers and other technocrats such as planners) embarked on a process that Macfarlane calls *disguised design* to industrialize the river and keep the tourists marvelling at the grandeur of the Falls. This was no simple trick. Certainly some industrial structures could be hidden from the gaze of casual visitors by placing them beyond the usual lines of viewer interest, up- or down-stream. Others could be turned, by cunning description, into spectacles themselves, as examples of the industrial sublime and evidence of humanity's growing capacity to bring nature into useful service.

But HEP generation required the massive diversion of water from above the Falls to run through tunnels and turbines before being returned to the river below them. As the demand for electrical power grew after the Second World War, legislated limits on the reduction of flow over the Falls (established by the Niagara Diversion Treaty of 1950) began to pinch both the capacity of the electricity supply and the quantity of water available for power generation. In broad terms, the amount of water going over the Falls during daylight hours is about half of the regular flow; night time flows are approximately half of those during the day. Yet the Falls could be reduced to a trickle if diversions were maximized. To address these circumstances, and keep the Falls flowing, engineers designed ingenious and extensive means of pumping water into a storage reservoir during the dark hours of low electricity demand, and then running that stored water through the penstocks to meet daytime power needs.

Alongside the development of such grand designs, engineers also devoted themselves (in Macfarlane's fine phrase) to protecting "Niagara Falls from itself." With the infrastructure of tourism (from viewpoints to hotels and promenades to parks) established hard by the cataracts and submerged engineering works above them, the Falls could hardly be allowed to continue, unfettered, their inexorable march upstream. Nor was it wise, in the minds of those whose eyes were fixed on the splendour of the view, to let the reduced flow dissipate in trickles around the edge of the cataract. So the Horseshoe Falls were made to better validate their name: their crestline shortened and their curve exaggerated. As Macfarlane notes, the waterfall was transformed into a version of its imagined self, "a unique form of hybrid envirotechnical infrastructure blending not only steel and concrete but water and rock, weeds and ice" (see p. 133).

This is a striking, perhaps even disconcerting, way of thinking about Niagara Falls. The details of their transformation, presented in the pages that follow, tend to provoke unease, if not outrage. How sadly we are misled; is there not something amiss in the manipulation of these iconic cataracts? But why should we think this way? On one level, the cunningly contrived, composite, contemporary landscape around the Falls bears comparison to some of England's most famous landscape gardens. Think, for example, of Lancelot (Capability) Brown's work at Blenheim Palace in Oxfordshire. There he transformed an earlier Baroque garden, with regimented parterres, mock fortifications, and a canalized river alongside the oldest woodland in Europe, into a picturesque park with a serpentine lake and clusters of trees planted in carefully selected locations. A large dam widened and deepened the river and fed a newly fabricated cascade. Sunken walls and ditches (ha-has) kept livestock from the house and kitchen garden while preserving the view. In the end, most landscapes reflect a mix of human actions and natural processes. Densely occupied central cities dominated by concrete and glass, the gardens of country estates, and remote areas designated as "wilderness" are all contrived to some degree.[19]

In the end, the challenge is neither to mark, nor simply deplore, transformations of the earth but to ask (as Macfarlane does) why and how they occur, and (much more difficult this) to weigh the balance of their implications. Here too, *Fixing Niagara Falls* offers food for thought. Much of what was done to the Falls and their surrounds was justified as action for the public good – preserving the spectacle for tourists; bringing heat, light, and jobs to the people, and so on. But expenditures on engineering and remedial works generally came from the public purse (or indirectly, from taxpayers' pockets) and the benefits of these investments (in the form of cheap power or commercial opportunities in the hospitality industry) accrued disproportionately (in the form of profits) to relatively few industrialists and entrepreneurs.

Arresting as the changes documented in these pages are, Macfarlane's success in teasing unexpected insights from unlikely sources also warrants remark. Here we find engineers forced to grapple with the effects of frazil (or slush) ice on their infrastructure, confronting "a dearth of factual data," realizing the imprecision of their knowledge, bracketing out certain conditions and factors as "Acts of God," and coming to live with uncertainty and approximations. So, seemingly opaque archival records, a black box of "technical discussions and engineering specifications," form the basis of a riff on "the technopolitics of disguised design" – or, more plainly, an examination of the ways in which uncertainty and politics forced technical experts to grapple with qualitative as well as quantitative evidence and led them, despite lingering reservations, to rely on physical scale models in planning engineering works. So too, we are here reminded that although Niagara is but a tiny dot on a map of North America, it is tied through the electricity grid to most of the continent. And that the Falls themselves can only be properly understood as part of a much larger hydrographic system affected by diversions into, and out of, the Great Lakes basin hundreds of kilometres to the west.

To some fair degree, such insights flow from Macfarlane's particular approach to his subject. A prolific student of rivers, boundary waters, diplomacy, and high modernism, he works the fertile ground between environmental history and the history of technology in developing what is now commonly known as an "envirotech" perspective. As Sarah Pritchard, a leading proponent of work of this sort, noted a few years ago, this realm of scholarly inquiry is fast-evolving and multi-faceted. One among its several strands focuses on "the physical (re)production of 'nature' and 'technology'" and argues that "'biological' organisms, 'natural' environments, and 'technological' objects are ... [all] envirotechnical entities ... [or] material hybrids of what is typically called 'nature and technology.'"[20] Hydroelectric power developments – which necessarily combine water, land, machines, and engineering knowledge in particular places, where local environmental conditions test the universal principles of both scientific understanding and engineering solutions, and regional, national and even international cultural, economic, social, and political circumstances inflect possible and actual outcomes – offer almost classic instances of such blending and blurring of the categories of everyday understanding.[21] For all the specific (and hitherto little-pondered) detail – about water diversions, political discussions, and the artifice involved in "maintaining" the twentieth-century falls of Niagara – to be found in the pages that follow, this book makes a concerted effort to drive understandings of hybridity and envirotech in new directions.

Ultimately, and perhaps most importantly, *Fixing Niagara Falls* also pushes readers to think more deeply about their own, and their society's, roles and responsibilities in shaping the world. For all the alterations to the Niagara River, and the changes made to the Falls, it is hard to find real villains in this piece. Electricity sparked enormous enthusiasm in the final decade or so of the nineteenth century. After decades of accelerating industrial production based, since the eighteenth century, in coal-fired, steam-powered, ever-larger, more capital-intensive factories, this new form of power seemed to augur a different future. Small electric motors driving individual machines could be set up almost anywhere, dispersing production across the landscape, allowing workers new freedom to shape the rhythms of their days, and eliminating the crowding, impoverishment, squalor, and vice manifest in coalfield industrial towns.

This vision was an attractive one set against the backdrop of all that had come before, and several turn-of-the-twentieth-century thinkers were intrigued by it. The essential problems of urban-industrial concentration, as contemporaries saw them, were well framed in Ebenezer Howard's *To-morrow! A Peaceful Path to Real Reform*, published in 1898. Contemplating the rapid urbanization of Britain, fuelled by the steady migration of people from the countryside, Howard recognized that there were pros and cons to both city and country life. Those who lived in the countryside enjoyed low rents, fresh air, bright sunshine, abundant water, and the beauties of nature, but they also had to work long hours for small returns in settings devoid of amusement and "society." Town-dwellers, by contrast, could take advantage of social opportunities, numerous places of amusement, well-lit streets, and relatively high wages (so long as they were employed); the downsides ranged from foul air and murky skies to high rents and prices, long hours of work and the "isolation of crowds." The splendour of the "palatial edifices" found in cities had to be balanced against the proliferation of slums and gin palaces there.

Conceiving of these positive and negative factors as the polarities of horseshoe magnets attracting people to, or repelling them from, town and country, Howard defied understandings of magnetism to envisage a third magnet. It was a real crowd-puller, stripped of the negative dimensions, and exhibiting only the positive aspects of town and country life. This town-country magnet showcased all the virtues – low rents, high wages, bright homes and gardens, lack of smoke and slums, natural beauty and social opportunity, freedom and cooperation – that, Howard argued, would accrue from the establishment of small, well-planned towns (later characterized as "Garden Cities") at some remove from the accumulating social and environmental ills of large centres such as London. Howard's proposal was by no means anti-industrial. Indeed, some have found a thoroughgoing "technological and machinist vision of the city" in his writings.[22] Certainly each carefully planned Garden City was to have manufacturing at its core. Local employment was vital as Howard sought to foster a productive, harmonious combination of decentralized industry and civic life in towns of 30,000 to 60,000 people. "The towns will be split up and scattered and will make innumerable small and compact residential and industrial centres, which rapid transport will bring close together," was how one of Howard's later collaborators summarized the idea.[23] Some contemporaries noted, skeptically, that industry had been planted in dozens of villages since the eighteenth century, many of which had grown into major industrial cities. But those industries had been fuelled by coal. Howard had a different future in mind. Unable to see beyond the railway for the conveyance of bulk goods, he nonetheless insisted that "the smoke fiend" would be "kept well within bounds." In the Garden City, all machinery would be "driven by electric energy, with the result that the cost of electricity for lighting and other purposes is greatly reduced."[24]

The decentralizing potential of electricity also found arresting public expression in the work of the Russian anarchist, scientist, and philosopher Peter Kropotkin. The first edition of his *Fields, Factories, and Workshops,* published in Britain in 1899, offered a trenchant critique of the social and economic consequences resulting from the wholehearted embrace, in the name of efficiency, of Adam Smith's doctrines of specialization and the "division of labour" – that lay behind the routinization of work in large factories.[25] The division and subdivision of tasks had proceeded so far that "the modern ideal of a workman" seemed to be someone "only capable of making all day long and for a whole life the same infinitesimal part of something." A different future was possible. By all means, wrote

Kropotkin, build factories and workshops, but disperse smaller establishments – "the countless variety of workshops and factories which are required to satisfy the infinite diversity of tastes among civilised men" – across the countryside. These could be "airy and hygienic, and consequently economical, factories in which human life is of more account than machinery and the making of extra profits," not places "in which children lose all the appearance of children in the atmosphere of an industrial hell." Rather than being driven to industrial toil by hunger, workers in these establishments would be "attracted by the desire of finding an activity suited to their tastes, and ... aided by the motor and the machine, ... [would] choose the branch of activity which best suits their inclinations."

Thirteen years later, a second edition of *Fields, Factories, and Workshops* celebrated the "new impulse" given to "the growth of an infinite variety of small enterprises by ... the distribution of electrical motive power."[26] In Manchester, in northern England, "the municipal supply of electrical power" provided small factory owners with "a cheap supply of motive power, exactly in the proportion required at a given time," and allowed them "to pay only for what is really consumed." Even more remarkably, in Oyonnax, a small French town on the River Ain well known for the production of combs in small workshops, the introduction of "electricity, generated by a waterfall," and distributed through the settlement had brought "into motion small motors of from onequarter [sic] to twelve horsepower," and encouraged 300 workers to leave "at once the small workshops and ... work in their houses." Kropotkin's enthusiasm for electricity extended well beyond its promise of decentralized production: his vision embraced a revolutionary new form of society, one in which the different capacities of human beings could be fully and variously realized, one in which – the subtitle of his book suggested – "Industry [was] Combined with Agriculture and Brain Work with Manual Work" and people found happiness, and riches enough, through "the work of their own hands and intelligence," without exploiting others. These, he concluded, were the bountiful horizons opened "to the unprejudiced mind" by his analysis – and the promise of electricity.

That the future delivered few of the outcomes envisaged by Howard and Kropotkin should not diminish the allure of their ideas at the time they were articulated. Rather, it should prompt us to wonder why these ideas bore so little fruit. Some point to the sheer magnitude of the Niagara River and Falls to explain the Ontario–New York State manifestation of this conundrum. The physical and engineering challenges involved in harnessing this flow for hydroelectricity generation required massive investments of labour, skill, technical know-how, and capital. Large-scale developments – diversion works, tunnels, and powerhouses – offered the only realistic way of utilizing the potential energy of the Falls. This deployment of enormous, capital-intensive, cutting-edge technologies to increase the electricity supply, and the associated costs of building high-voltage transmission systems, have been characterized as "hard-path" development, because the magnitude of the works creates an inertial effect, locking-in dependence on existing infrastructure and reducing flexibility with respect to future energy options.

There are forceful echoes of Amory Lovins's 1976 distinction between hard- and soft-path energy futures here.[27] But they hinder as much as they help when thinking about why the brave new world of small-scale production powered by electricity (envisaged by turn-of-the-twentieth-century futurists) never properly came to pass. Compelling as Lovin's case for soft-path energy systems that minimize fossil fuel consumption continues to be, it skirts some crucial issues. Written in the white heat of concern about the expansion of nuclear (and to lesser extent oil, gas, and coal-fired) electricity generation, his arguments essentially ignore both large-scale HEP production and the potential for dispersed consumption of centrally generated electricity. Certainly conservation (demand reduction), and the use of "soft" (in the sense of "flexible, resilient, sustainable, and benign") technologies such as wind and solar generation were sensible proposals in 1976 (and are even more urgent today).[28] True, the costs of long-distance electricity transmission are significant: Lovins suggested that about half of all domestic electricity bills went to pay system overheads, from "transmission lines [and] transformers ... [to] meters and people to read them." No doubt the smaller scale, local generation of electricity will increase. But massive HEP schemes continue to be built, and there are advantages (as well as drawbacks) inherent in the existence of large scale energy "grids." The question remains: Why were so few of the anticipated demand-side benefits of widely distributed electricity not realized after 1900?[29]

American social philosopher and historian of technology Lewis Mumford offered one possible answer. A humanist romantic who valued many aspects of medieval life and deplored the human, social, and environmental costs of the Industrial Revolution, Mumford embraced electricity's potential. Like Kropotkin, he believed it would usher in a new epoch in the life of humankind (Mumford called it the "neotechnic era") in which production would be freed from the tyranny of coal, and labourers would be able to establish small, viable factories. Early in the 1930s, however, he lamented that the enormous transformative potential of small-scale electrically powered machinery had not been realized: just as a mineral might be leached from a rock and replaced by another that adopts its form to create a pseudomorph, the new technology had been incorporated into pre-existing frameworks. Most of the power generated on the New York side of the Niagara River went to large industrial plants: electrochemical and electrometallurgical establishments, chemical plants, steel and aluminum mills, oil refineries, and so on. "We have merely used our new machines and energies to further processes which were begun under the auspices of capitalist and military enterprises: we have not yet utilized them to conquer these forms of enterprise and subdue them to more vital and human purposes," wrote Mumford in *Technics and Civilization*.[30] Older forms of technology had constrained new developments, and vested political and financial interests had worked to "maintain, renew, and stabilize the structure of the old order." It seemed clear, to Mumford at least, that "the neotechnic refinement of the machine, without a coordinate development of higher social purposes," had neutered the potential for transformation inherent in the new technology. The hope it stimulated, for the development of radically new forms of economy and society, had been stillborn.[31]

These musings seem, to me, to have a great deal of relevance in the third decade of the third millennium, as the world worries about global warming, and the climate crisis deepens. Almost half a century after Amory Lovins warned that then-current commitments to producing electricity depended on "future technologies whose time has passed," human society remains heavily dependent on those self-same technologies.[32] Coal is still mined in enormous quantities and converted, inefficiently, into electricity that is used, in inefficient turn, for such purposes as heating and cooling. Oil and gas wells have proliferated – the US alone has about a million active wells – and world production of sweet, light crude oil continued to increase until 2005. "Unconventional" production has expanded since that date, but returns on energy invested have fallen as inhospitable environments require new extraction techniques.[33] The ecosystemic consequences of coal, oil, and gas combustion are clear in increasing concentrations of CO_2 in the atmosphere. Meanwhile, methane, an even more potent "greenhouse gas" than CO_2, escapes largely unremarked on, and unregulated from, oil and gas production facilities.[34] Scientists' concerns about tipping points, and increasingly forceful public protests about the implications of climate change, place human society at the edge of a precipice.

New technologies, such as wind and solar systems, and dispersed, small-scale, environmentally benign run-of-the-river hydroelectricity generators promise to prevent or mitigate the fall, and energy conservation strategies can help to moderate its threat. Yet, regarded as "safety harnesses" rather than as transformers, as supplements and stand-ins rather than as substitutes and replacements, none of these technologies have been pursued or adopted with the alacrity required. We would do well, at this juncture, to think hard about the unrequited dreams of those early twentieth-century intellectuals who hoped that the distributed, democratic potential of electrification would lead to new and better forms of work and life and to ponder why they went so largely unrealized. It was neither accident nor oversight that turned neotechnic means to paleotechnic ends in the twentieth century. Powerful interests with much invested in older forms and systems of production shaped the political, economic, and ultimately moral choices that led to the deployment of new and potentially progressive technologies in reactionary ways.

In this age of hyper-connectivity, of widely dispersed and powerful computing and communication technologies, and impending environmental crisis we have both the means and the imperative to radically reshape our ways of being in the world. *Fixing Niagara Falls* is not – and was never intended as – a manual for such reconstruction, but writ large the story unfolded in these pages demonstrates that the course of change is neither predestined nor beyond contestation. It reminds us that humans determine the ways in which technologies are used and that it is we who must, in times of crisis, seek to create "fresh integrations of work and art and life."[35] The need is urgent. Let us not waste this chance by presuming, collectively, that we face an insurmountable opportunity.[36]

Acknowledgments

Books such as this one wouldn't be possible without winning the academic lottery a few times – from supportive supervisors, mentors, and colleagues; to the necessary resources, time, and funding; I've been lucky and privileged in many ways. *Fixing Niagara Falls* had its roots in my book on the St. Lawrence River, and thus I owe a great deal to Serge Durflinger, Norman Hillmer, and Nancy Langston. I also need to recognize earlier academic mentors, such as John Courtney and Brett Fairbairn. Many other colleagues have supported my Niagara research through encouragement, advice, or exchanges of ideas. Andrew Watson, Maurits Ertsen, Christine Keiner, Jeremy Mouat, Andrea Gaynor, James Hull, Amy Slaton, Barbara Hahn, Colin Duncan, and D.C. Jackson read and provided valuable feedback on parts of this book at its various stages and iterations. Murray Clamen, Marty Reuss, Lynne Heasley, and Rob Link read the whole manuscript in draft form, and their incisive comments undoubtedly made this a better book. I'm very grateful for the anonymous peer reviewers the Press lined up: they were generous, wise, and the model of what productive peer review can look like. I thank Colin Duncan for his advice and his indexing skills. I had the opportunity to present on Niagara Falls in various forums where I received valuable feedback: ASEH, SHOT, CSTHA, CHA, ACSUS, ICEHO, IWHA, Borders in Globalization, Quelques Arpents de Neige, McMaster University, Niagara University, Brock University, Burning Books in Buffalo, the "Estimated Truths" workshop at the Max Planck Institute, Osher Lifelong Learning Institute in Kalamazoo, and the speaker series in the departments of History and Geography at Western Michigan University.

For about as long as I've been working on Niagara Falls, I've been involved with the Network in Canadian History and Environment (NiCHE), an amazing scholarly community that has offered a supportive environment, many friendships, and an online space in which I could workshop many of my Niagara ideas. I've been lucky enough to serve as an executive member of the International Water History Association (IWHA), and I thank the folks there I've worked with and learned from. Several other organizations and outlets also gave me the opportunity to share my Niagara work in blog or article form: Rachel Carson Center's *Arcadia,* FLOW for Water, *Active History, Washington Post, Slate,* and *Buffalo News.* Some concepts deployed in this book came out of co-authored articles with Andrew Watson and Peter Kitay. I have previously published some of my Niagara findings in academic journals, and I thank them (and the various editors and reviewers involved with these articles) for permission to reprint some parts and ideas in this book: Daniel Macfarlane, "Nature Empowered: Hydraulic Models and the Engineering of Niagara Falls," *Technology and Culture* 61, 1 (January 2020); Daniel Macfarlane, "Negotiating Niagara Falls: US-Canada Environmental and Energy Diplomacy," *Diplomatic History* 43, 5 (November 2019); Daniel Macfarlane, "Saving Niagara From Itself: The Campaign to Preserve and Enhance the American Falls, 1965–1975," *Environment and History* 24, 4 (November 2019); Daniel Macfarlane, "'A Completely Man-Made and Artificial Cataract': The Transnational Manipulation of Niagara Falls," *Environmental History* 18, 4 (October 2013).

I was fortunate to receive financial support for my Niagara research from a number of sources: a SSHRC postdoctoral fellowship at Carleton University; a Banting postdoctoral fellowship at Michigan Tech University; a Fulbright Visiting Research Chair at Michigan State University's Canadian Studies Center; a Visiting Scholar position in Carleton University's School of Canadian Studies; various NiCHE grants; a travel grant from the Eisenhower Archives; and ASPP funding. This book was mostly written at Western Michigan University, my academic home since 2014. I want to express my profound thanks to my department, the Institute of the Environment and Sustainability, and my colleagues and students. WMU provided financial support through various internal awards and grants, particularly FRACAA, CDDA, and SFSA. WMU's Jason Glatz once again provided stellar maps and images. I'm especially appreciative of the continued support of Lynne Heasley, my collaborator and office neighbour.

I thank the archivists and staff at Library and Archives Canada, National Archives and Records Administration (NARA) II, Truman Presidential Archives, Eisenhower Presidential Archives, Buffalo History Museum, Niagara Falls Public Library (NY), Niagara Falls Public Library (Ontario), Queen Victoria Niagara Falls Park Commission, Syracuse University Library, and the Archives of Ontario. Much of my research was carried out in the archives of the International Joint Commission, and I need to thank the staff of the Canadian Section of the IJC, particularly Jeff Laberge. The New York Power Authority (NYPA), formerly the Power Authority of the State of New York (PASNY), provided scanned records of some internal documents regarding the Robert Moses Niagara Power Plant and related infrastructure. Ontario Power Generation (OPG) provided records related to the Hydro-Electric Power Commission of Ontario (HEPCO for short, but often referred to as "Ontario Hydro") and its Niagara undertakings. The views and opinions expressed in this book are the author's own and do not necessarily reflect opinions or positions of OPG or NYPA (having supplied these records). I should note that the Ontario Hydro records from OPG were generally not catalogued or organized by file groups; thus, though they are a treasure trove of information, it was not possible in most cases to provide precise cataloguing or archival information such as box, volume, or file numbers.

The UBC Press staff expertly guided me once again through all the hoops: Randy Schmidt handed me off to James McNevin who, as editor, ushered the manuscript to the finish line, where Production Manager Megan Brand, Copy Editor Frank Chow, Cover Designer Will Brown, and the rest of the Press team took over. My thanks to the editor of the Nature | History | Society series, Graeme Wynn, whose editorial advice again saved me from myself on many occasions.

As much as I may have gotten lucky in the academic game, I've been even luckier outside of it. A number of friends, many going back to my years in Saskatoon and Ottawa, have humoured my droning on about manipulated waterscapes and such. My biggest debts are to my family, primarily for their love and encouragement, but also for all the invisible labour that makes a book like this possible. My father, Bill Macfarlane, passed away in 2016, and his attitude during his declining health was an inspiration; this book is dedicated to him. It was on the return leg of a family visit to my dad's hometown in Ontario that I first visited Niagara

Falls. He always supported my academic pursuits, as has my mom, Becky Macfarlane. She has watched our kids on numerous occasions over the years and continues to be a go-to proofreader. My in-laws Bob and Vivian Thomson (and their extended family) spent considerable time with our kids, which allowed me to work on this book, and they have provided a home-away-from-home in Ottawa. I appreciate the support of many other relatives, including my brothers Tim and Eric and their families; as well as my Grandma, Erna Reimer, who passed away in 2019; and my Grandpa, Art Reimer, who continues to take an interest in the history of controlling water.

Most importantly, I am humbled and blessed by the love and support of my caring and patient wife, Jen, and our amazing kids, Elizabeth and Lucas. Because of this book, we went to Niagara Falls so frequently that the railings along the Niagara Gorge served as our equivalent of those growth charts parents mark up on a kitchen door jamb: in each successive set of photos at the waterfall, the kids were a little bigger. There are surely worse places to drag one's family on research trips, and I hope they cherish the memories of our times together at the Falls as much as I do.

Abbreviations

AC	alternating current
AFIB	American Falls International Board
AFL	American Federation of Labor
ALCOA	Aluminum Company of America
AOC	Area of Concern
BWT	Boundary Waters Treaty
CERCLA	Comprehensive Environmental Response, Compensation, and Liability Act
DC	direct current
FPC	Federal Power Commission
HEPCO	Hydro-Electric Power Commission of Ontario (Ontario Hydro)
Hz	Hertz
IJC	International Joint Commission
INFEB	International Niagara Falls Engineering Board
IWC	International Waterways Commission
kW	Kilowatt
kWh	Kilowatt Hour
NYPA	New York Power Authority
OPG	Ontario Power Generation
PASNY	Power Authority of the State of New York

POPs	persistent organic pollutants
RAP	Remedial Action Plan
SINB	Special International Niagara Board
USACE	United States Army Corps of Engineers
WES	Waterways Experiment Station

Fixing Niagara Falls

INTRODUCTION

Characterizing Niagara

> *If you wish to see this place in its grandeur, hasten. If you delay, your Niagara will have been spoiled for you. Already the forest round about is being cleared. The Romans are putting steeples on the Pantheon. I don't give the Americans ten years to establish a saw or flour mill at the base of the Cataract.*
>
> *Alexis de Tocqueville, 1831*

MY FIRST EXPERIENCE at Niagara Falls was on the return leg of a long family vacation. It was the summer before my last year of high school. I don't remember most of that trip to Niagara, aside from a few foggy snippets. I vaguely recall going on the *Maid of the Mist,* though perhaps I am confusing photographs with memories. But I definitely remember experiencing a general sense of awe. Niagara Falls obviously made enough of an impression on me that I was like fertile soil when, in the course of doing research on other aspects of the Great Lakes–St. Lawrence system more than a decade after that family excursion, I discovered that Niagara Falls had been heavily manipulated.

In fact, one could almost say that Niagara Falls is *fake.*

It might be jarring to hear such a statement. After all, Niagara Falls is the world's most famous waterfall. Niagara was *the* epitome of the natural sublime. Though the meaning of "the sublime" has changed over time, it was classically defined by Edmund Burke in a 1757 treatise as natural features that combined beauty, awe, and terror. "Sublime" was a favourite word in eighteenth- and nineteenth-century accounts of visiting the Falls. Though Niagara is no longer the quintessential example of the sublime, it still has ineffable qualities and persists as an icon – or cliché – that has attained the status of a common reference point. We compare things to Niagara Falls to convey a sense of size, magnitude, and grandeur. Though it is neither the tallest, widest, nor largest waterfall by volume, it is the only cataract that combines all three elements in such impressive proportions.[1]

Furthermore, until the middle of the nineteenth century, Niagara was *thought* to be the largest waterfall in the world. Its location, along with the fact that Niagara Falls has been so heavily marketed over the years, further helps explain its hold on the world's imagination. Indeed, other notable waterfalls – Victoria, Iguazu, Angel, and so on – are not nearly as close to large population centres or so easily accessible (tourists don't even need to get out of their cars to see Niagara Falls). Most of the world's other large waterfalls aren't in the northern hemisphere and, unlike Niagara, don't turn into an icy wonderland in the winter.[2] Reading between the lines, what further distinguished Niagara Falls for many was racial and cultural chauvinism: the waterfall came to be controlled by cultures that believed they knew how best to appreciate and appropriate its liquid wealth so that it wasn't squandered by simply running to the sea.

The *genius loci* of Niagara Falls is widely recognized. It became a symbol of an entire continent and an entire cultural inheritance, in a way that isn't true of any other waterfall.[3] Some have speculated that, among the various artistic media, including prints, paintings, lithographs, maps, aquatints, engravings, and photographs, Niagara was the most commonly represented image from the American continent during the nineteenth century.[4] Though the waterfall is split between two nations, and the more spectacular Horseshoe Falls is predominantly in Canada, until the twentieth century Niagara was more prominently associated with the United States. The waterfall fed the fledgling republic's conception of its abundance, strength, and limitless possibilities. The perceived taming of Niagara's might gave the nation an exuberant confidence in its energy supplies, power over nature, and manifest destiny. Consequently, Niagara Falls is often cited as the birthplace of hydro power. While this role has been exaggerated, Niagara Falls is inextricably linked to the birth of hydroelectric generation and distribution on a large scale. The availability of energy made the Niagara frontier the ideal host for the Aluminum Company of America (ALCOA) and many important chemical industries, as well as the victim of the environmental impacts of all these industries, such as the toxic pollution of Love Canal (Niagara Falls is of course linked with the word "love" in other ways: it was one of the most well-known honeymoon destinations in North America). Despite its association with untamed nature, Niagara is the most industrialized and commercialized of the globe's tourist waterfalls.

Arguments and Approaches

Many excellent books have been written about Niagara Falls.[5] A number of studies consider its status as the repository of the sublime, and the fading of this sublimity, from cultural, social, and artistic perspectives.[6] Then there is the long legacy of preservation, park developments, and landscape design: New York's State Reservation at Niagara is the oldest state park in the United States, while the Queen Victoria Park in Ontario is one of the most famous public parks in Canada.[7] Several studies consider imagined utopias at Niagara Falls.[8] Many look at Niagara as a cultural touchstone, as a site of the carnivalesque and kitsch – from barrels to tightropes to wax museums – including tourism and honeymooners.[9] Given its geographic and spatial location, Niagara has attracted its share of scholars interested in borderlands.[10] Others examine it as a site of technological prowess, from its bridges to electricity generation to transmission networks.[11] Still others delve into the industrial and chemical factories that took advantage of the cheap power – the flip side of which is the Niagara frontier as a space of deindustrialization, rust belt, and toxins.[12]

Why the need for this book then? My answer is that little attention has been paid to the waterfall itself, especially in the post–First World War era. Scholars have addressed in some detail what transpired between the 1870s and the 1910s but have largely ignored the rest of the twentieth century, which is precisely the period on which this book concentrates. It was a time when massive public and state-sponsored hydro-power plants were built at Niagara Falls, and when both Canada and the United States sought to remake the waterfall in order to preserve tourism while accommodating power developments. Most existing studies treat the Niagara torrents either as a backdrop, in front of which impressive things were built, or a blank screen on which social and cultural mores are projected, much like the coloured lights that shine on the waterfall at night. This book seeks to foreground the waterfall while showing that it is a tangled blend of nature and culture. In the process, the terms "waterscape" and "fallscape" are used to refer to the interface of the fluid, terrestrial, and infrastructural elements that together constitute the defined space, or microgeography, which we call "Niagara Falls" (see Figure 0.2).

Above all else, most previous works about the history of Niagara Falls seem unaware of – or, if they are aware, then unconcerned with – the radical reshaping of the physical contours of the waterfall that occurred in the twentieth century.[13] The Niagara Falls of today bears only a partial resemblance to its former self: it is smaller, it has a different shape and location, and much less water plunges over its lip. There is a good chance that the Falls used to feel and sound, maybe even smell, different from how they do today. The physical manipulation of the waterfall, including the politics and diplomacy that enabled engineering alterations, is the chief concern of this book. Niagara's modern history is defined by the tension, or contradiction, between power and beauty.[14] The 1950 Niagara Diversion Treaty between the United States and Canada is the hinge on which this study pivots. As a result of this treaty, between half and three-quarters of the water that would otherwise plunge over Niagara Falls is instead sent through huge tunnels that feed enormous hydroelectric stations some five miles downstream. To visually mask the impact of diverting all that water, the United States and Canada cooperatively reengineered the cataracts, particularly the Horseshoe Falls. The two governments sought to improve, rationalize, preserve, and enhance the waterfall – that is, they sought to *fix* Niagara Falls, to rectify those aspects of this border waterway that, from an anthropocentric perspective, were not efficient.

Since various governments were responsible for the remaking of Niagara Falls – two federal, one state, one provincial, and multiple municipal and binational institutions – I am primarily interested in the public hydroelectric generating stations and the remedial works constructed by governmental bodies or agencies. Therefore, relatively little time is spent discussing privately developed hydro-power works, all but one of which were out of operation or publicly controlled by the later 1950s anyway. The role of the state is stressed because the combination of governments and industrial capitalism was the prime historical manipulator of Niagara's landscape and non-human nature; these were the agents that determined that the highest uses of Niagara were energy and tourism.[15] That said, it is important to recognize that there were differences of opinion and conflicts not only between, but within, the different levels of government in both nations.[16]

This book blends environmental and technological history, an approach commonly called *envirotech*, with an emphasis on water and energy history. Since the deepest channel of the Niagara River demarcates the international boundary between the United States and Canada, this is an inherently transborder book, as well as a contribution to international, political, and borderlands history. I pay equal attention to both sides of Niagara Falls: the United States and Canada, the state of New York and the province of Ontario. In order to do so, I examined archival files from many governmental bodies and institutions in both Canada and the United States:

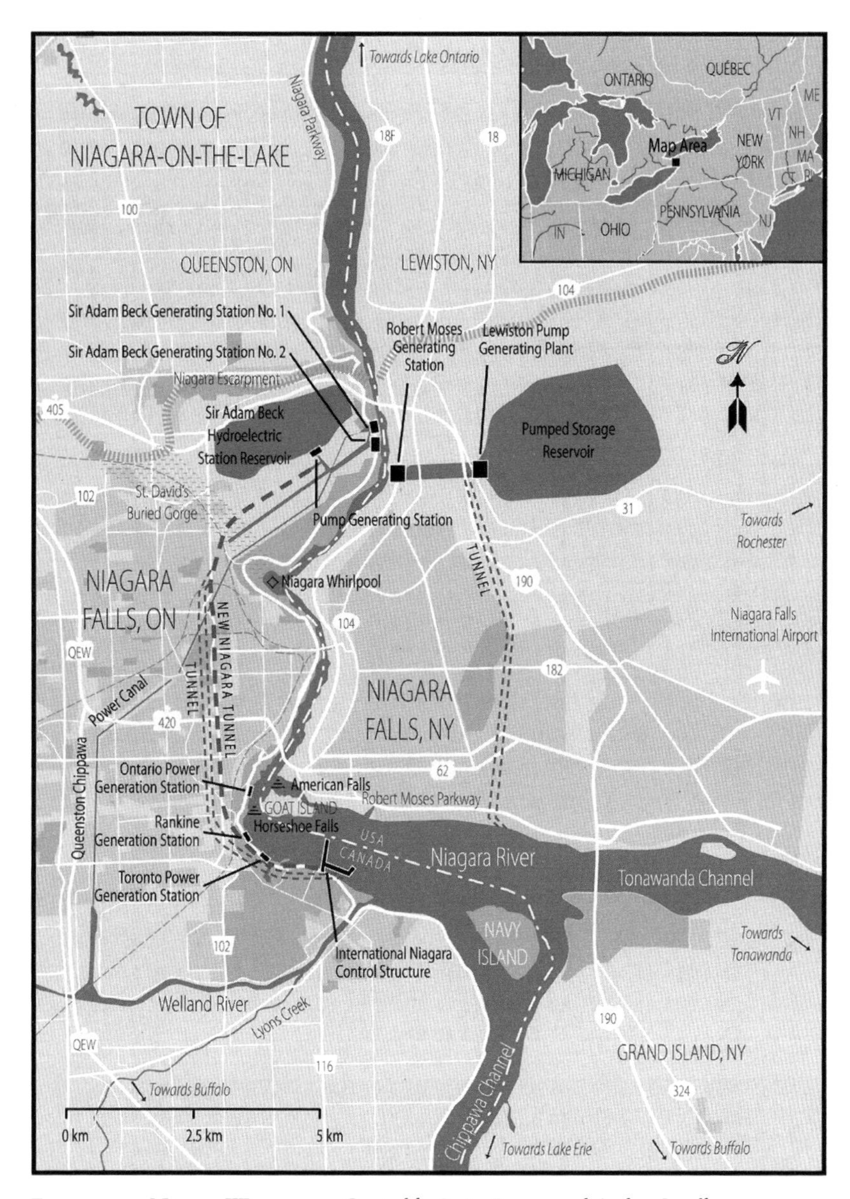

FIGURE 0.2 Niagara Waterscape. *Created by Rajiv Rawat and Anders Sandberg, based on a map by Daniel Macfarlane*

various branches of the two federal governments and diplomatic services, presidential libraries and archives, local Niagara libraries and collections, and the relevant Province of Ontario and State of New York holdings. I was able to access the records of the bilateral International Joint Commission, which may have been the most revealing of the archives I consulted. I also received access to some archival sources from the two public power entities that develop electricity at Niagara Falls: the Power Authority of the State of New York (PASNY), which is now known as the New York Power Authority (NYPA), and the Hydro-Electric Power Commission of Ontario (HEPCO). In this book I will refer to HEPCO as "Ontario Hydro," the long-standing nickname that was adopted as the commission's official title in the 1970s.[17]

I was fortunate to be able to draw on an extremely deep global literature on water, rivers, and hydroelectricity. Underpinning this study is the notion that nature and the infrastructures we create by blending nature and technologies exhibit types of agency and historical causation.[18] Rivers and waterfalls are historical actors. Rivers are shaped by humans, but they also shape human history. Water provides both opportunities and constraints; it opens up many possibilities while simultaneously limiting many others; it inspires dreams and frustrates ambitions; it provides life and takes life. A river can serve as a major power source, transportation corridor, nurturing source for agriculture, quenching font of drinking water, sustainer of fish and fowl, artistic inspiration, and nationalist or regional repository of identity. But it is also a receptacle of waste and pollution, wrecker of ships, conduit of disease, and flood hazard. The embedded energy in water, which humans try to capture in various forms, thwarts as many plans as it enables.

Various debates about what are deemed to be the greatest public goods run like threads through the history of Niagara Falls. Though statist developments and regulations were often undertaken in the name of the collective good, they also raised concerns among the general populace that taxpayers were subsidizing cheap electricity for industrialists and the infrastructure necessary for private profit. Such concerns were well-founded. Although the availability of cheap electricity helped revolutionize modern living standards in North America, it did not usher in the democratic utopia predicted by many, and only some of the savings trickled down to the average person. Most of the electricity produced from Niagara water on the New York side was sold in bulk to industry rather than to local domestic consumers (granted, Ontario's Niagara power was added to the province-wide electricity grid).

As this book will show, the "preservation" of Niagara had multivalent and sometimes contradictory meanings. Ironically, the act of preserving Niagara Falls usually seemed to be synonymous with changing it. This book will show that, though widely considered to be one of the continent's natural icons, Niagara Falls is in fact quite *un*natural.[19] The Falls are as artificial as they are natural – a type of "organic machine" in the now-classic formulation.[20] The Niagara waterscape has become a part of the built environment, an exemplar of the technological sublime.[21] Technocrats concealed the industrialization of Niagara's waterscape by helping the Falls continue to resemble their past appearance – a process labelled here as *disguised design* – so as to maintain and boost tourism levels. In the minds of those diverting water and designing remedial works, they were merely protecting Niagara Falls from itself. These "remedial works" – a term that encompassed the suite of various engineering interventions and control structures that included excavations, fills, reclamations, weirs, and dams – were labelled as such precisely because they were meant to correct something considered faulty. Instead of a messy, receding waterfall, an idealized and synthetic version was sculpted and frozen in place. The copious technical discussions and engineering specifications in the archival records serve as a sort of black box from which I can tease out the technopolitics of disguised design – for example, how Niagara experts dealt with uncertainty and politics, came to rely on hydraulic scale models, and sought to reconcile qualitative and quantitative factors.[22]

The growth and reach of infrastructure is a defining feature of modern life, and Niagara Falls is no exception. After 1945, the North American electrical grid tied Niagara to most of the continent energy-wise. Given that the utility and actual shape of Niagara Falls have been predicated on energy considerations, I suggest that the Falls are profitably understood as an energy landscape or sacrifice zone (even if it seems anathema to lump such a picturesque landmark in with the likes of denuded tar sands or strip-mined coal mountains). Niagara Falls was also linked to another wide-ranging and geographically diffuse envirotechnical system: diversions in and out of the Great Lakes basin hundreds of miles to the west affected water levels in the Niagara River and thus production at Niagara generating stations. In a way, this meant that the entire Great Lakes–St. Lawrence basin – as well as other major watersheds, such as the Mississippi and Hudson Bay, whose levels were affected by these same diversions – became a scaled-up and interconnected infrastructural system to develop Niagara hydroelectricity.

But Niagara Falls itself also became infrastructure. Thomas Zeller helpfully defines infrastructure as "large, state-sponsored, transformationist projects which mobilise environmental and technological resources for the attainment of specific goals."[23] Indeed, one of the central claims advanced in this book is that the Niagara waterfall and river were purposefully transformed into principal parts of a larger terraqueous infrastructure, a hybrid envirotechnical system that was submerged and concealed by a flowing facade.[24] Even the seemingly natural features that constitute Niagara Falls (water, ice, rock, and weeds) were intentionally enrolled as working parts of the infrastructure.[25]

While the American and Canadian nation-states were busy shaping Niagara Falls, they were themselves being shaped by the Falls. Numerous scholars have addressed how states and societies have been structured and changed by the ways they relate to water resources, such as Donald Worster's "hydraulic society" in the western United States.[26] At Niagara, government control of water as well as hydro power profoundly influenced North American state-building and socio-democratic politics. The *idea* of Niagara Falls was politically potent, especially for the ways in which Niagara fostered socio-technical imaginaries of perpetual economic growth and abundance. In more tangible ways, Niagara Falls was central to the evolution of federal and state/provincial policies in a number of areas, such as conservation and parkland, federalism, water rights, and electricity regulation. And its policy influence stretched beyond domestic issues: Niagara Falls was one of the foremost issues in the environmental and energy diplomacy that underpins so much of the modern Canada-US relationship.[27] For example, it was a key factor in the creation of the 1909 Boundary Waters Treaty and the International Joint Commission, as well as in the evolution of continental electricity exports and imports, while Niagara negotiations in the mid-twentieth century involved some of the earliest examples of North American subnational diplomacy.[28]

Though Niagara Falls is an important shared cultural landscape, it has different cultural meanings and hydrosocial relations on each side of the border.[29] Both countries evinced distinct forms of hydraulic nationalism, which springs from the juxtaposition of national identity, technology, and major water basins.[30] In Ontario, and in Canada as a whole, hydroelectricity has been intimately intertwined with political identity to a much greater extent than in the United States, in large part due to Niagara.[31] As one engineering journal put it in 1953, Canada is "hydro-conscious."[32] Hydroelectric development was so attractive in Canada not only because the country was amply endowed with viable sites but because "white coal" reduced Canadian reliance on American sources of energy.[33] Residents of

Figure 0.3 Aerial view of Niagara Falls in 2018

Ontario and several other Canadian provinces today still refer to their domestic electricity invoice as a "hydro bill."

Setting the Stage

To help situate the reader, I would like to provide an outline of both the rest of this book and Niagara's geophysical properties. First, let us turn to the Niagara River, whose green waters are the by-product of dissolved salts and "rock flour," primarily limestone (an estimated sixty tons of dissolved minerals are swept over Niagara Falls every minute).[34] Carrying the flow and energy of the upper Great Lakes basin, these waters hydrologically connect Lake Erie to Lake Ontario. The Great Lakes–St. Lawrence basin holds about 84 percent of North American – and over 90 percent of American – surface freshwater, and the Niagara is one of the ten largest rivers by volume on the continent. Measured by discharge, the Niagara River is the second-biggest river in both New York State and Ontario. In fact, the volume of the Niagara River is roughly ten times greater than that of New York State's most prominent river, the Hudson. But the Niagara is only thirty-six miles long, which is not very lengthy compared with other large continental waterways. It has a watershed of approximately 264,000 square miles but few tributaries. As a result, the Niagara River can be considered a "strait."

The river drops 326 feet in total, most of that in an eight-mile span between Chippawa and Queenston, Ontario. About half of that drop is at Niagara Falls proper, with a fall of another 140 feet in the rapids above and below the Falls. The mean flow of the Niagara River at the waterfalls, without any diversions, is officially 202,000 cubic feet per second (cfs), the equivalent of roughly 13,000 bathtubs of water. It appears that this volume was much higher prior to the twentieth century, though without question some of the difference was the result of varying natural causes and rudimentary measurement techniques and technologies. To illustrate, an 1841 estimate pegged the flow of the river at 374,000 cfs, while New York State engineer and surveyor John Bogart estimated in 1890 that its flow was 275,000 cfs.[35] Nonetheless, compared with mountain- or precipitation-fed rivers, the flow is very uniform and steady within a given year. The Niagara River's volume does fluctuate from year to year, tied to oscillations in Great Lakes water levels driven by natural precipitation, ice cover, and evaporation, as well as anthropogenic interventions such as diversions and engineering works. The Great Lake that most directly determines the flow in the Niagara River is of course Erie, with an underwater rock ledge at the head of the Niagara River influencing the rate of water inflow.

After beginning at Buffalo, the Niagara River widens out, moving relatively slowly between low banks for eighteen miles to the head of the rapids (or cascades) above the Falls and splitting into two channels to go around Grand Island and Navy Island. The four miles of river from the lower end of Grand Island to the head of the cascades, opposite the southern end of Goat Island, is known as the Chippawa–Grass Island Pool. This

Figure 0.4 Historical erosion at the Horseshoe Falls (lines indicate crestline at a given year; aerial image is from 2017). *Map by Jason Glatz (Western Michigan University Libraries) and Daniel Macfarlane*

pool transitions into the upper rapids – a drop of about fifty feet over a distance of one mile.

Niagara Falls consists of three cataracts (see Figure 0.3). The American Falls carries about 10 percent of the river's flow. The minuscule Bridal Veil Falls – which could be considered part of the American Falls – drops between Luna Island and Goat Island. Roughly 90 percent of the Niagara River's water goes over the magnificent Horseshoe Falls (also known as the Canadian Falls), which straddles the international boundary. The *horseshoe* moniker is apt not only because of this waterfall's shape but since, like the hydro power from the flowing water, it evokes notions of brute force – horsepower – that just need to be harnessed. The American Falls, entirely in US territory along with the Bridal Veil Falls and Goat Island, has a much smaller crestline but is taller (by approximately ten feet) than the Horseshoe Falls; the fallen boulders (talus) at the base of the American Falls, however, reduce the sheer drop by roughly half. The cross-section of the bed of the river here is not horizontal: the Niagara River slopes gradually toward the western (Canadian) side, which results in gravity sending water toward the Horseshoe Falls.

The famous drop of Niagara Falls comes from the level change of the Niagara Escarpment, or "the mountain," as it is colloquially called by locals. The escarpment is a distinct topographical feature running through the Great Lakes basin. The location where the Niagara River originally plunged over the escarpment was the birthplace of Niagara Falls. The waterway has relentlessly eroded its way more than seven miles south since glacial ice receded about 12,500 years ago (note that this book will generally employ the imperial system of measurements since that is what was used for the majority of the time period covered). That is, the feature that we call "Niagara Falls" – the place where the water drops dramatically from the upper river to the lower river – is constantly migrating (see Figure 0.4). As the waterfall slowly worked its way back, it cut down through much older sedimentary rock layers – the product of clay, mud, sand, and shell sediments left behind by saltwater seas – compressed by earth's geomorphic processes. Looking at the sides of the Niagara Gorge downstream from the Falls, these different types of sedimentary rock form visually apparent bands, a sort of geological layer cake.[36] These various types of rock and shale react differently to the erosive force of water, depending on variables such as composition, thickness, fracture lines, and hydrostatic pressure.

The American, Bridal Veil, and Horseshoe Falls plunge dramatically into the narrow but elongated Maid of the Mist Pool. The pool leads to the first stretch of the perilous lower river rapids. In the Whirlpool Rapids, the water reaches speeds of over thirty miles per hour, making it one of the fastest rivers in North America. The rocky bottom produces phalanxes of standing waves. The river trends northwest until it arrives at the Whirlpool, where it performs a counter-clockwise rotation then exits, making a hard northeast turn. The channel runs through another chasm and some smaller rapids, past the enormous hydro-power stations and a gap in the imposing cliff of the Niagara Escarpment. The Niagara River then widens, settles, and slows, almost as if it is catching its breath, spent and exhausted from its journey up to this point. For the rest of its course, there are no rapids, gorges, or islands. At its mouth, the river deposits sand and sediment as it debouches into Lake Ontario.

The escarpment is responsible for the region's different microclimates: the fruit belt running to Lake Ontario and a snow belt to the south. The Niagara Peninsula is part of the Carolinian life zone and its sylvan biodiversity, particularly at that latitude, was renowned: the father of North American botany, Asa Gray, declared that Goat Island contained the greatest variety of plant species on the continent.[37] It should be noted, however, that nineteenth-century pronouncements about this rich biodiversity may have been exaggerated.[38] Contemporary ecologists are therefore unsure about the proper baselines with which to evaluate long-term changes to the biodiversity of the area, though clearly that diversity has been adversely affected by human activities.

Chapter 1 of this book begins with a selective history of the Niagara region, running from the deep past up to the early twentieth century. Readers well versed in Niagara lore will find little on tightrope walkers, barrels, and bazaars, but when it comes to those events, processes, and innovations connected to this book's central themes, my discussion becomes more detailed and nuanced. It includes the growth of industry directly at and along the margins of the waterfall, as well as the ensuing public outcry about the visual impact on the Niagara landscape that eventually led to the creation of Niagara parks in the late nineteenth century. We then turn to the fin-de-siècle diversions and hydroelectric complexes that followed hot on the heels of the preservation movement. Chapters 2 and 3 cover the first half of the twentieth century. By the early twentieth century, hydroelectric plants dotted both the New York and Ontario shorelines above and below the Falls, and alterations to the cataracts had already been made to facilitate diversions. The proliferating and sometimes contradictory calls for further preservation of Niagara Falls found expression in various types of legislation and international agreements, which are detailed in Chapter 2. During the interwar years, covered in Chapter 3, Canada and the United States undertook binational studies on Niagara remedial works, as well as some failed international agreements, in attempts to mitigate the scenic impact of the continually increasing water diversions. Seeking to legally enshrine higher diversion levels, the United States and Canada signed the Niagara Diversion Treaty in 1950.

Chapter 4 focuses on the construction of the massive public hydroelectric stations built at Niagara Falls after the Second World War. The first of these was the Sir Adam Beck Generating Station No. 2, completed by Ontario Hydro in the early 1950s not far from the Niagara Escarpment. After much delay, in 1961 the Power Authority of the State of New York opened its own hydro megaproject across the gorge, the Robert Moses Niagara Power Plant. In addition to detailing the creation of these generating plants, Chapter 4 addresses the larger infrastructures they required around the region, including diversion works and reservoirs. Chapter 5 covers the international implementation of the 1950 treaty. Remedial works, built by Ontario Hydro and the US Army Corps of Engineers, involved the installation of structures and a range of physical reconfigurations to the river and waterfall. Engineers sought to produce a pleasing "curtain of water" over an unbroken crestline, with the appropriate colour and, in response to tourist complaints about getting wet, not too much spray or mist. The overarching goal was to achieve a sufficient "impression of volume" to captivate tourists and obscure the fact that at least half the water, which would otherwise have gone over the waterfall, was diverted for power production. The International Control Structure, a gated dam that is part of the International Niagara Control Works, was built above the Falls to apportion the flow of water, and it was extended in the early 1960s. Chapter 6 explains the 1965–75 campaign to preserve and enhance the American Falls. Concerned about rockfalls and the resulting talus at the base of the smaller of the two main Niagara waterfalls, local interests pushed higher-level governments and the International Joint Commission to examine whether this talus could be removed and the American Falls improved. However, the experts eventually decided, with public support, to let nature take its course and not significantly re-engineer the American Falls.

The Conclusion offers some final thoughts about Niagara's modern history. Other major hydroelectric projects of the twentieth century obliterated and dominated the rivers they remade, but Niagara was an exception: as Niagara Falls was turned into a tap, it was being changed to have it appear more like its past self. Measuring by water volume, since the 1950s the real waterfall has been downstream in the penstocks of the enormous hydro-power stations. Since Niagara Falls has been so extensively manipulated, it can be thought of as a simulacrum: an imitation of something that existed in the past.

1

Harnessing Niagara: Developments up to the Twentieth Century

Among the many natural curiosities which this country affords, the cataract of Niagara is infinitely the greatest.

Andrew Elicott, 1789

The Edward Dean Adams Power Plant, opened in 1895 on the New York side of Niagara Falls, was an electricity pioneer. Despite its heritage significance, when the Robert Moses Niagara Power Plant came online in the early 1960s, the buildings that made up the Adams station, which by that time had long been redundant and outdated, were mostly torn down. Today, all that is left standing of this historic plant, the symbol of fin-de-siècle American progress, is one building: the transformer station. The impressive archway from the entrance to generating station no. 1 of the Adams plant still exists, albeit in a disembodied form: it was taken apart and reassembled on Goat Island, towering over a walkway that leads from the main parking lot to the Cave of the Winds tour, framing a statue of Nikola Tesla. Built in the Beaux Arts fashion with locally quarried and rough-finished limestone, like the two generating stations that flanked it, the Adams Power Plant Transformer House is registered as a US National Historic Landmark (see Figure 1.1). According to a 1978 landmark designation document, "this 1½ story building is presently utilized by a local chemical firm to house its frequency converters. It is well maintained."[1]

But when I went to photograph this last vestige of the Adams plant, "well maintained" was not the first adjective that came to mind. The transformer station is in a nondescript location just off Buffalo Avenue, a brisk walk from the waterfall. Surrounded by other industrial buildings and a municipal wastewater treatment plant, it is now privately owned

FIGURE 1.1 Adams Power Plant Transformer House today

and, as of early 2020, was for sale. There appears to be a dump or junkyard out back, and the last time I visited, an RV had been added to the menagerie of abandoned vehicles. The roof has seen better days. The inlet canal that brought water from the Niagara River to the generating stations, and then into the discharge tunnel, was filled in – and in its place seems to be some sort of water treatment lagoon. It is a poor monument to a complex that was hailed as enabling one of the greatest advances in electricity. In a profound way, however, it symbolizes how much of the history of Niagara Falls is hidden beneath the surface.

Many Niagara scholars have analyzed the late nineteenth-century period, and more has been written about the history of Niagara Falls during the Progressive Era than any other epoch. In this chapter, I first provide some backstory on the pre-twentieth-century human history of the Niagara frontier, highlighting aspects that will help readers better understand subsequent chapters. We then turn to the earliest hydroelectric stations built at Niagara Falls. I survey the rapid pace of technological change, political and diplomatic developments, and societal ferment in these decades, delving into more detail when necessary for establishing the context in which public hydro-power projects developed. Weeding through the various power stations and their often misleading names can be tricky, and though entire books have been written about these individual stations, this chapter provides a selective history that foregrounds those aspects that are most relevant to the themes of this book.

Onguiaahra

First Peoples began occupying the Niagara region around 9,000 BCE. The Niagara River was a trading crossroads and the Falls were an important spiritual and physical resource that supported Indigenous lifeways. In addition to fish, meat was gathered from animals that went over the Falls. Oral and archaeological evidence indicates that Haudenosaunee (Iroquoian) groups developed agricultural villages nearby that featured palisaded longhouses.[2] By the early seventeenth century, the Neutrals were the predominant Indigenous group in the area, though there is evidence that their habitation of the region went back further.[3]

Early European explorers, such as Jacques Cartier and Samuel de Champlain, were told about the Falls. The latter relayed that it was called *Onguiaahra* or some variation thereof, which may have referred to the Neutrals, or may have meant "thunder of the waters" or "neck" in reference to the river as a strait connecting two lakes. At any rate, various understandings of this word came to be interpreted as *Niagara.* By the middle of the seventeenth century, the Neutrals had been almost completely wiped out by disease and conflict; the Seneca and Mississauga nations, and then the Tuscarora, eventually occupied the Niagara frontier. Though several other Europeans had made it close to Niagara Falls, the first to produce a record of seeing the waterfall was Father Louis Hennepin, who witnessed the Falls in 1678 as part of an exploratory party. Hennepin distorted and exaggerated the scale of the waterfall, and in the following decades his depictions spread around Europe and the world. Niagara Falls became a symbol of – a sort of stand-in for – all of North America and its wilderness.

Forts were built. French and British settlers trickled into the area. Around the midpoint of the eighteenth century, the first recorded use of the Falls for power took place: a small ditch was dug to power a sawmill.[4] Travel accounts by first-time visitors to the Falls usually commented on its size – many taking issue with Hennepin's assertions – and often noted that the mist and the thunder of the waterfall could be apprehended from miles away. Other common concerns in such travelogues included debates over whether birds could survive flying near the Falls, whether it was rarefied or condensed air that made it hard to breathe behind the waterfall, and whether congealed spray or foam created white rock-like substances.[5] The US-British treaties in the aftermath of the American Revolution established the Niagara River as the international boundary line. The First

Nations groups in the area were forced to leave or sell most of their territory. Niagara then became a focal point in the War of 1812: the British/neo-Canadians and their various Native allies held off the Americans, and the Treaty of Ghent, along with subsequent agreements, established the international boundary as cutting through the Horseshoe Falls.

Thus, force and the power of the state imposed a process of settler colonialism by which the Indigenous inhabitants of the region were moved or disenfranchised so that European settlers could appropriate the land for their purposes.[6] The Niagara Corridor was at first valuable as a transportation and portage route, and the Falls offered many nascent economic development possibilities. For this reason, moving local peoples away from the river margins was one of the first priorities of settling the area and developing agriculture, and then tourism and industry. Leaving Niagara Falls to Indigenous Peoples was considered the same as letting its waters pour unfettered over the precipice – in either case, the potential power was squandered. Indigenous conceptions of Niagara Falls, and their embodied ways of knowing its water and ice, needed to be replaced by quantifiable notions of Niagara Falls as a commodity that could be properly exploited by an industrializing society in which private property rights were sacrosanct. Much of this was done "legally" – though these are legal fictions when they are structurally designed to extinguish Indigenous claims, rights, and modes of living.

From a settler perspective, however, the displacement of Indigenous Peoples brought peace and borders, which brought stability, which in turn brought commerce and transportation improvements: the Erie and Welland Canals to circumvent Niagara Falls, then later railways. The Erie and Oswego Canals were key links in, and became key attractions along, the fashionable "Northern Tour" to Niagara. The number of visitors to the Falls rose significantly in the mid-nineteenth century. The first tourist spectacles soon followed. Establishments catering to tourism cropped up on both sides of the river, and Niagara enjoyed an extended period as the prime tourist landmark on the continent.

Communities near Niagara Falls developed in both New York State and Ontario, connected by ferry and then bridges (see Figures 1.2 and 1.3).[7] Not long afterward, Frederick Church crafted his famous portrait of Niagara Falls, and soon the Falls were illuminated by lights for the first time.[8] Daredevils such as the Great Blondin crossed the gorge by tightrope or challenged the rapids in different contraptions. About 40,000 people visited annually. As Karen Dubinsky puts it, "Niagara became famous for being famous."[9] In a case of what has been termed "allowably indigenous" –

FIGURE 1.2 View from the American side in the nineteenth century

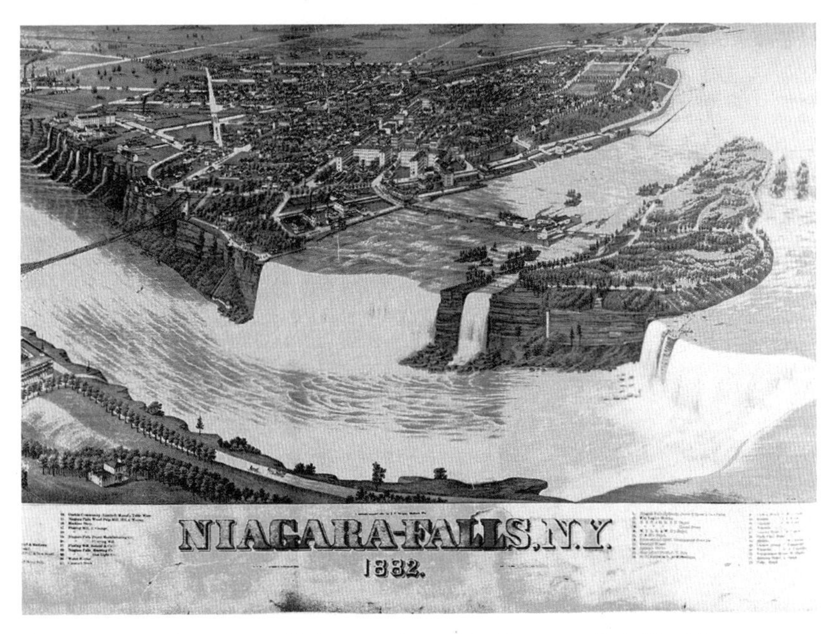

FIGURE 1.3 Bird's-eye view of Niagara Falls in 1882. *Niagara Falls Public Library, New York*

in reference to those enriching aspects of Native cultures that settler society permits to be visible – First Peoples were part of the mythical Niagara histories peddled to tourists, since they represented untamed wilderness.[10] Entrepreneurs in both countries – Canada became a nation in 1867 – built stairs and ladders to take curious onlookers to the base of the Falls and behind. Although the strip of land along the Falls and the gorge was supposed to be government-controlled, private interests installed fences and obstacles so that curious onlookers would have to pay to use their facilities (see Figure 1.4).

At the start of the nineteenth century, New York State had auctioned off a strip of land bordering the Niagara River. This strip included the riverbed and riparian rights to the water. Niagara offered hydraulic power opportunities that exceeded the capability of the imagination. At this point, hydraulic power was obtained along the shoreline of the upper river just by utilizing the drop of the rapids above the Falls.[11] Various factories and industries congregated at these rapids, constructing an intersecting network of diversion works, dams, canals, and millraces. These industries later spread to the downstream gorge, which was traversed by a number

FIGURE 1.4 At Prospect Point in the late nineteenth century. *Niagara Falls Public Library, New York*

of new bridges that associated Niagara with pushing the technological envelope.[12] In the 1840s, developers had purchased property for a hydraulic canal, although digging would not begin until the following decade. The scheme fell into insolvency but was resurrected. In 1861, a hydraulic canal was completed, taking water from above the cataract to the High Bank area downstream from the American Falls, where the declivity was 210 feet to the lower Niagara River. This canal initially attracted few users, but things changed when Jacob Schoellkopf's Niagara Falls Hydraulic Power and Manufacturing Company purchased the canal in 1879. Within a few years, seven mills were taking hydraulic power from the canal.

Free Niagara

New developments began to divert water from the upper river rapids and return it to the lower river gorge. This was more easily achieved on the east (New York) side of the waterway since the river turns sharply to the northeast after the waterfall, making for a short diversion route from the American Falls rapids to the east bank of the river below the Falls. By the late 1870s, the American side of the Niagara River was in an "advanced stage of visual blight."[13] In the words of author Archer Butler Hulbert, "the spectator ... cannot help seeing this mass of incongruous and ugly structures extending along the whole course of the Rapids and to the brink of the Falls. Of course, under these circumstances the Rapids are degraded into a mill race, and the Fall itself seems to be lacking a water-wheel."[14] The tailraces of the many mills and factories poured mini-torrents down the gorge face, forming a wall of small waterfalls (see Figures 1.5 and 1.6). Above the Falls, the shoreline and islands in the channel leading to the American Falls were thick with sundry commercial buildings, warehouses, icehouses, mills, hotels, and so on. At the waterfall proper, there was a jumble of vendors, carts, and confidence men, and visitors had to pay proprietors just for the privilege of viewing the falling water. In reference to all the fees and gratuities, one tourist lamented: "I could hardly divest my mind of the idea that I was not 'doing' Niagara, but that Niagara was 'doing' me."[15]

Niagara's reputation as the top resort destination was in decided decline, and its status as the icon of American nature was being surpassed.[16] A preservationist movement to "free Niagara" by creating parkland arose on both sides of the border. But it was noticeably stronger to the south, where a wide-ranging alliance of middle-class and elite reformers sought public

FIGURE 1.5 Industry along the High Bank in the nineteenth century. *Niagara Falls Public Library, New York*

FIGURE 1.6 Industrial ruins along the High Bank today

ownership of Niagara Falls. While most participants in the "free Niagara" movement wanted to protect the cataract from rapacious industrialists and ugly encroachments, these sorts of preservationist anxieties were also, consciously or not, bound up in efforts to combat hostile forces that threatened not just the waterfall but the ideals and goals of white North America, particularly those of the upper classes.[17] Governments on both sides of the border cracked down on tourism purveyors considered to be crooks and rip-off artists.[18] Certainly many were. But the ethnicity of some of the tourism operators (e.g., Jews and African Americans) motivated at least some of the charlatan characterizations. Thus, preserving Niagara was a discursive framing that partly relied on mobilizing racial and class fears: Niagara was not so much "freed" but rather "simply placed in different, more culturally acceptable hands."[19]

Even claiming that Niagara Falls should be publicly controlled was not as altruistic as it might appear. Creating some sort of Niagara preserve that was free for everyone essentially subsidized the cost of visiting Niagara Falls for those who could actually afford to vacation there – at the time, these were mostly those with financial means. For the hoi polloi who could partake, the experience was paternalistically meant to be a civilizing lesson of sorts. And this preservationist movement was even more directly colonial in the sense that it continued long-standing assumptions about what constituted the appropriate and efficient uses of the setting, and whom these uses were intended to benefit (i.e., industry and tourism took precedence over other uses and cultures).

A group of American cultural elites seized on the idea of a public park. They were led by Frederick Church, the Hudson River School painter who had given Niagara its most famous artistic rendering, and Frederick Law Olmsted. The "free Niagara" campaign grew quickly, channelled through a Niagara Falls Association composed of prominent citizens and backed by widespread popular sentiments. This included a petition that, in Pierre Berton's estimation, featured the signatures of more "illustrious and distinguished persons" than any other comparable effort.[20] Within a decade, the campaign achieved stunning success.

Other early state and national parks were often leftover lands that had little economic or agricultural value until railways ran nearby. By comparison, Niagara was already a veritable American shrine. By no means, however, were all Americans disenchanted with what was happening at the Falls. Visitors to Niagara were almost as likely to praise the harnessing of Niagara and to call for further development as they were to lament its despoilment. Some industrial capital sectors supported the preservation of Niagara Falls because they saw an opportunity to further their economic interests and investments. Many thought that Niagara would be *most* attractive when all of its water was put toward producing power. For example, Sir William Thomson, later Lord Kelvin, remarked that

> a long while must elapse before the whole of the volume of water now passing over Niagara Falls could possibly be utilized for the production of power, but if the whole of the water were so utilized and if the lofty cliffs over which the waters now tumble, were bare, think what would then be their aspect! The face of the precipice would be covered with aquatic plants giving in summer a splendor of color which with all their watery magnificence the Falls do not now possess, while the pool below would have a quiet beauty instead of its present turbulence.[21]

Despite considerably less industrial development, the Canadian side was not in a much better state than its cross-river counterpart when the "free Niagara" movement got underway. The chain reserve kept a strip sixty-six feet wide in the hands of the Crown. This precluded the types of industrial development that flourished in New York, as did much of the topography (for example, a ridge runs along the Canadian side, set back a bit from the waterfall, and much of this area was cedar swamp). However, hotel operators and other tourist purveyors had long encroached on the chain reserve, with or without permission, creating a carnival atmosphere of con artists, extortionists, and low-brow attractions. Moreover, upriver from the Falls, mills occupied parts of the river bank and various nearshore islands.

Members of the "free Niagara" entourage met with Governor General Lord Dufferin to get Canadian cooperation. They sought his help because of his central role in saving heritage landscapes in Quebec City. Dufferin became the first to officially propose the creation of an international Niagara park, which attracted many powerful proponents in North America and abroad. In New York State, a survey of the prospective Niagara parkland was quickly authorized and undertaken: it recommended a small, publicly owned reservation around the cataract. But gubernatorial changes and resistance from other quarters in the state led to several years of unsuccessful efforts in the New York legislature.[22] Finally, in 1883, with the Niagara Falls Association as the motive force and supported by a massive letter-writing campaign, a bill was passed to create a state reservation at Niagara.

Opponents fought a rearguard action, however, and it took until 1885 for the legislature to appropriate the necessary funds for Niagara land acquisition. In the end, the state spent $1.4 million to acquire the requisite property through eminent domain. The new reserve totalled 412 acres, three-quarters of which was underwater. The unsubmerged portion took the form of a mile-long strip, ranging in width from 100 to 800 feet, along the Niagara River, as well as Goat Island and other islands in the American Falls channel. On July 15, 1885, some 75,000 people gathered to celebrate New York's State Reservation at Niagara, as it was officially titled.[23] This was the first time that an American state "had used public money to expropriate property purely for aesthetic reasons."[24] Testifying to the blurring of the lines between a "reservation" and a "park," it eventually inherited the title of the country's oldest state park. But the actual reservation still needed to be created. There was considerable debate about its form and appearance: Should it evoke sylvan wilderness, or be in the English garden tradition? Or should it be a park replete with modern amenities? Regardless, the first step was to remove all the existing industry and related buildings, a procedure that left the land constituting the new reservation disfigured and denuded.

It was not until 1887 that the reservation commissioners hired Frederick Law Olmsted and his partner, Calvert Vaux, to design the area. The pair aspired to "restore the primacy of nature" while simultaneously accommodating throngs of tourists.[25] Olmsted wished to avoid emphasizing the terror and awe associated with Niagara's sublime past, aiming instead to evince spiritual regeneration and moral uplift to the visitor through verdant peacefulness. He favoured the upper rapids and was apprehensive about the waterfall drawing too much attention away from the "picturesque" surroundings he planned to create. Drawing on their past projects, such as Central Park in New York City and parks in nearby Buffalo, the reservation was intended to look as natural as possible and free of obvious human intrusions such as buildings and monuments. But it was also a manufactured naturalness that needed constant maintenance, and the reservation operators would later introduce park and commercial features that seemed incongruous with a reserve or preserve. Nonetheless, when completed, the State Reservation was widely acclaimed and gave Niagara a new lease on life as a prime place to see and be seen.

The genesis of the Canadian Niagara park has for many years been told "as a case of the virtuous people and the wise political leaders successfully overcoming the social abuses of commercial and corporate interests."[26] As in the American preservationist push, however, there were mixed motives at play. Gerald Killan has shown that, though the Canadian park movement did originate in concerns about preserving or restoring the natural state of the Falls, "the initial impetus of the nature lovers soon gave way to the influence of private park promoters. These businessmen viewed Niagara Falls as a resource to be conserved and wisely exploited as a tourist attraction for their personal profit."[27]

As early as 1880, the Canadian federal government had passed an act allowing it to expropriate land for the park, but questions about whether the provincial government had jurisdiction over the chain reserve stalled matters for several years, with each level of government trying to get the other to take responsibility. Entrepreneurs spurred on by potential profit opportunities, particularly railroad interests, helped bring matters to a head.[28] After several different commissions and inquiries and suggestions of a privately funded park, in 1885 the Ontario government passed an act to create a parks commission that would investigate and make recommendations for the future of the prospective park.

The commission endorsed provincial control of the park, which should include land between the the ridge and river. This included the edge of the Horseshoe Falls and the territory running from the Dufferin Islands to a spot close to today's Rainbow Bridge, as well as several islands – a total of 154 acres above water and 317 acres below. In 1887, the Ontario legislature, led by Premier Oliver Mowat, half-heartedly passed another act formally instituting the Queen Victoria Niagara Falls Park along the recommended lines. The new parks commission began the process of acquiring the necessary property and removing most of the existing buildings along the front while other new facilities were installed.[29] Park officials undertook landscaping, though there was little effort to naturalize the area compared with New York's State Reservation at Niagara. Be that as it may, when the new Queen Victoria Niagara Falls Park was officially opened in 1888, it was considered the "first expressly 'natural' park in British North America."[30]

Both of the new park systems were free to the public, a major change from the days of paying for the privilege of a view of the waterfall. The public responded by visiting in record numbers. Within a decade, the Queen Victoria Niagara Falls Park had expanded several times, to about 675 acres.[31] The State Reservation could not easily do the same, hemmed in as it was by industry. But it had an important advantage: it received operating funds from the state government, whereas its Ontario counterpart had to raise its own financial support in order to be self-sustaining. The result was that, in order to fund itself without charging entrance fees, the Queen Victoria

Niagara Falls Park allowed commercial concerns and hydroelectric franchises within the park boundaries in exchange for annual rental fees.[32] New York's State Reservation did not allow any power developers within its boundaries. This was a reversal of sorts, for it would now be Canadian, rather than American, powerhouses in tourist sightlines.

War of the Currents

The last decades of the nineteenth century were a period of incredible technological ferment and cultural anxiety concerning the continent's most recognizable natural icon. Those preoccupied by the landscape of the future focused their attention on the seemingly limitless energy of Niagara. Some plans were rational and realistic and, enabled by a rapid succession of socio-technological advancements, led to the proliferation of hydroelectric stations and industrial enterprises. Other schemes were, as both William Irwin and Patrick McGreevy have discussed, much more utopian.[33] Such unfinished undertakings were not benign, however. Although razor blade tycoon King Camp Gillette's envisioned Metropolis at Niagara, for example, never came to pass, a canal started but then abandoned by one William T. Love would later become the site of America's most famous toxic crisis. Technological advancements also stoked dystopian anxieties: Would the sacred waterfall be completely drained and commercialized?

Many businessmen had supported the creation of the State Reservation at Niagara. Some did so for "enlightened" reasons, in the parlance of the times, while some seemed to sense that the creation of the parkland might be, from an industrialist's perspective, a sort of reverse sacrifice zone: a small area near the cataract would be protected in exchange for the right to develop the surrounding area. With industry shunted to the margins of the Niagara tourist zone, the waterfall was preserved for tourists, while locals would come to bear the cost of living in a toxic landscape.

Although the State of New York had created the reservation, it seemingly had no compunction about authorizing developments that extracted water from the Niagara River. The year after creating the reservation, the state legislature was already granting charters to multiple companies that allowed them to divert water right outside the park boundaries. By 1894, eight companies had power charters on the American side. The state charged the companies nothing, and most had no restrictions on the amount of water they could divert, save for technological and capacity limits. There were rumors that the existing franchises, both real and on paper, could cumulatively deplete the entire river.[34]

These franchises generally took advantage of the new power form, hydroelectricity, which had been emerging as a viable alternative to hydraulic power while the State Reservation was in the process of being legally established. Hydro-power innovations, and electricity in general, were prime contributors to what has been termed the second industrial revolution.[35] Niagara's first hydroelectric generator – and one of the first in the world – was installed in 1881 by Jakob Schoellkopf, owner of the Niagara Falls Hydraulic Power and Manufacturing Company. The generator was housed in one of the mills at the High Bank taking water from the hydraulic canal, and produced direct current with one of Charles Brush's dynamos. Schoellkopf built another generating station close to the first installation, this one at the base of the gorge below the High Bank. Up to that point, other turbines and water wheels had not been able to use the full drop of water down the gorge wall, but the technology had advanced enough that Schoellkopf could take advantage of the head from the forebay of the hydraulic canal.

Other interests had also explored water conduits and power station configurations. Thomas Evershed, a New York engineer, developed a plan for a tunnel that would take water from above the American Falls and discharge it at the gorge.[36] However, this was initially intended to be a hydraulic, not a hydro-power, enterprise, featuring many small intakes for factories. Formed in 1886, the Niagara Falls Power Company took over the Evershed tunnel idea. This company was the engineering arm of the Cataract Construction Company, a holding firm that had big-name financial backers such as J.P. Morgan and John Jacob Astor. It was a model of Gilded Age interlocking directorships, conflicts of interest, and vertical integration that would have made John D. Rockefeller proud and trust-busting reformers furious.

By the early 1890s, the tunnel was well advanced and a central station design (rather than the multiple intakes and wheel pits in the original Evershed tunnel scheme) had been adopted. The company was still not absolutely certain about what generation and distribution form to choose for this new development – in addition to electricity, mechanical power and compressed air were options on the table. To attract the requisite technology, and attempt to get it as cheaply as possible, the Niagara Falls Power Company formed the International Niagara Falls Commission to hold an international competition. This commission was headed by Lord

Kelvin, arguably the world's most eminent theoretical physicist and the mind responsible for the absolute temperature scale and the second law of thermodynamics. The commission consulted with experts around the world, and toured water control facilities in far-flung places such as Switzerland.[37] A range of entries were submitted to the competition in various categories, including alternating current (AC) and direct current (DC) proposals, as well as non-electrical methods. Some prizes were awarded, but none provided a complete and workable power production and distribution method that satisfied the contest organizers.

As the types of entries suggest, this competition coincided with the so-called war of the currents between AC and DC. That battle has, of course, been well documented, and we need not rehash it at length here.[38] Thomas Edison, who had created a DC distribution station in New York City in 1882, followed by a number of other stations in other locales, was the major advocate of that form. But the new developers were banking on the need to send power a longer distance (such as to Buffalo) to make it profitable, since it was reasonably, though ultimately incorrectly, believed that Niagara Falls could not attract enough industrial customers.[39] DC could not be distributed long distances, which, along with being cheaper, was the advantage of polyphase AC induction motors whose voltage could be stepped up and down. After seeing Nikola Tesla and George Westinghouse's display of polyphase AC at the World's Columbian Exposition in Chicago in 1893, and in light of the fact that a polyphase generation and transmission system was installed in California that same year, the International Niagara Falls Commission recommended its method. (Steven Lubar argues that the decision to use AC was based as much on cultural criteria as material: that the technology used at the Falls should match their natural grandeur.[40]) Westinghouse was awarded the contract for the generators. The generating station was named after Edward D. Adams, president of the Cataract Construction Company.[41]

Polyphase AC combined with the revolutionary central station model meant that all the power would be produced at one location by large generators and then transmitted at high voltage over longer distances to multiple recipients. The design of the Adams plant would divert water about 1.5 miles above the American Falls, through a 1,200-foot intake canal at Port Day that served as a reservoir. From there, water would drop through penstocks to the bottom of a powerhouse, generating electrical energy. After leaving the turbines, the water would then flow northwest through the newly constructed 1.25-mile Evershed tunnel – itself heralded as a

FIGURE 1.7 Adams Power Plant under construction. *Niagara Falls Public Library, New York*

major achievement – under the City of Niagara Falls, pouring into the bottom of the gorge about one-quarter of a mile downstream from the American Falls.[42]

The first power came online in August 1895. When completed, the Adams plant's ten turbines produced 50,000 horsepower (around 37,000 kilowatts) of two-phase electricity, utilizing a head of 135 feet. It was the first large-scale AC generating and polyphase transmission plant in the United States, and it deployed technology that could transform high-voltage power into the various currents (including DC) needed by customers and sectors (such as industrial and lighting). Twenty-five-cycle frequency was selected as a compromise, and it became the electrical industry standard, though later 60-cycle frequency would be provided to domestic customers. In 1896, the Niagara Falls Power Company began sending three-phase AC (11,000 volts) the twenty-two miles to Buffalo, though most of the power remained in the local vicinity for industries like aluminum. All this, and the central station approach, made it an "electrical wonder of the world." The powerhouse and adjoining building were spared no expense and famed architect Stanford White gave them classical treatment (see Figures 1.7 and 1.8).[43] The Adams generating station soon became

FIGURE 1.8 Inside a generating station at the Adams Power Plant. *Niagara Falls Public Library, New York*

a regular stop for tourists. The company quickly added a second powerhouse, built between 1899 and 1904, doubling the plant's electrical power capacity.

Industry remained along the Niagara River, but outside the boundaries of the State Reservation, in two main areas. The first was at the High Bank downstream from the Falls. The second area was just upstream of the

American Falls in the Buffalo Avenue area, where Port Day took water out of the river. Though factories and mills no longer encroached so closely on the waterfall, these two areas grew to become some of the most concentrated industrial zones in the country, particularly in the electrochemical and electrometallurgical fields, which were well established before the First World War.[44] "Electro-chemistry is essentially a child of Niagara," proclaimed one commentator in 1912.[45] A bit further afield, in places like Buffalo and Lackawanna, heavy industry also set up, churning out products like steel and iron, as well as cereal, beer, and lumber. Cheap power was crucial, but so were factors such as easy transportation options and the path dependencies that resulted from initial investments (such as economies of scale, interdependencies, and vertical and horizontal integrations of processes, products, firms, and industries).

With the ideal sites for a hydro development on the New York side already snapped up, at least until there were further technological advances, American hydro financiers looked across the river for more opportunities. The Canadian Niagara Power Company had been granted the right to develop power in Ontario in 1892.[46] This American-controlled company failed to begin construction, however, in large part because the Adams plant design had so disrupted the industry. The International Railway Plant actually produced the first hydro power on the Canadian side in 1893.[47] This small plant conducted water through a canal just above the Falls and discharged the water through a tunnel that exited at the face of the cliff below the Horseshoe Falls. The electricity was used to operate the International Railway Company's trolley line, and power sent across the river for the trolley reputedly represented the world's first-ever international electricity interconnection. In 1898, the DeCew Falls hydroelectric station became operational. Taking water from the Welland Canal just west of Niagara Falls, it sent electricity thirty-five miles to Hamilton, Ontario.

Canadian Niagara Power's original charter was revoked in 1899, and the rights to develop power from water were divided between this company and several other concerns.[48] In the first decade of the twentieth century, Ontario's Niagara parks commission granted other charters. Three hydroelectric stations were subsequently completed on the Canadian side by the Electrical Development Company, Canadian Niagara Power, and the Ontario Power Company. The proliferation of hydroelectric companies was part and parcel of a larger continent-wide explosion in electricity: between 1902 and 1912 hydro-power generation mushroomed by 464 percent in the United States and supplied one-third of the output of central electrical stations.[49] The Edward Dean Adams Power Plant alone was responsible for one-tenth of all the electrical power generated in the United States. This growth helped drive the shift from using electricity only for lighting to also powering machines. More widely available electricity in turn brought with it a transformation of both workplace and household labour, with the attendant range of social consequences.

The first large power station at Niagara Falls on the Ontario side was the Canadian Niagara Power Generating Station, later known as the Rankine Generating Station, which featured the largest generators in the world when it came online in 1905.[50] It was modelled on the Adams plant and was a subsidiary of the same parent concern, the Niagara Falls Power Company. This plant is near the brink of the Horseshoe Falls, and discharged its water to the lower river through a tunnel. As the company directors planned from the beginning, the bulk of the electricity was exported to the United States.[51] Another new station, the Ontario Power Plant, was owned by the Ontario Power Company; however, the name is misleading, for this was an American company and it signed export contracts to send the electricity across the river.[52] The Ontario Power Plant was placed in the gorge right below the Falls along the waterline, and brought water from the upper river via underground conduits. The Toronto Power Plant, owned by the Electrical Development Corporation (EDC), was sited along the upper cascades. The EDC was the only Canadian concern among the new plants on the Ontario side, though a group of Toronto robber barons controlled it and sent most of its power to that city.[53] The architecture of all these stations, like those on the American side, was along classical lines to project strength, solidity, and power.

Thus, within the span of only a decade, a spate of power stations had sprung up around Niagara Falls (see Figure 1.9). The various Niagara hydroelectric plants on both sides of the border could cumulatively produce well over half a million horsepower. While that may not seem like much by modern standards, or even by the standards of the mid-twentieth century, this figure dwarfed the hydraulic power available from places like Lowell, Massachusetts, which just a few decades earlier was seen as a revolutionary energy source because it could produce the equivalent of up to 20,000 horsepower. It is important to realize that, before certain technological and scale advances enabled industry to tap nearly the full potential of generating and dam sites, hydro power from Niagara was popularly perceived as inexhaustible. Niagara's hydropower was so attractive not only because it seemed limitless but because it was so clean compared with burning coal, oil, or biomass. According to historian H.V. Nelles,

Figure 1.9 View of the Horseshoe Falls in the early twentieth century (note the Rankine generating station in the upper left). *Niagara Falls Public Library, New York*

hydroelectricity "existed in harmony with the rational and the romantic world" and "could resolve the paradox of ugliness that had blighted nineteenth-century industrialism; it could create factories *and* natural beauty."[54] As a direct result of the new power and jobs, the sister cities of Niagara Falls expanded rapidly: between 1890 and 1910, the population on the Ontario side of the waterfall almost tripled, while the size of its New York counterpart grew sixfold.

Conclusion

Robert Belfield writes that the Niagara Falls Power Company had introduced a technological system – the universal electric power system, which means that the resulting power could be used at different voltages (AC and DC) and for different applications (lights, small motors, large motors, and so on) – that was so successful that earlier hydraulic power approaches at Niagara immediately became obsolete.[55] The Adams plant technology was transferred to or imitated by subsequent power developments on both sides of Niagara and beyond.[56] The fascination with the new technology and energy form was further enhanced by the fact that it was derived from a place that still held a magical aura in the public consciousness. This was an exciting era when new electricity-related technology was developed or advanced every few years. However, the wide acceptance of hydroelectricity, and the central plant model, should not be taken as inevitable developments. As with all energy transitions, there were many contingencies, the supply preceded the demand, and infrastructures and technologies had to be provided or improved to convince producers and consumers that they should adopt the new source of power.[57]

Niagara became the synecdoche for a host of advances related to hydroelectricity, and the symbol of what electricity derived from water could offer. After all, by 1905, all of the generating stations in service at Niagara Falls on both sides of the border cumulatively produced the same amount of electrical energy as the rest of the United States put together. But Niagara's electrical "firstness" does get exaggerated. This is likely because of American technological exceptionalism narratives, but also the tendency when discussing the evolution of technologies to collapse inventions into oversimplified stories that privilege certain events, companies, and individuals (for example, framing Niagara Falls as the birthplace of hydroelectricity was good for Westinghouse and bad for General Electric), ignore the many necessary but incremental small advances, and then reify this narrative into concrete truth.

Niagara Falls and electricity became intertwined in the minds of so many because both were exemplars of unlimited power, danger, and utopian possibilities; quite simply, it makes for a compelling story if the greatest natural wonder of the world brought forth what some thought of as the greatest invention of modern times. Nonetheless, hundreds of places in Canada, the United States, and the United Kingdom produced hydro power before the Adams station. Nor did the Adams station involve the first transmission line, as both AC and DC electricity had already been transmitted various distances elsewhere.[58]

These qualifications need not reduce Niagara's importance to the evolution of hydrotechnology. In much the same way that Niagara's physical impressiveness stems not from being the biggest in the world by any one measurement but from the magnitude of its combination of numerous factors, so too does the history of Niagara as the cradle of hydroelectricity reflect its combination of several important advances on a larger scale, even if it cannot claim to have been the first in all of them. Niagara's first

AC station generated far more power than its predecessors and contemporaries. And because of its volume, and the technological capacity of late nineteenth-century generators, Niagara Falls was seen as a boundless source of energy close to urban centres and transportation networks. The Adams station had tapped only a portion of Niagara's potential, which far exceeded other water-power sites. To be sure, future generating stations at Niagara would be even larger, with several taking their turn at bearing the mantle of largest in the world.

Therefore, even though the story is more complicated and nuanced than is often presented, Niagara does deserve the title as the *main* locale where hydroelectric generation and transmission from a large central station was proven. Because of Niagara's potential, location, and symbolism, the establishment of power generation there "took on a larger-than-life role in the future of the country" and substantially conditioned North American attitudes toward electricity. Hydro power was an exceptional and "clean" energy source – compared with fossil fuels – that could be sent virtually anywhere and that represented a "utopian and progressive force for the future."[59] Looking back from the twenty-first century, it is difficult to appreciate how profound a psychological impact this energetic abundance had on North American society.

2
Saving Niagara: Innovation and Change in the Early Twentieth Century

The dynamos and galleries of the Niagara Falls Power Company impressed me far more profoundly than the Cave of the Winds; they are indeed, to my mind, greater and more beautiful than the accidental eddying of air beside a downpour ... I think the huge social and industrial process of America will win in this conflict, and at last swallow up Niagara altogether.

H.G. Wells, 1906

ANYONE VISITING the Canadian side of Niagara Falls should take a fifteen-minute walk upstream from Table Rock, the viewing area at the western flank of the Horseshoe Falls. Paralleling the Niagara Parkway, the path takes you over the forebay of the Rankine generating station. The original designers would be pleased to hear that this old powerhouse still conveys a sense of stability and power. The same is true of the Toronto Power Generating Station, fronted with classical columns and friezes, just a little further along. Keep going, and you can either turn south into the peaceful Dufferin Islands recreation area or continue east to a pedestrian bridge leading across the intake for the Ontario Power generating station. While there are many small islands and rock outcroppings dotting the rapids, some permanent and some ephemeral, the attentive observer will notice a low wall in the water, running parallel to the shoreline. This training wall leads to a dam that looks like a set of enormous concrete gates stretched halfway across the Niagara River: the International Control Structure. Behind it, visible on the horizon, are what appear to be two oversized garage doors, several storeys high, standing alone on the shoreline.

These are gates for the tunnels taking water to the Adam Beck generating stations.

These structures are all supposed to be here. Some we have already encountered; others, like the control structure, will be addressed in future chapters. They have all become fixtures of the Niagara waterscape. So has the old iron scow – but that was completely accidental.

Out in the rapids, not far from the brink of the Horseshoe Falls, sits the rusting hulk of the scow (see Figure 2.1). Back in August 1918, two men were working aboard this dredging barge when it broke free from a tugboat. As it raced out of control through the churning rapids, the pair managed to open the dredging doors on the bottom, and the vessel grounded on a rock shoal near the Toronto Power Generating Station. But now they were marooned past the point where any rescue vessel could reach them, perilously close to plunging to an almost certain death. After an epic rescue that took until the next day – interrupted by a thunderstorm and involving Niagara legend Red Hill and a breaches buoy shot to the scow from the roof of the nearby powerhouse – both men were miraculously brought ashore alive.

For the next century, the scuttled scow was a resolute sentinel, unmoved while the river and waterfall changed around it. However, videos captured by drone flights in the twenty-first century showed that the vessel had disintegrated considerably, giving rise to local speculation that it might not hold on much longer. Perhaps sensing this, in the summer of 2018 Niagara park officials organized a centenary celebration and unveiled a new heritage display on the shore. Then, during a big storm on Halloween night of 2019, the scow moved about one hundred feet further downstream before, further bent out of shape, it again lodged on the riverbed. Whether the iron scow will remain in place for another century, or soon crumble or tumble over the Horseshoe Falls, is anybody's guess.

But going back to its First World War–era origins, the initial grounding of the scow effectively capped a several-decades stretch during which the Niagara fallscape was significantly changed. This chapter details the manifold alterations made between the late 1890s and early 1920s, including new power stations and diversions as well as adjustments to the waterfall, along with widespread concerns that erosion was causing the Falls to, as popular parlance put it, commit "suicide." However, to fully understand those changes, we must also cover the institutional and international wrangling in this period that determined how the United States and Canada would share and divide the resources of the Niagara River. This requires tackling an alphabet soup of acts, agencies, and acronyms, but I implore

FIGURE 2.1 The iron scow in 2018 (former Toronto Power Generating Station in foreground, International Control Structure in background)

the reader to stick with it since the devil is often in the details, and so much of Niagara's later history was influenced by choices and arrangements made in this era.

Niagara Hydropolitics

In much the same way that fossil fuels have profoundly shaped socio-democratic politics in those nations that produce and consume them, Niagara Falls was key to the development of *hydro democracy.* At Niagara Falls, the powers and interests of two North American states overlap and collide, throwing their similarities and differences into sharper relief. In both countries, Niagara Falls played a role in the national evolution of political and jurisdictional debates about water power. Hydro democracy was especially apparent in Canada, and Ontario specifically, where governmental control of generating stations and diversions at Niagara Falls gave hydro-power considerations an outsized role in electoral politics, political economy and natural resource development, transborder relations, and federalism.[1] This is true of New York State too, albeit to a lesser extent. The prolific waters of the Niagara – as well as the Great Lakes–St. Lawrence basin in general – fostered the belief that energy abundance could produce egalitarian societies (or at least equality within those segments of society deemed deserving because of ethnicity, class, and gender) and perpetual economic growth. Cheap energy would facilitate a growth economy based at first on industrial expansion and networked services, including lighting and streetcars, and later on consumer and household consumption.

In the 1890s, the United States and Canada entered into talks about some sort of international agreement to govern shared waters. This included the Deep Waterways Commission, which proposed a dam where Lake Erie flows into the Niagara River.[2] In the following years, the United States Board of Engineers also recommended a dam across the Niagara River, in order to raise Lake Erie by three feet and counteract the lowering throughout the Great Lakes basin caused by the Chicago Diversion.[3] This diversion was made possible by the Chicago Sanitary and Ship Canal, which took water out of Lake Michigan, thus lowering water levels in the Great Lakes basin. It became a thorn in the side of US-Canada relations.[4] Other proposals for a dam in the lower Niagara River were presented, potentially in conjunction with a power tunnel from the Whirlpool to the Devil's Hole to channel water to a hydro-power station. For example, the Thompson-Porter Cataract Company outlined a plan for a dam near today's Moses and Beck power stations that would raise the lower river and flood out many features of the lower gorge. The company claimed this would avoid "disturbing the Horseshoe and American Falls" while generating 2 million horsepower.[5] Foreshadowing the remedial works and diversion regime created in the 1950s, Thomas H. Norton proposed an enormous dam across the Niagara River about a half-mile upstream from the waterfall that could divert all of the water for power production, except for the eight hours each day when Niagara Falls would be turned back on.[6] Others suggested various iterations of an "intermittent waterfall," such as schemes that would allow water to flow over Niagara Falls only on Sundays.[7]

The United States was embroiled in protracted and intertwined debates concerning private versus public power, state versus federal control of power, and unbridled economic expansion versus conservation or preservation of resources.[8] Although the creation of the State Reservation at Niagara had pushed industry away from the waterfall, the New York legislature continued to grant charters for prospective hydroelectric developments that would take water from above the Falls. But taking away too much water could be just as disastrous from a scenic perspective as putting up intrusive buildings. Andrew H. Green and Charles M. Dow, successive presidents of the Commissioners of the State Reservation at Niagara and leaders of the public movement to protect Niagara Falls, urged a freeze on New York State's water diversion permits.[9]

However, the annual reports of the State Reservation reveal that the likes of Dow and Green were often at odds with other commissioners, many of whom had financial holdings in the power industry – e.g., some were directors of the Niagara Falls Power Company – and used their positions on the commission to further their commercial interests. These blatant conflicts of interest were not always readily apparent to the general public. Some industrialists wanted to increase diversions, but others with a stake in the Niagara power industry hoped to freeze them in order to protect their existing monopoly or thwart rival companies' access to water. And some pushed for, or even helped craft, diversion legislation that would lightly regulate their industry; this early form of what is today called "corporate capture" led to watered-down laws that would assuage concerns but pre-empt more stringent policies. The reservation commissioners requested that the federal government seek an international agreement concerning the Falls and attempted to have the New York legislature adopt diversion limits. Other politicians and bureaucrats also called for an agreement with Canada.[10]

It was about this time that J. Horace McFarland, a figure underappreciated by historians of the North American conservation movement, began his rise to prominence.[11] He was president of the newly formed American Civic Association, and he emerged as the leader of a national "save Niagara" crusade. In the September 1905 issue of *Ladies' Home Journal,* McFarland penned a muckraking piece with the title "Shall We Make a Coal-Pile of Niagara?" No single article lamenting the imminent destruction of Niagara Falls expressed greater outrage or so successfully aroused the ire of the American public.[12] McFarland met with President Theodore Roosevelt, who took a very close and personal interest in Niagara affairs, and gained his support.[13] In his 1905 message to Congress, Roosevelt called for the preservation of the Falls and scouted for a cross-border accord.[14] In the meantime, McFarland lobbied Congress, wrote many pieces in newspapers and magazines, and toured the lecture circuit.[15] Though he campaigned hard in Ontario as well, "the idea of saving Niagara Falls at the expense of curtailing power development" was, in his estimation, unpopular in that "fuel-deprived and industrially under-developed province."[16]

As congressional debate about Niagara Falls indicated, and putting aside New York's decades of indiscriminately granting Niagara water rights, the US government and influential segments of the public did appear to be more concerned than Canadians about preserving the appearance of Niagara Falls. True, aesthetic protection had some economic motivations for the tourist industry in the surrounding communities, not to mention the millions of dollars in sunk costs that had been spent creating the Niagara reservation.[17] But many American industrialists were vocal about preserving the Falls – whether out of real concern or self-interest – and without their support the preservationist efforts would likely have come to naught. There was also a less apparent motivation: if Canada could not absorb all the Niagara hydro power produced within its borders, and limitations were put on the amount that it could transmit to the United States, then the stations on the Ontario side could not increase their generation since there would be no one to whom they could sell the power. This meant that more water would flow over the waterfall, which would be good for tourism, while the New York generating stations maintained their output. Other interests, such as coal, also favoured the curtailment of diversions so that there would be less energy competition.

But congressional debates about Niagara were complicated by nationwide debates about federal versus state jurisdiction over water power. Would, and should, federal authority override the water rights already granted by New York? Such concerns took on even greater importance since the state government flirted with opening up all water-power rights to private interests. This was all further complicated by the fact that the Niagara River was a border water, which some argued brought it under federal purview. It was deemed a navigable waterway, also theoretically making it the responsibility of the national government – except that the river was clearly not navigable at the Falls, where diversions took place.

The chairman of the House Rivers and Harbors Committee, Ohio representative Theodore Burton, introduced Niagara legislation in 1906.[18] The Burton bill, as it was called, proposed to prohibit additional diversions of water from the Niagara River and the four Great Lakes above it, as well as prevent any additional electricity imports from the Canadian side of Niagara. The bill also stressed the need for international cooperation to preserve the Falls, which would give the federal government jurisdiction over diversion rights in the non-navigable portion of a navigable river. While Burton himself was opposed to diversions because of his connection with navigational interests, particularly the Lake Carriers' Association, some powerful corporate interests aligned against the proposed bill.[19]

The Burton Act was passed at the end of June 1906, but political pressure had resulted in significant amendments to the legislation.[20] Both Burton and McFarland were displeased with its final form, which retreated from stronger prohibitions on diversions above Niagara Falls as a result of lobbying from power companies, and increased the caps on the diversions that would be allowed at the Falls. These diversions were subject to permits from the secretary of war, who was authorized to grant revocable permits for Niagara diversions not exceeding 15,600 cubic feet per second (cfs) on the condition that this amount not interfere with the scenic grandeur of the Falls or the navigability of the river. The amount of electricity that could be imported from Canada was raised, compared with previous versions of the bill, to 160,000 horsepower by irrevocable permits and another 350,000 horsepower in revocable permits. This added up to close to the total amount of electricity generated on the Ontario side of Niagara Falls. The bill also called for a treaty with Canada, via Great Britain, to regulate Niagara diversions and the maintenance of the cataract's visual splendour. In the belief that such a treaty was imminent, the Burton Act was originally authorized to run for only three years. But when a US-Canada border waters accord was signed in 1909, its Niagara measures were more lenient. Consequently, joint resolutions were passed to extend the Burton Act several times – the first by two years, until 1911, after which it was twice extended for an additional year. It finally expired in March 1913.[21]

The two American-owned powerhouses on the Canadian side were allowed to export power over the border so long as they sold half the electricity in Ontario – but only when it was required in the Canadian province, which was not the case in the first decade of the twentieth century. As a consequence, the Ontario powerhouses exported most of the electrical energy across the river to their parent utilities. Under the authority of the Burton Act, Secretary of War William Howard Taft issued licences to the three Niagara power stations operating in Ontario,[22] but the two American-owned companies were permitted to export almost 40 percent more power than the Canadian-owned EDC station. Limiting electricity importation from Canada was an indirect way of limiting water diversions on the Canadian side, since any restriction placed on the electricity brought into the United States would proportionately limit the amount of water diverted in Ontario for power. These limitations on export privileges from Canada were also intended to give the United States the upper hand in bargaining for a boundary waters accord.

After repeated calls for transborder action, the International Waterways Commission (IWC) was finally constituted. The US federal government was especially concerned about diversions for hydroelectricity, at Niagara Falls in particular.[23] The Canadians were more worried about diversion impacts on navigation at places *other* than Niagara Falls, and coveted an international agreement with a broad purview – not only geographically but in terms of establishing general rules that would apply to diversions from all border waters. Canada's seeming disinterest in protecting Niagara was partially motivated by a desire to hinder the Americans from taking control of power production on both sides of the river. The Canadians were also less concerned because it would take much larger volumes of water diversions to seriously impinge on the Horseshoe Falls, which was mostly in Canadian territory, compared with the much smaller American Falls.

Niagara diversions became the Roosevelt administration's top IWC priority.[24] The two sides batted back and forth various proposals for diversion limitations at Niagara, eventually issuing a joint report that promoted partial Niagara preservation. The commission had decided that about one-quarter of the Niagara River's total volume – which it estimated at 222,000 cfs – could be diverted without injuring the scenic effect. However, it admitted that one-quarter was essentially an educated guess and political compromise, and its report concluded that the total should be apportioned as follows: a maximum diversion volume of 36,000 cfs on the Canadian side, which was the sum of the cumulative intake capacities of existing power stations there, and 18,500 cfs on the American side.[25] The bases for these numbers were the recognition that much of the power was exported to the US anyway, even though a larger share of the water fell in Canadian territory, and the volume of the Chicago Diversion, which was subtracted from the American share. These amounts would sufficiently serve the existing and soon-to-be completed power stations. Canadian representatives did not think there was much point fighting for more than 36,000 cfs until there was a substantial Ontario market for it.

While the Americans had convinced the Canadians to set Niagara diversion limits, the latter had convinced the former, using Niagara Falls as a bargaining chip, to enshrine such measures in a comprehensive accord setting joint rules for *all* border waters. Several years of further discussions produced several different drafts. The resulting 1909 Boundary Waters Treaty (BWT) was a compromise.[26] Securing the agreement was a significant coup for Canada, since the much more powerful United States agreed to a commission within which the two countries were equal.[27] The treaty notably granted equal navigation access to shared waters, and adopted regulations concerning water diversions and changes to water levels. Essentially, any changes in the level of a border water needed agreement through the International Joint Commission (IJC), or a special agreement between the federal governments outside the IJC (though the subsequent construction or maintenance of any structures needed IJC approval). The treaty established the IJC as a six-member body with an equal number of Canadian and US appointees, who are technically independent from the government that appointed them.

The Boundary Waters Treaty mostly incorporated the Niagara recommendations of the IWC. Article V of the treaty dealt specifically with Niagara Falls. It aimed to avoid disturbing the existing vested power rights at Niagara, and to avoid hurting Lake Erie and Niagara River levels. Limits on the amount of Niagara power that could be exported from each country were left on the cutting room floor. Canada could divert no more than 36,000 cfs from the river above Niagara Falls, and the United States 20,000 cfs, which was slightly more than the 18,500 cfs in the IWC's recommendations.[28]

The volume had been upped to 20,000 cfs in the final flurry of negotiations for the 1909 treaty, though the archival record does not provide a clear explanation as to why. During the negotiations that produced the BWT, the importance different diplomats placed on the various factors (higher flow volume on Ontario side; existing stations and infrastructure; existing electricity exports; Chicago Diversion) was inconsistent. Likely, the final number was based on a combination of all these variables, though which one was more important is difficult to determine and may have shifted or been contested, even within both national camps. Secretary of State Elihu Root, who was deeply involved in the crafting of the BWT and believed that Canada would have been perfectly happy to dry up the Falls for power, seemed to suggest in testimony before the Senate Foreign Relations Committee that the additional 1,500 cfs was to put the US allotment at "round numbers, so our limit is higher than we want" so as to allow for future growth.[29] He also insisted that no reference be made to the Chicago Diversion trade-off in the final text of the treaty.[30]

Thus, the murky rationale for the final apportionment was as much about power politics as geography. Some commentators criticized the unequal distribution of Niagara water in the treaty as a "sellout," while others decried the aesthetic injury that would result from the diversions permitted under the treaty. In any case, this distribution of Niagara water in the BWT would prompt further Niagara negotiations for the next half-century, with each side at various times grumbling that the basis of the initial Niagara balance had been invalidated, or that it should have gotten more. While it is difficult to determine which nation fared better in the long term, Canadian negotiators admitted behind the scenes that the United States might have gotten the better end of the deal in the short term because so much of the hydro power was exported to that country. The American power companies that set up shop on Ontario's Niagara shore had made no effort to erect electricity distribution facilities in Canada, and all signs indicated that they did not intend to live up to the promise to keep half of their product in Ontario. Along with worries that the Toronto utility trusts (which owned the Electrical Development Company's power station) might grab a monopoly on Ontario's share of Niagara power, this anxiety galvanized a public power movement that led to the creation of an organization that would become one of the most important players in the modern Niagara saga: the Hydro-Electric Power Commission of Ontario, often referred to simply as "Ontario Hydro."

The rise of Ontario Hydro, the source of the "people's power," as the utility's slogan proclaimed, was a distinct development in North American energy history. Created in 1906, after much political debate and posturing it began distributing electricity in 1910.[31] This quasi-Crown corporation, which combined elements of a government department and a municipally owned cooperative, was spawned from a Progressive Era emphasis on state intervention, antimonopolism, and the rhetoric of "fairness." It promised cheap electricity – "power at cost" – for the underdeveloped middle class and small manufacturers, rather than rich industrialists and haute bourgeois in Toronto and across the border in the United States.[32] As *Saturday Night* magazine phrased it, Ontario Hydro championed "popular rights as opposed to monopolistic privilege."[33] At the same time, as Ontario Hydro expanded its reach in the coming decades, it would often conflate the public interest with its institutional self-interest and self-preservation.

With Ontario Hydro in place, Canada's Niagara electricity was increasingly public, while the American was private. But the battles about the future of Niagara Falls had led to more government intervention in natural resource management in the United States.[34] Moreover, progressive and conservationist desires to regulate industry and environmental damage were arguably stronger in New York than in Ontario, and the state's water policies yo-yoed back and forth on the public power question.[35] That no equivalent to Ontario Hydro was created in New York until 1931 had as much to do with pragmatic factors – such as the state's existing spatial distribution of and access to electricity, differing riparian rights, and Ontario's lack of coal supplies – as it did with ideology concerning public versus private ownership.

Some of the same concerns that spurred the formation of Ontario Hydro were linked to the Canadian formation of the National Commission of Conservation as well as the passage of the federal Fluid and Electricity

Export Act of 1907. Though the federal government, led by Wilfrid Laurier's Liberals, promoted a fairly laissez-faire approach to electricity exports, this act was a partial rejoinder to the export discrimination under the Burton Act. Moreover, it represented the federal Liberals' attempt to curb the growing public power movement in Ontario, and reassert some federal jurisdiction in the area of utilities and water rights in response to recent Niagara developments.[36] This 1907 act required Canadian power exports to secure an annual licence, gave Parliament the authority to levy an export duty on hydroelectricity, disallowed hydro power from being sold at a lower price in the United States, and featured a recall clause that allowed exports to be quickly revoked if the power was required in Canada. Granted, the act was ambiguous about how to determine much of this, and some of these powers were never really exercised or enforced.[37] The federal government's authority to regulate power exports therefore remained unsettled, and as of 1910 about one-third of the total electricity generated in Canada was sent across the border.[38]

Even if Canadian power export policy was not fixed, by the start of the First World War transboundary governance of northern North American border waters had been given a firm foundation, one that lasts to this day. The Boundary Waters Treaty proved to be a major step for the Canadian-American relationship in general, and even more so for Niagara Falls and other border flows. Though this treaty contained the seeds of environmental protection and cross-border cooperation, it was an agreement that at its core was as much about mutual exploitation, and the two nations were now free to aggressively pursue both their desire to wrest maximum power from Niagara and, as the next section details, to adjust the waterfall accordingly. At the same time, both countries would soon challenge the treaty's apportionment of Niagara water, leading to more rounds of diplomatic discussion (see Figure 2.2).

Niagara's Suicide

While politicians and engineers had talked and debated, erosion continued to gnaw away at the waterfall. The pace of erosion caused widespread concern that Niagara Falls was committing "suicide."[39] American commentators argued that Niagara was a national symbol and its potential demise should concern the whole country, though there continued to be differences of opinion about whether the state or federal government

FIGURE 2.2 Schoellkopf power station. *Niagara Falls Public Library, New York*

should exercise responsibility. Canadian authorities, too, promoted some type of intervention.[40]

The rate of recession of the Horseshoe Falls varied – two, five, or eight feet per year – depending on which study one consulted and which area of the cataract was being discussed, since the middle eroded much more quickly than the edges. During the nineteenth century, the smooth "U" curve of the Horseshoe Falls transitioned into more of a "V." In 1888, the annual report of the State Reservation at Niagara reported, over a span of less than fifteen years, "a noticeable alteration in the outline [of the Horseshoe Falls]: a decided recession has taken place."[41] Then a sudden rockfall in 1889, and another in 1905, created a notch in the centre of the half-ellipse that served to pull more water in that direction. Water was concentrating in the middle of the horseshoe, leaving the grey rock flanks progressively denuded. In fact, the dewatered flank ruined the Ontario excursion behind the falling water known as "going under the Falls," and authorities elected to build scenic tunnel portals instead.[42]

During the decades covered in this chapter, the landscape of Goat Island was altered significantly.[43] Tourism and past agricultural use – it was named

Goat Island for a reason – had changed the island's flora and fauna. The island was also perennially being reshaped and shrunk by erosion. To check this loss, in the 1880s the State Reservation installed a 1,500-foot buttress of timber and stone along the Goat Island shore.[44] During the creation of the reservation any unsightly buildings had been eliminated, and the shoreline graded and landscaped. Wing dams and diversion weirs were removed, hydraulic raceways were filled in, as were parts of the riverbank, with stone riprapping added to better resist erosion. Trees were planted. The areas around Goat Island, particularly Bath Island, had been heavily industrialized; now they went through extensive rehabilitation in order to look natural. A new bridge crossed to Goat Island. To the north of the American Falls, the park staff of the new reservation also went to great lengths to "naturalize" and green the area. The famed observation tower at Terrapin Rocks on the western edge of Goat Island, which was first erected in 1833, had been deemed an eyesore and removed so that it would not compete with other vantage points.

On the Canadian side, Table Rock was the foremost place to view the Falls. "The Table Rock is a very large flat rock projecting from the bank and overhanging its base very much," recorded one visitor in 1787; "this being the nearest part to the Great Fall, you are of course almost stunned with its noise and perfectly wet with continual mist arising."[45] But erosion was slowly but surely shrinking Table Rock, which was notoriously unstable: rockfalls were known to have occurred there in 1818, 1828, 1829, 1850, 1853, 1876, and 1897.[46]

At the start of the twentieth century, the average flow of the Niagara River was commonly accepted as around 222,000–224,000 cfs.[47] This is worth pointing out because within a few decades experts widely agreed that the average flow of the river was about 200,000 cfs. The reasons for this volume decrease are not entirely clear. Surely some of it was that the river volume had temporarily declined, partly in accord with variable natural supply in the Great Lakes basin as well as water manipulations upstream in the Great Lakes system, and partly from increased diversions from the Niagara River. Inaccurate measurement procedures may have been a factor too; the limited means of gauging a turbulent river, and constant diversions and alterations to the aquatic geography, mean that important baseline information has been lost to time. It is likely, however, that some of the difference was a revisionist history, intentional or subliminal, aimed at obscuring the role that power diversions had played in decreasing the flow of the river.[48]

As part of the United States Lake Survey, the US Army Corps of Engineers spent several years analyzing the Niagara River. These investigations provided a much clearer picture of the hydrology and bathymetry, uncovering many of the basic physical facts about the river and waterfall. According to Corps of Engineers estimates, diversions from the Chippawa–Grass Island Pool lowered the upper Niagara River by 5.5 inches at the head of the rapids above the Falls, and Lake Erie by a little more than an inch, which in turn slightly lowered the lakes above it. The Corps of Engineers concluded that, as of June 1911, the Horseshoe Falls flow had decreased 15–20 percent, and the flanks had been lowered 0.29 feet at the Goat Island end and 0.72 feet at the Canadian side.[49]

Submitting his report, the Chief of Engineers hit on many of the points that others had made about what accounted for Niagara's impressive nature:

> The "scenic grandeur of Niagara Falls" appears ... to be dependent on opinion and sentiment, and it seems almost absurd to demonstrate, by physical measurement of any kind, what the effect ... of any diversion will be upon the Falls, considered solely as a spectacle. If, however, it be conceded that the "scenic grandeur" of the Falls is dependent largely, if not exclusively, upon the awe with which they impress the spectator, and that this sensation is due to the irresistible power of their enormous volume of flow and upon the height of the fall, then even grandeur is susceptible of measurement, since reduction in volume and height will measurably, if not sensibly, affect the Falls as a spectacle. Moreover, the effect produced by the Falls is intimately connected with unity of sensation, and this is seriously disturbed by breaks in the crestlines, which follow reduction in volume.[50]

The development of Niagara Falls produced disparate reactions. Industrial interests framed themselves, disingenuously, as protectors of the Falls. They pointed out that it was the flowing water that caused the erosion. By diverting water around the waterfall, they were reducing erosion and saving the great cataract. Or so they claimed. Many members of the public joined industrialists in the view that it would be better to let the Falls run dry, perhaps turning them on for tourists during the weekend, like some sort of amusement ride (see Figure 2.3).

Other significant physical modifications were made to the wider fallscape in this period. Because of the many hydro-power developments, the area all around the Falls was a perpetual construction zone for a quarter-century. This left the parkland on the Canadian side of the river quite disturbed,

FIGURE 2.3 Ice bridge. *Niagara Falls Public Library, New York*

and the parks commission fretting over how to restore it. The EDC power station was built on reclaimed land, in a spot that was previously submerged under eight to twenty-four feet of water.[51] Power operations had submerged the main island at the Dufferin Islands, an embayment about a half-mile above the Horseshoe Falls along the west bank of the upper river.[52] Excavated materials were used there to extend the islands, a series of cascades was formed in the channel, "and with a view to the restoration of natural effects a group of small islands is being added to those now in existence."[53]

Ice formation was a pre-eminent concern at the Niagara power stations. Ice that forms on the river or comes from Lake Erie is hazardous for a hydro station. Because of the northern location, the Niagara River became a working laboratory for major scientific studies of ice formation and control, with some of the earliest research taking place here.[54] While designing the Adams power plant in the 1890s, the planners had scoured the globe for existing knowledge about how ice affects water wheels and other hydraulic ventures. Planners considered various methods of dealing with ice: guard grills for intakes, racks heated by steam, ice sluices, alteration of the water velocity, and volume of the intake canals to affect ice formation, and so on. In subsequent winters, and at other Niagara power stations, derricks, cranes, dynamite, dredging, dikes, ice runs, and ice-breaking tugs were used to address the gelid buildup.[55]

Trial and error remained the name of the game, however, since Niagara's frozen water still befuddled engineers, and damage and shutdown of the generators were common. The infamous ice bridge incident of 1912 brought the point home in tragic fashion. When the Falls froze over, people would congregate on the ice mountain – the "ice bridge" – for festivities. In 1912, the ice broke away and floated downstream, taking several people to their untimely deaths.

A range of other initiatives combined to further alter the appearance of the Niagara waterscape. Secretary of War Taft led a committee that secured a number of changes in New York territory, such as removal of unsightly structures, tailraces, and rubbish. Design changes were made to powerhouse no. 3 of the Hydraulic Power Company of Niagara Falls: the penstocks, for example, were cloaked in rubble masonry to harmonize with the colour and texture of the surrounding cliffs.[56] Pollution in the Niagara River and other boundary waters was also gaining attention. IJC researchers stated that "the pollution below the falls is gross," but a cross-border agreement on water quality would have to wait for half a century.[57]

In conjunction with cofferdams for power station construction, the erosion of the Falls had impinged on the ability of intake works to draw water for the Town of Niagara Falls, Ontario.[58] The town hired Isham Randolph, the chief engineer for the Sanitary District of Chicago, to build a submerged dam to raise the water levels at the intakes, located along the shore just above the Horseshoe Falls. Randolph created a concrete column fifty feet high, and then tipped it into the water; unfortunately, the falling structure landed on large boulders, which prevented it from achieving its full purpose. However, it reputedly still raised the water level at the intake about 10.5 inches.[59] This slightly counteracted the unsightly masses of rock at the exposed flanks of Table Rock. Other schemes were proposed for submerged dams in order to push water to the shallower flanks, thereby diminishing recession and improving appearance. Using surplus material from excavations for the new generating station tunnels, the Queen Victoria Niagara Falls Park reclaimed about 400 feet of the foreshore along the Canadian side upriver from the Falls, raising it to the same level as the adjoining shoreline. Doing so reduced the length of the lip of the Horseshoe Falls by about one-sixth.[60]

This reclamation was among the most significant changes to the Niagara waterscape in this period. Yet it went virtually unremarked on, both then and in the years since, even though the various alterations had indelibly revamped the Niagara Falls scene over the span of a generation. The 1912 report of the Commissioners of the Queen Victoria Niagara Falls Park likewise contained an offhand comment about the extent to which they had altered the littoral zone of the Niagara River: since the commission's inception, about fifty acres had been filled in along the shoreline or turned into islands.[61] Furthermore, Cedar Island, a large crescent-shaped mass of land joined to the mainland by a pedestrian bridge, ceased to be an island when the lagoon between it and the Canadian shore was filled in by material from power tunnel excavations. Other nearshore islands disappeared in similar fashion.

Temples of Energy

Each country could opt out of the Boundary Waters Treaty after five years (with one year's notice required). Ongoing deliberations made it clear that no one was taking the continuation of the treaty and the IJC for granted. Indeed, there was serious discussion of ending the treaty, in part because of its Niagara provisions. After the Burton Act expired in 1913, the two American hydroelectric power companies at Niagara, the Hydraulic Power Company of Niagara Falls and the Niagara Falls Power Company, diverted 6,500 cfs and 8,600 cfs, respectively. This was the situation when the Great War broke out. During that war, all the Niagara water that could be utilized was made available for power diversions, and the existing stations often exceeded their legal diversion limits.[62]

Even before the United States formally entered the war, the War Department requisitioned all Niagara hydro power and directed it toward war-related uses.[63] Many New York establishments in Niagara Falls and Buffalo depended on the electricity imported from Canada to make vital wartime products such as chlorine, phosphorus, carborundum, aluminum, silicon, chromium, and magnesium. In many cases, the Niagara-Buffalo district was responsible for all, or most, American supplies of these electricity-intensive processes and products. Beginning in January 1917, with the United States on the brink of entering the world conflict, the

War Department was authorized to license the use of Niagara River waters, and the secretary of war issued revocable permits, which cumulatively went up to the 20,000 cfs limit of the Boundary Waters Treaty. Under government pressure, the Niagara Falls Power Company and the Hydraulic Power Company of Niagara Falls merged (along with the Cliff Electrical Distributing Company), retaining the name of the former and bringing the two main New York power stations at Niagara under unified control. In 1914, the Niagara Falls Power Company had replaced its existing plant in the gorge with another new generating station, no. 3, featuring the largest generators in the world. In order to more efficiently use the increased wartime water allocations, and at the request of the federal government, a new powerhouse, no. 3-B, was added in 1918.[64] Total capacity now stood at 100,000 horsepower. These stations soon took water up to the 20,000 cfs ceiling, though heavy ice cover in the Niagara River sometimes curtailed hydro-power production, such as in the winter of 1917–18.[65]

The Canadian federal government, too, moved to control Niagara electricity in wartime, though the object of its concern was most often exports across the border. There was already strong domestic Canadian resistance to electricity exports to the United States before 1914, especially in Ontario. This opposition reached fever pitch during the war, resulting in the so-called Repatriation Crisis and various studies into the nature of Canadian electrical development and exports, such as the Drayton Report.[66] Adam Beck, the chairman of Ontario Hydro, pressured the federal government to stop all power exports from Niagara to the United States, partly as a means of repudiating the export contracts that Ontario Hydro had inherited when it purchased one of the private generating stations.[67] Canadian authorities resisted, but cut electricity exports to the United States from a maximum of 112,500 horsepower to about 91,000 horsepower. This became a point of friction in Canada-US relations – with the United States threatening to cut off coal exports to Canada in retaliation – as it already was for federal-Ontario relations.[68] But meetings between American and Canadian federal officials led to a "mutually helpful understanding" about the Niagara electricity situation that kept export restrictions to a minimum.

During the war, the US government increased its role in regulating electricity not just at Niagara, but across the country. Private interests reluctantly went along, but pushed back after the war ended, as exemplified by the battle over hydroelectric developments at Muscle Shoals and the Hoover Dam.[69] While renewing domestic Niagara withdrawal permits,

Congress had also asked for a comprehensive investigation of water diversions there.[70] The ensuing report, released after the war, contended that it was feasible to divert 80,000 cfs from Niagara (plus develop more power along the lower river) and to build a submerged dam or other remedial works to spread the reduced flow more evenly across the crest of the waterfall.[71] The wartime diversion situation stateside continued until June 1920, when the Federal Water Power Act and a revised Federal Power Commission (FPC) came into effect.[72] The FPC was made up of the heads of the War, Interior, and Agriculture departments. Its purpose was to replace the previous piecemeal way of approving hydro-power projects – the new commission could grant fifty-year licences – and systematically preserve and strategically develop the nation's water-power resources. But this commission consistently favoured development over preservation, private over public power, and hydro power over multi-purpose water developments.[73] This, along with the commission suffering from a lack of funds and personnel, led one scholar to note that "from the beginning, the FPC was ineffective as a regulatory agency."[74]

The 1920 Federal Water Power Act permanently moved the American Niagara diversion limits to those set by the Boundary Waters Treaty. Moreover, the Federal Power Commission was now responsible for apportioning US diversion rights from the Niagara River. In 1921, the FPC granted its very first licence, which went to the Niagara Falls Power Company so that it could use the full portion of the US 20,000 cfs share of Niagara water (except for 500 cfs). This company then established interconnections with other distribution systems in New York State, forming one of the earliest substantial electrical grid systems in the country. It enlarged its big power station in the gorge, as well as the hydraulic canal that ran through the city above. Since the capacity of the canal was limited, the company added a mile-long tunnel. Powerhouse no. 3-C came online in 1925. This 210,000-horsepower expansion made the tripartite complex, by this time known as the Schoellkopf Generating Station, the largest in existence (taking the crown from the recently completed Queenston–Chippawa development, described below). No longer on the cutting edge of efficiency just several decades after its celebrated opening, the historic Adams power station was relegated to "standby" status.[75]

By throwing off their self-imposed diversion shackles, both nations had turned a corner in their attitudes about the tension between beauty and power. One of the war's lessons was the importance of hydroelectricity from Niagara Falls and the need to obtain even more. Between 1917 and

1920, electrical generation in the United States went up by 42 percent, with a total installed hydroelectric capacity of 9 million horsepower.[76] There was no drop-off in demand after the Treaty of Versailles as commercial establishments and residential areas quickly absorbed the additional electricity. Canada's electrical generating capacity had gone up only 3 percent during the war, but the country made up for its lack of wartime expansion by almost tripling its kilowatt output in the postwar decade.[77] In 1925, the Canadian government enacted a minor duty on electricity sold to the United States. This marked the start of a gradual shift in which the American and Canadian grids became more interconnected and "electricity trade between the two countries changed from unidirectional firm power sales from Canada to the United States to interruptible power sales in both directions."[78]

Though Ontario Hydro had imitated existing technology at first, it broke new ground when it created its Niagara distribution system, particularly in the realm of high-voltage transmission networks.[79] According to Robert Belfield, the evolution and diffusion of the "Niagara system" pioneered by the Niagara Falls Power Company accounted for the timing and structure of Ontario Hydro's infrastructural evolution – and, in fact, made the public power movement feasible by offering standardized designs.[80] Within a few years of its establishment, Ontario Hydro began to build and acquire power stations, leaving the confines of Southern Ontario.[81] By the First World War, Ontario Hydro had the largest transmission system in the world, and it was still in the process of making interconnections across the province; by 1930, it would be the world's largest generator and distributor of electricity.

Ontario Hydro chairman Adam Beck had used the war situation as an opportunity to undermine the private generating companies operating at Niagara.[82] According to Christopher Armstrong, "Beck procured from the Ontario cabinet an order-in-council in June 1914 formally dividing up the 36,000 cubic feet per second of water which Canada could divert at the Falls for power purposes."[83] This division resulted in some unallocated water, which Beck pointed to as grounds for new generating facilities. Washington briefly objected, citing previous plans for a power dam in the lower Niagara River. But this amount fell within Canada's rights under the BWT. Other acts, instigated by Beck, were passed in Ontario that further gave Ontario Hydro an advantage over its private competition. Two new provincial statutes, the Ontario Niagara Development Act and the Water Power Regulation Act, collectively altered regulations for the public development of Niagara power in the province.[84] The private power concerns were opposed, but a New Year's Day 1917 plebiscite open to the concerned municipalities gave Ontario Hydro a mandate to start serving as an energy producer rather than just a distributor.

In 1917, Ontario Hydro purchased the Ontario Power Company, and then proceeded to install another conduit and two generating units in the generating station, enabling it to generate over 200,000 horsepower.[85] Ontario's total generation from Niagara was now 425,000 horsepower, which, in the estimation of the Queen Victoria Niagara Falls Park Commission, required a diversion of about 31,000 cfs.[86] With Ontario Hydro's 1922 purchase of Toronto Power's generating station, only the Rankine plant was left in private ownership on the Canadian side. And Ontario Hydro was already making efforts to provide power to rural areas.[87] This helps explain why hydro power represented 97 percent of the electricity produced in Canada as of 1920; in the United States, by comparison, until the end of the Second World War hydro power would account for only about one-third of the electricity produced in the United States.

Ontario Hydro had been considering the construction of a major power plant at Niagara even before the First World War. In early 1917, it broke ground on its new Queenston–Chippawa station, just south of the Niagara Escarpment. This publicly owned "temple of energy" took almost five years to complete.[88] Its placement allowed for the utilization of most of the 327-foot drop, or head, between Lakes Erie and Ontario, thus developing electricity more efficiently (in other words, the new Ontario Hydro station could produce two to three times more power than the Ontario Power Company's station and its head of 180 feet).[89] Ontario Hydro dug the nine-mile long Chippawa canal to carry water to the power station.[90] To supply the new canal, it installed intake works that drew water from the upper Niagara River into the Welland River, which was reversed and enlarged.[91] The new station and its connected infrastructure was built on land transferred from the parks commission and purchased from the International Railway Company and other property owners.[92]

The park commissioners insisted that Ontario Hydro build its new station in a way that would harmonize with its natural surroundings. This it did, and landscaped the area as well. Lead engineers Frederick Gaby and Harry Acres designed the reinforced concrete and steel station. The original budget was about $24 million, the final bill $84 million. The overrun was due to a variety of factors. The war caused manpower problems, for example, though at the height of construction over 10,000 workers toiled. Whereas the early twentieth-century power plants had been built mostly by animal and human muscle power, Ontario Hydro took advantage of many new electric construction techniques, such as the world's largest electrically operated shovels, and an electric railway. Many techniques for the canal and powerhouse, and the connected distribution system, marked important advances in hydraulic science and engineering, including the turbines, the control valves, and the canal control gate. "Building, maintaining and upgrading that infrastructure put Ontario Hydro's engineers in the forefront of this key technology of the second industrial revolution" and, particularly for concrete, "made Hydro a key player in the development of technical standards in Canada."[93] The new generating station itself did not represent any radically new overall design but was mostly a larger version of existing Niagara central stations. With dimensions of 560 feet long and 180 feet high, the equivalent of a nine-storey building, it was said to be the tallest commercial structure in the British Empire.[94]

Model studies, which would be essential parts of planning a few decades later, played a small and contested role. For the power canal, Ontario Hydro had engaged a University of Toronto professor to undertake hydraulic studies, including a model of the Niagara waterscape in miniature (a scale of 1:20) placed in the Dufferin Islands. The model was used for two summers to test five different types of water intake, with ice represented by floating bottles weighted with water and wood.[95] Ontario Hydro engineers deemed the model's performance "very satisfactory" for it behaved as predicted by their theories. Facing cost overruns, they brought in several outside engineering consultants, including famed hydraulic engineer Hugh L. Cooper. Almost two decades earlier, he had been integral to planning the Toronto Power Generating Station; over the course of his career, he was also involved in the design of a dam on the Mississippi River, the Wilson Dam at Muscle Shoals, and others abroad such as the Aswan Dam in Egypt and another across the Dnieper River in Soviet Ukraine.[96] When presented with Ontario Hydro's studies, however, Cooper demurred and called into question the model results "which in no sense represent working conditions."[97] He critiqued other aspects of the design for the Queenston–Chippawa station, and argued that, rather than a canal abstracting water from above the Falls, the lower river should have instead been dammed for power production.[98]

The Dufferin Islands had themselves already been considerably reconstituted using spoil from other projects, including new islands, shoreline changes, and artificial cascades. Whereas old maps show Dufferin as a side channel, or embayment, by now it was essentially a controlled water feature in a rock garden. Ontario Hydro dumped spoil and excess material in other parts of the river and surrounding landscape. Foreshore areas were extended into the river, and the dry riverbed at the western margin of the Horseshoe Falls was raised to the general level of the parkland, where low swampy places were filled in. The International Joint Commission later surmised that these changed the river levels and distribution of river flow, though the precise amount is impossible to determine without detailed records.[99]

The Queenston–Chippawa station went far over budget in large part because of the imperious practices of Chairman Adam Beck. The Gregory Commission was appointed to investigate Beck's practices since he was spending public money and there were worries that Ontario Hydro was becoming an unaccountable empire unto itself.[100] Thus, when the first generator at the Queenston–Chippawa development became operational in the last few days of 1921, it was under a cloud. A year later, four generators with a capacity of 240,000 horsepower produced 25-cycle power.[101] The total capacity of the plant would eventually be 650,000 horsepower.[102]

By 1923, generating stations on both sides of the border had a cumulative capacity that exceeded the legal limitations. Some would have to run at less than capacity or be relegated to standby status. Many industrialists, bureaucrats, and government agencies, including the Ontario parks commission, thought that more water could be abstracted for power, especially if compensating works were built, without detracting from what the chief of the US Army Corps of Engineers had termed the "unity of sensation." One option was to revise the 1909 Boundary Waters Treaty, specifically the Niagara clause. To get around existing limitations to a certain extent, the American power companies persuaded their government to interpret "Article V of BWT to mean 20 percent above the treaty figure for short periods but so that total generation for the day doesn't average more than treaty limitation."[103] Despite its dubious legality – under this principle, companies could have theoretically diverted the entirety of the Niagara River for a short period and still have met the "average" requirement in the longer term – Ontario eagerly adopted this interpretation. It was soon taking around 40,000 cfs on certain days, while the Americans routinely took 5,000 cfs above their Niagara allotment.

Conclusion

Niagara Falls at the turn of the twentieth century was a roiling cauldron of innovation and technological change. Transportation and other infrastructural improvements at Niagara, such as elevators, had made it much easier for the average person to visit the natural spectacle. In 1901, an estimated three million people, far above the normal one million visitors, came to the Niagara reservation, drawn by the Pan-American Exposition in Buffalo. Despite this growth, William Irwin writes that a steady stream of events, such as "middle-aged schoolmarm" Annie Taylor going over the Falls in a barrel, resulted in the "vulgarization of Niagara."[104]

Since Niagara Falls formed the border, developing it led to a great deal of legislation and diplomatic talks that oscillated between protecting and exploiting it. At length, the various federal, state, and provincial governments discussed Niagara internally and with each other. Niagara Falls played a central role in the legal regimes that evolved to help govern not only the border and natural resources but the two nations themselves. Niagara Falls symbolically shaped political developments in North America while political developments literally shaped Niagara Falls. To illustrate, bilateral Niagara diplomacy – technically trilateral, since Great Britain still handled much of Canada's foreign affairs on paper, if not increasingly in practice – paved the way toward smoother relations between the two North American nations and "helped foster an Anglo-American epistemic community."[105]

The Niagara waterscape was modified significantly in the years leading up to and including the First World War. By the end of the first decade of the twentieth century, both sides of Niagara Falls were flush with power-producing stations that took significant quantities of water away from the cataract. By the end of the second decade, an enormous power development had begun the process of replacing the original Ontario stations. We do not know the cumulative impacts of all the shoreline fills, island removals, diversion channels, and wing dams since so many were not formally approved or recorded. Turning former "wilderness" into an industrial and tourist zone, and then turning some of those industrial sites into parkland, meant that many aspects of the former fallscape have been lost.[106] That said, like the rest of the natural world, the Niagara River is not a static entity, frozen in time – it would have been changing itself anyway. The water of the Niagara River is a powerful force that, given time, can both submerge its own past and reclaim what humankind builds. The result of this collision is an ever-changing new normal that blends human and non-human nature.

3
Negotiating Niagara: Environmental Diplomacy and the 1950 Treaty

Well, the principle seems the same. The water still keeps falling over.

Winston Churchill, 1943

NOWADAYS, TERRAPIN POINT is a sloping triangle of manicured grass, traversed by stairs and paths, jutting out from verdant Goat Island at the eastern edge of the mighty Horseshoe Falls (see Figure 3.1). Soaked tourists crowd the railing, buffeted by spray. They pose for pictures and selfies, often theorizing about whether they could survive going over in some sort of contraption. Truth be told, I myself have often conjectured about the survivability of a plunge over the precipice. The falling water has a hypnotizing effect, mesmerizing in its pull. Imperceptibly and unconsciously, I find myself leaning a bit over the railing. Then the reality of the drop, the water's crushing power, and the rocks below induces a sensation of vertigo, snapping me out of my trance. Niagara has not lost its ability to instill feelings of the sublime – awe and terror commingled.

Paeans and panegyrics to Niagara's sublimity used to be regularly recorded in diaries, guestbooks, and postcards. Many of these reflections have survived to the present day. To give but one example, Helen Keller, whose sensory experience of the waterfall did not include sight, limned Niagara thus: "I wish I could describe the cataract as it is, its beauty and awful grandeur, and the fearful and irresistible plunge of its waters over the brow of the precipice. One feels helpless and overwhelmed in the presence of such a vast force."[1] At the time of Keller's observations, however, the Horseshoe Falls was wider than it is today, and a great deal more water tumbled over the rim. Terrapin Point did not even exist as a solid

FIGURE 3.1 Contemporary view from Terrapin Point

land mass until the 1950s. Before then, it was part of the cataract and known as Terrapin Rocks: water interspersed with a series of boulders and shoals (which had the appearance of turtle shells, hence the eponymous name), on which walkways and viewing stations had been built.

All these changes to the waterscape were designed to be hidden, but engineering alterations needed to be authorized by various governments, and this drawn-out political process was also hidden from the view of the general public. This chapter considers the political and engineering negotiations between 1920 and 1950 that laid the groundwork for Niagara's ultimate reordering. By the early 1920s, Ontario Hydro was operating its showpiece Queenston–Chippawa generating station, joining new generating units opened in New York State. Technological prowess and political willingness kept increasing the amount of horsepower wrung from Niagara. Experts optimistically believed they could figure out a way to get this energy while also preserving the scenic integrity of the waterfall – and thus the tourist industry as well. In other words: Could they divert even greater quantities of water but reshape Niagara Falls to disguise what they were doing? Several more rounds of environmental diplomacy would be necessary, however, before the grandiose schemes to fully rationalize Niagara Falls were realized. In 1929, 1932, and 1941, Canada and the United States signed accords that dealt with Niagara Falls, but none of them were ratified by the US Senate. Finally, the two countries signed – and then enacted – the 1950 Niagara Diversion Treaty.

Hydro Diplomacy

Despite the emphasis on power expansion at Niagara, there was plenty of alarm about aesthetics. Transnational engineering boards staffed with, or drawing from, various types of expertise sought to protect the waterfall's appeal while simultaneously increasing diversions for power, all while counteracting the continual erosion of the Horseshoe Falls. This required seemingly endless political negotiations, and relied on quantitative techniques to address what were often qualitative considerations.

In the span of a generation, erosion had deepened the middle of the channel, which drew water away from the flanks of the Horseshoe Falls. As a result, the flanks had only thin rivulets going over, and sometimes they went completely dry. The dried fringes attracted curious and daring onlookers, which created a potentially perilous situation. To rectify this, as part of a rehabilitation of the edge of the Horseshoe Falls in 1920, the Queen Victoria Niagara Falls Parks Commission removed a section of the rock flank so that water could flow over it. Part of the motivation for this was to return flow to the front of the portals for Ontario's scenic tunnels in the chasm below.[2]

By the summer of 1923, both national governments had agreed to an International Niagara Board of Control and appointed representatives.[3] The board's purpose was to coordinate the monitoring of the diverting entities.[4] This body began the process of compiling standardized information on the diversion situation and the efficiency of the existing generating stations. In 1925, it produced a report concluding that neither country exceeded in the aggregate its daily diversion limits. Further study indicated that since 1764 the rate of erosion at the southern apex of the Horseshoe Falls had been 5.1 feet per year; since 1917, the rate of erosion at the same place had been 7 feet.[5]

Ontario indubitably wanted higher diversions, as the province was having difficulty meeting power demands. Jurisdiction over streambeds and water power remained the pivotal dispute between the federal government and the province. Howard Ferguson, the Ontario premier, asked for stepped-up diversions in a letter to the federal minister of the interior, arguing that the limitations on Niagara water withdrawals in the Boundary Waters Treaty no longer reflected the amount that could be safely diverted without injury to other interests.[6] For its part, the United States was claiming that any diversion proportions should now be equal. A separate Niagara agreement or convention, permitting each country to increase its allotted water under the oversight of the new board of control, was attractive to many officials.

The Department of the Interior and Ontario Hydro created a Niagara Research Committee to prepare data to support Canada's case, and the Geological Survey of Canada undertook an assessment of Niagara Falls.[7] The United States engaged in similar studies. For example, Samuel S. Wyer, an associate in mineral technology at the Smithsonian Institution, authored an extensive study on the Niagara situation. Granted, the study's obvious biases – it was a thinly veiled attack on Ontario Hydro and public power since the author represented the private power lobby – clashed with the ostensibly objective institution that employed him and published the report.[8] Nevertheless, Wyer rhapsodized about why Niagara was the "most sublime" waterfall in the world.[9] A section was dedicated to proving that the American Falls were better than the Horseshoe Falls since the mist and spray of the latter spoiled the view, and "the abnormal concentration of water in the V-shaped notch, rather than its uniform distribution over the crestline leaves many bare spots and gives the crestline a ragged appearance." Conversely, the greater beauty of the American Falls was the result of "its unbroken crestline, generous supply of a thin sheet of water, uniform distribution and clear vision because of the absence of spray."[10] Wyer called for remedial works – a submerged dike, or artificial islands, above the cataract – to stop erosion. After all, the Niagara parks agencies were in the midst of improving facilities for the approximately 2 million tourists who visited both sides annually. For example, work started on a new Table Rock House, the parkway along the river was lengthened, and improved Falls illumination lights were installed in 1925.[11] The latter, in the words of historian David E. Nye, merged "electricity and one of the most powerful natural symbols in a new version of the electric sublime."[12]

Stoked by pressure from the various Niagara preservationist groups, American officials emphasized to Canada that they did not want to consider additional power development without safeguarding the scenic values. The two governments eventually tasked the International Niagara Board of Control with studying both diversions and scenic beauty.[13] An additional member from each side was added to the existing board, creating the Special International Niagara Board (SINB). The American inclusion of J. Horace McFarland indicated that Washington was serious about scenic values. The terms of reference asked the board to consider the extent to which erosion affected the Falls, whether any works or changes could remedy this, and what minimum flow of water would preserve the scenic beauty.

The SINB produced an interim report in December 1927. The report began by noting a base assumption: that the American Falls were more beautiful than the Horseshoe Falls because they had a more uniform curtain of water. The larger cataract should thus be apportioned to present a more consistent and less broken flow over the brink. Many thought that the Horseshoe Falls were "in danger of committing suicide" because of erosion, and that the "proposed thinning" would help check the erosion and also allow more water to be taken for power purposes. However, the SINB declared that the Falls' scenic value was not in serious and imminent danger if proper controls were put in place. The board attempted to debunk the "suicide myth," contending that the cataracts had always been receding at a consistent rate. Nonetheless, alterations could still be beneficial. After all, with lower flow volumes going over the Falls came the risk of white water replacing the "beautiful green" colour, which had traditionally been one of the principal attractions of the crescent-shaped waterfall and an interesting contrast to the whitish American Falls. But the board warned that heavy-handed intervention in the rapids above the Falls would destroy "the broad expanse of cascades which constitutes another of the chief charms of Niagara."

The comments about colour resulted from the unique methods the board employed to study the subjective and objective features of Niagara. Its approach included photographs and aerial surveys, and also a special Niagara telecolorimeter developed by Mr. A.B. Clark, an American colour physicist. The telecolorimeter, which resembled a road surveying device, "determined the cause and relation of the deep green appearance of the Horseshoe Fall, which is such an important feature of the great cataract[,]" to test for the desired "greenish-blue" shade. That specific hue, which resulted from deeper water going over the lip, was considered superior to the whitish colour resulting from a thin flow over the precipice. At least, the Canadian engineers considered it superior; the American engineers preferred the virtues of a white but well-distributed wall of water.[14] Additionally, as part of the SINB's investigations, the chief engineer of the Niagara Falls Power Company, John Lyell Harper, author of the 1919 book *The Suicide of the Horseshoe Fall,* tested various types of compensating works on a scale model he had built for this purpose. Reputedly the "largest working model ever built,"[15] it was outdoors and about four feet high and one hundred feet square. However, the SINB did not feel "that the specific performance of the works of this character and magnitude in the turbulent water of the upper rapids can be successfully predicted upon experiments on even the most accurately constructed small scale models." That said, the model demonstrations had "shown that remedial works which will greatly improve the scenic value of the Falls are eminently practicable."[16]

The board had been exploring the placement of weirs (also called submerged dams, barrages, or artificial cascades) in various places in the Chippawa–Grass Island Pool, which fed the rapids leading to the Falls.[17] It recommended the use of weirs to strategically divert water from the middle part of the Falls to the edges in order to improve the appearance of the crestline by distributing the water more evenly along its length. Removal of exposed boulders and shoals, along with strategic excavations and submerged deflecting weirs irregularly set at diagonal angles across the current, would intercept the heavier flow in the deep channel and push it toward the flanks. A submerged rock-filled weir, jutting out approximately one-half mile from the Canadian shore in the Chippawa–Grass Island Pool, would send more water to the American Falls, as ongoing diversions had thinned the rapids above it. The SINB thought that eventually a total of 100,000 cubic feet per second (cfs) could be withdrawn without tarnishing the look of the Falls, provided appropriate compensating works were put in place. In the meantime, it proposed that during the non-tourist season power companies be permitted to divert another 10,000 cfs from each side.

The Niagara Falls Power Company, for its part, had started looking at an American version of the Queenston–Chippawa development, across the gorge at Lewiston. (This company was also on the cusp of a major reorganization, along with other power companies, into the Niagara Hudson Power Corporation.) The Niagara Falls Power Company and Ontario Hydro offered to build the remedial works recommended by the SINB in exchange for receiving a combined additional 10,000 cfs in "experimental" diversions for power. "The dominating element of the Niagara problem," wrote Ontario Hydro officials, "is the discovery of a safe relation or division as between what may be called 'scenic beauty water' and 'power water.'"[18] The SINB concurred. The diplomatic services of the two countries began the work of reconciling drafts and hashing out treaty language.[19]

On January 2, 1929, both countries signed the Niagara Convention and Protocol (formally the Convention and Protocol between Canada and the United States Regarding the Niagara Falls and the Niagara River). This

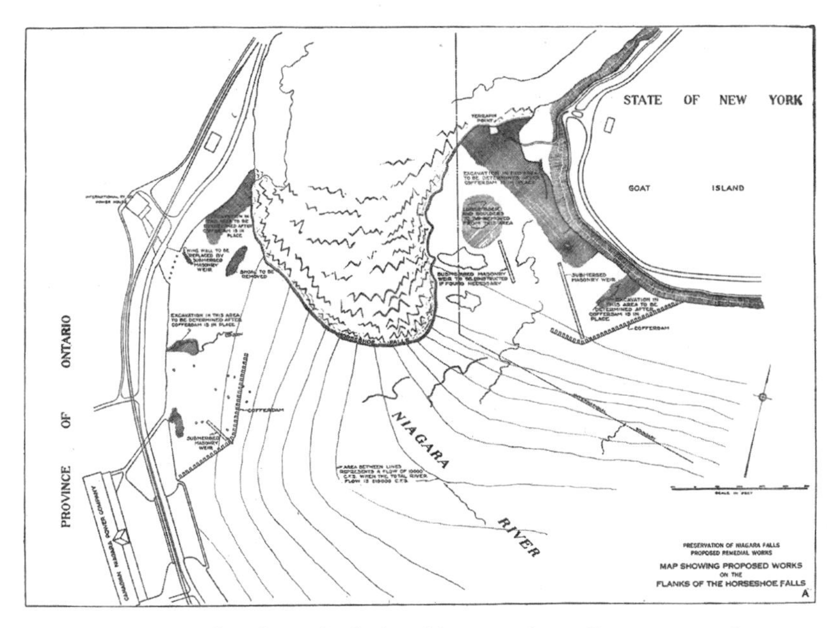

FIGURE 3.2 Proposed works on the flanks of the Horseshoe Falls. *International Joint Commission*

brief treaty had two key aspects: 1) construction of remedial works, and 2) the amendment of Article V of the Boundary Waters Treaty in order to permit additional temporary diversions of water. The agreement provided for remedial works to ensure unbroken crestlines at all times on both the Canadian and American Falls and an enhancement of their scenic beauty. Article II authorized "experimental" daily diversions of 10,000 cfs on both sides during the non-tourist season (between October 1 and March 31). These would continue for a seven-year period, after which they could be made permanent. The attached protocol noted that the construction of the remedial works should follow the Special International Niagara Board's previous recommendations, and be under that board's supervision.

Also appended to the convention were the Ontario Hydro and Niagara Falls Power Company's plans for the remedial works. Intended to be as inconspicuous as possible, the plans included a submerged rock-filled weir in the Chippawa–Grass Island Pool, raising the level there by a foot and pushing water over to the American Falls. Barrages, which could be progressively built as necessary, would be installed in the rapids above the Horseshoe Falls near the Canadian shoreline and Goat Island. A number of rocks and shoals, and the riverbed abutting both sides of the flanks of the Horseshoe Falls, would be excavated so that they would be lower, allowing more water to flow over them (see Figure 3.2).

This would all completely "reclothe" the flanks of the cataract and reduce erosion by deflecting water away from the centre of the waterfall. The additional seasonal diversions by the utilities would pay for these control works, which were projected to run somewhere between one-half and three-quarters of a million dollars – though it should be noted that this cost was sure to be far less than the profits that the utilities would reap from the expanded power diversions.

After some delay, however, the convention ran into trouble in the US Senate Foreign Relations Committee.[20] The main opposition stemmed from the public and private ownership elements. Namely, that the American share of the remedial works would be constructed by the Niagara Falls Power Company, which would receive the right to the additional temporary diversions. This was seen as an egregious concession to private power interests, with the beautification aspects intended mostly as a panacea.[21] The critics were not far off the mark, as the private utilities were a driving force, though not the only one, behind the 1929 convention. In February 1931, the Senate Foreign Relations Committee unanimously came down against the ratification of the 1929 Niagara Convention and Protocol on the grounds that power development and scenic preservation should be considered separately. But the main problem was that the treaty would give a private power company undue advantages. A survey of newspaper responses to the decision suggested strong sentiment for preserving the Falls, though not for the benefit of a private power company.[22]

It did not help matters that New York governor Franklin D. Roosevelt was miffed that his state's government was not consulted about the convention. He asserted New York's sovereignty over the American share of the Niagara River's water.[23] During the 1920s, Roosevelt's gubernatorial predecessor had made several attempts at a public power utility, but the Power Authority of the State of New York did not come into being until Roosevelt took office.[24] Though he was pragmatic about power issues, Roosevelt preferred publicly financed and government-controlled hydroelectricity – with Ontario Hydro as a key inspiration and model – which was distributed as cheaply as possible to aid rural areas.[25]

Despite the convention's failure, the SINB pushed on toward a final report. Released in spring 1931 and titled *Preservation and Improvement of the Scenic Beauty of the Niagara Falls and Rapids,* the report analyzed every aspect of Niagara Falls aesthetics and the potential emotional responses of tourists. Noting that the visual appeal of the Falls stemmed from diverse factors and perspectives, the report considered what exactly constituted "scenic grandeur." For example, was it the height, width, volume, colour, or lines that made Niagara such a spectacle? The report's sections on colour are fascinating. As a result of telecolorimeter studies, the engineers determined that "the optimum stream velocity to insure maximum persistence of the color curtain for the conservation minimum depth of 5 feet would be no greater than 15 feet per second."[26] Their respective water colour proclivities were likely tied to the shade exhibited by their respective cataracts: the report suggested that the American Falls were more appealing than the Horseshoe Falls, in part because of the excessive mist and spray at the latter, which forced onlookers to seek shelter when the wind was blowing in certain directions. Apparently folks didn't want to be inconvenienced by a wet waterfall.

All factors considered, the report concluded that a sufficiently distributed volume of flow, or at least the "impression of volume" – which was largely the result of an unbroken crestline or "curtain of water" – was more important than the actual volume of water dropping over the waterfall. The riverbed above the Falls should be manipulated in order to apportion the necessary volume of water to create the desired effect. Remedial works, in the form of submerged weirs and excavations, would achieve this purpose while allowing for increased power diversions of 20,000 cfs, bringing the total allowable to 70,000 cfs. This report, along with the 1929 treaty, would serve as the conceptual basis for attempts in the following decades to control the Falls of Niagara.

In the wake of the failed treaty, there were other prospects for Niagara power advocates, such as an international accord encompassing all boundary water issues, or at least those in the Great Lakes–St. Lawrence basin, with a St. Lawrence Seaway as the centrepiece. The Great Lakes–St. Lawrence Deep Waterway Treaty was signed on July 18, 1932. It outlined a twenty-seven-foot deep waterway, or seaway, as well as hydroelectric development in the St. Lawrence River. The treaty also dealt with a range of boundary water issues in the Great Lakes–St. Lawrence basin, including remedial works at Niagara Falls and other Great Lakes diversions, such as those from Chicago and the Albany River basin (the Long Lac and Ogoki diversions).[27]

After Congress resumed in autumn 1932, the Senate Foreign Relations Committee reported the treaty out favourably. The recently created Power Authority of the State of New York and the US Army Corps of Engineers (USACE) arrived at an arrangement outlining their division of responsibilities. Over the course of 1933, the treaty was debated and studied, though Roosevelt, now president, did not formally introduce it until 1934. The treaty had in the meantime become ensnared in US domestic politics and in the conflicts between regions and interest groups that had long plagued the project in the United States.[28] When the time came for a vote in March 1934, a majority of the Senate, but not the two-thirds majority required for ratification, approved the treaty.

Back to Niagara

Proponents continued to try to pass the Great Lakes–St. Lawrence Deep Waterway Treaty, or suggest amendments to make other members of Congress more receptive to it. But the prospects were not bright, and the rest of the 1930s was taken up by negotiations and bickering between the two federal governments and New York and Ontario. Only the outbreak of the Second World War broke the logjam, starting a chain of agreements that eventually culminated in the remaking of Niagara Falls.

In the mid-1930s, however, Ontario was growing impatient. Mitchell "Mitch" Hepburn had won the premiership of Ontario in the 1934 election with a "back to Niagara" campaign – Ontario was running short of power, and though it had other generation options and would exploit many of them, Niagara Falls was the simplest and quickest means of additional power production. The campaign had stressed Ontario Hydro's purported mismanagement and malfeasance, and immediately after his victory Hepburn cleaned house in Ontario Hydro's upper management and passed a new Power Commission Act. He soon wrote to Prime Minister Robert Bennett, pushing for movement on an international Niagara accord that would permit increased water diversions. He was also irked by Ontario Hydro's need to continue honouring the export contracts it had inherited when it took over the American-owned stations at Niagara Falls. Additionally, the previous Ontario government had signed electricity import contracts with Quebec utilities that had become an albatross around the provincial neck.[29]

Hydroelectricity exerted an influence on Ontario's politics and statist evolution even beyond its statistical significance, contributing to a provincial "hydro myth."[30] Granted, hydro power was the source of about 90 percent of the electricity generated in Canada in the 1930s, led by Ontario. But industry and manufacturing consumed the majority of the electricity, while household consumption (especially outside urban areas) remained low, relative to the US, until about the midpoint of the twentieth century. Indeed, average Canadian households relied on the organic energy regime (e.g., coal and wood) much longer than Americans.[31]

Roosevelt did not like the power aspects of the 1929 convention, and in any event seemed more concerned about the scenic works. This reflected the greater urgency in the United States over preservation of the beauty of the Falls, which the Canadian government admitted.[32] The president wanted to address Niagara Falls as part of revised St. Lawrence Seaway discussions: subsume Niagara within a wider agreement to replace the unratified 1929 and 1932 accords.[33]

Hepburn's Liberal government faced obstacles in its attempts to tear up or modify the Quebec electricity contracts, making it difficult for Ontario to determine whether it would likely be facing an electricity shortage or surplus.[34] Ontario Hydro officials indicated that Niagara power productivity could be supplemented by additional water from the Ogoki and Long Lac diversions, which Ontario had first formally proposed in 1925 and which were included in the unratified 1932 treaty.[35] The plan was to reverse these two rivers so that they flowed south into Lake Superior instead of north to the Albany River system and eventually to Hudson Bay. The additional water would benefit lake and shipping levels all around the Great Lakes basin, the Long Lac reversal could assist in the transportation of pulpwood, and hydro-power stations could be built into these rejigged northern Ontario waterways. But the main attraction was that the diversions replaced the water decanted from the Great Lakes–St. Lawrence basin by the Chicago Diversion, and the extra volume of water put into the basin could be used at downstream Canadian power stations, particularly at Niagara.

Hepburn hatched a plan that would supply Ontario with more power from Niagara, supplemented by the diversions into Lake Superior and new stations on the Ottawa River, allowing him to take the electricity purchased from Quebec and export it for profit to upstate New York State.[36] Since the Ogoki–Long Lac diversions altered the levels of a border water, Ontario technically needed American consent. Hepburn wanted a separate Niagara treaty, not a wider St. Lawrence–Great Lakes agreement, whereas FDR wanted the latter. With Prime Minister Mackenzie King's concurrence, Roosevelt signalled that he would refuse to approve the exportation of additional Ontario electricity to the United States in order to bring Hepburn to the bargaining table. Hepburn apparently had financial interests in not just the Quebec power companies but also private power companies in the United States, including the Niagara Hudson Power Company, which produced all the Niagara power on the New York side and imported power from Ontario. These conflicts of interest are primarily what inspired Hepburn's strategy, including his opposition to comprehensive and public developments on the St. Lawrence, until the Second World War broke the stalemate.[37]

Even though Niagara negotiations did not get far during the 1930s, the Niagara environment was far from static during the decade. Since the mid-1920s, the commissioners of New York's State Reservation at Niagara had added 2,000 acres to the original reservation, including the Whirlpool, Devil's Hole, Lewiston Heights, Beaver Island, and Buckhorn Island state parks.[38] More dramatically, several large rock slides took place in the 1930s. Just before dinnertime on January 17, 1931, folks in the vicinity of the Falls heard a low rumble. Many felt a slight shock. Seventy-five thousand tons of rock had fallen from the centre of the American Falls, just below Robinson Island, extending for a distance of 280 feet along the face of the waterfall. On August 13, 1934, the Horseshoe Falls shed 45,000 tonnes of rock in a slide on the Goat Island side. In December that same year, a crack was discovered in a large overhang directly in front of the Table Rock House. This remaining fragment of Table Rock was a popular viewing site that included a parking lot.[39] In 1935, the Niagara Parks Commission (the new name for the Queen Victoria Niagara Falls Park Commission as of 1927) decided to blast off the offending section of rock. This process was not only a tourist draw, as thousands of people showed up to witness it, but, according to the parks commission, "actually improved the view of the Horseshoe Falls from Table Rock" (see Figure 3.3).[40]

The expansion of Ontario's hydroelectric facilities during the First World War had chewed up the parkland on the Canadian side immediately above, below, and abutting the Horseshoe Falls. Water conduits and trenches disfigured the grounds, but the utilities had resisted for many years the park commission's calls to rectify the situation. Finally, by the early 1930s, the various water conveyances had been hidden and the area landscaped.[41] Ontario parks officials made a number of improvements, such as construction of the Oakes Garden Theatre and the Clifton Gate Memorial Arch.[42] The Canadian Niagara Power Company constructed a gathering weir at the north end of their water intake in the 1930s.[43] This improved the company's water head for the generating station while cutting down on ice problems, and was also designed to improve the scenic effect in line with the SINB recommendations.[44] The weir pushed water toward the Canadian shore and provided "a better curtain of water over the Canadian end of the Horseshoe Falls."[45] Additionally, during construction of the weir, certain rocks and obstructions that interfered with the

FIGURE 3.3 Contemporary view of Table Rock

uniform distribution of water were removed, as was a waterworks pump house.

Ice jams had been a motivating factor for the Canadian Niagara Power Company's changes, but they were not alone. The Ontario Power Company's station in the gorge was beset by the so-called ice bridge that formed most winters at the base of the Falls. However, that station got off easy compared with the Honeymoon (Falls View) Bridge, which in January 1938 collapsed because of ice buildup below the Falls. (This was only the most recent bridge collapse at Niagara, but the first and last in the twentieth century.[46]) Ground was broken for the replacement structure, the Rainbow Bridge, in May 1940. It was said to be the longest hingeless arch bridge in the world, and designers placed its piers out of reach of the river ice.[47] Fully mastering ice would remain a chimera, however. It was one of the few elements of the Niagara environment, and the one form of water that engineers did not think they could fully control.

The cataclysm of the Second World War, beginning in 1939, demanded more and more electricity.[48] Premier Hepburn was suddenly open to the St. Lawrence project, provided Ontario could export power and complete its proposed diversions north of Lake Superior.[49] Ontario Hydro kept pressing for piecemeal development and looked elsewhere in the province for hydro-power possibilities. Since upstate New York was clamoring for more power exports from Canada, Roosevelt permitted Ontario to divert an extra 5,000 cfs at its Queenston–Chippawa plant and raise its electricity exports. This was conceivably the real reason for Hepburn's reversal. On both sides of the Niagara River, governments made infrastructural improvements to generating stations and their intakes in order to maximize capacity. Desperate for more electricity, the United States sought to create an enlarged power interchange with Canada, and Ontario specifically.[50] In fact, both the United States and Canada led the globe in expanding hydroelectric capacity during the war.

The extra 5,000 cfs of production at Queenston–Chippawa would come from the Ogoki–Long Lac diversions, which the United States agreed to in the fall of 1940. The reversal of Long Lac would be quick and simple because of the previous work done there, while the Ogoki diversion took about two years of construction. Over the course of the war, further diplomatic notes were exchanged, allowing Canada to increase the size of these Albany River basin diversions and increase power diversions in the Niagara peninsula.[51] On completion, Ogoki and Long Lac together encompassed the largest diversions *into* the Great Lakes basin.[52]

A comprehensive Great Lakes–St. Lawrence agreement remained the end goal, however, particularly for Roosevelt. The president had decided that a new accord should take the form of an executive agreement rather than a treaty, since the former needed the concurrence of only a majority of both houses of Congress, compared with a two-thirds Senate majority for the latter. On March 19, 1941, the two countries signed the Great Lakes–St. Lawrence Basin Agreement. The central goal was the construction of a St. Lawrence deep waterway and power development; in addition, Article IX provided for the construction of remedial works in the Niagara River, mostly in line with the recommendations of the 1929 convention.[53] Unlike the 1932 treaty, Ontario would now be responsible for the cost of Niagara remedial works, up to a total of $838,000. Another difference was the aforementioned substitution of a permanent 5,000 cfs diversion per side once the remedial works were finished, rather than a temporary diversion of 10,000 cfs each that could be discontinued after seven years.

As the ink dried on the 1941 agreement, Ontario and New York officials moved to enact its Niagara provisions. They recommended that New York could immediately take an additional 5,000 cfs, and Ontario 3,000 cfs, to be sent to idle generators to benefit war production. Once they ascertained that this removal of 8,000 cfs had no adverse scenic impact on the Falls, New York could take another 7,500 cfs.

The two countries tentatively agreed to move ahead with part of the remedial works referred to in Article IX: a submerged stone-filled weir to raise the level of the Chippawa–Grass Island Pool by a foot, more or less.[54] Although this barrier would ostensibly help the scenic effect, both New York and Ontario wanted it chiefly because it would aid power production by raising water levels and preventing ice blockages at the generating station intakes.[55] Other remedial works could be considered after the weir was completed. The United States also indicated that it would not object to Ontario Hydro's reduction by half of its cross-border export of nearly 80,000 horsepower to the Union Carbide and Carbon Corporation.

However, the 1941 executive agreement had not been ratified in the United States, and it would suffer the same fate as its predecessors. The agreement was attached to an omnibus bill, but before the House of Representatives could approve it, Japan attacked Pearl Harbor. With the United States joining the war, Congress deferred the bill until the end of the global conflict.

The two nations kept moving ahead with the Niagara provisions. Through diplomatic notes, they agreed to ignore the limits on Niagara River diversions so that all generating stations on the Niagara frontier could be used to their full capacity for the duration of the war.[56] Further notes were exchanged in 1944 that permitted Ontario to divert an additional 4,000 cfs, though the US Senate, through some oversight, never formally ratified this arrangement. Most, if not all, of the resulting Ontario power was exported to the United States.[57] During the war, diversions rose well above the limits set by the Boundary Waters Treaty, probably to 56,500 cfs for Canada and 32,500 cfs for the United States, though the net diversions were about equal, the equivalent of 41,500 cfs each, after taking electricity exports into account.[58] Undoubtedly, both nations would have approved more if necessary, but these water volume caps reflected the full capacity of the existing generating stations.[59] Ontario Hydro was also allowed to divert 2,500 cfs from the Welland Ship Canal (technically allotting it to the Albany River diversion contributions), which was used for extra production at its upgraded DeCew plant west of Niagara Falls.[60]

In 1942, work started on the new weir, a low underwater dam that allows water to flow over the top. The weir would jut out from the Canadian mainland into the middle of the river, a mile upstream from the Falls. It was jointly designed and built by Ontario Hydro and the US Army Corps of Engineers under the supervision of a construction subcommittee. A University of Toronto hydraulic laboratory built and ran tests of a model of the weir, but the Corps of Engineers deemed its results "unreliable," at least in part because of organizational and national engineering chauvinism.[61] An artificial island and causeway were needed to begin construction: a cableway was strung from a tower on this new island to the Canadian shore, using a kite and piano wire. Materials were then moved along the cableway and deposited in the appropriate location. The weir constricted the channel, which caused the water level upstream to rise and redistribute the flow toward the flanks of the river. Indeed, the bared fringes, which had been for decades identified as prime culprits in robbing Niagara of its beauty, were even more naked because of recent low levels on the Great Lakes. On that note, officials hoped that the weir would help limit the impact of diversions on Lake Erie's level to about three-fourths of an inch.[62]

The weir was 90 percent complete by the end of 1942, but was not fully finished until 1947 since planners wanted to delay so that they could evaluate the weir's impacts in light of changing Great Lakes levels and the seasonal growth of aquatic weeds.[63] The weir reportedly raised the water level in the Grass Island Pool a bit less than one foot. The power utilities gushed that the weir resulted in better power production (for example, Ontario Hydro claimed it enabled the production of over 16,000 extra horsepower) and improved the appeal of the American Falls by roughly doubling its flow of 5,000 cfs.[64]

However, different appraisals concluded that the weir essentially just offset the additional wartime diversions. For example, a 1945 report said that the weir raised water levels by 0.7 feet, whereas the increased power diversions lowered water levels by 0.73 feet.[65] Niagara park officials pointed to different negative scenic effects caused by the new weir. Because of a higher flow rate in the channel leading to the American Falls, banks eroded to such an extent that some trees washed away and other low areas flooded, submerging or threatening paths and roads.[66] Moreover, after redistributing some of the river flow from the west to the east side of the river, the parks commission argued that the weir resulted in a small decrease in flow over the Horseshoe Falls, compounding what was already lost through diversions. In subsequent years, the weir's effectiveness declined, though this may have stemmed from higher lake levels as well as ice chopping off part of the weir's crest.[67]

In conjunction with the weir installation, engineers considered other remedial works, such as more submerged weirs directly above the waterfall, or a gated control structure at the outlet of Lake Erie, but could not come to a consensus. The US Army Corps of Engineers considered the desirability of these interventions "doubtful" because of cost and intrusion on the tourist view. Instead, it averred, shallow excavations would have been a preferable way of achieving the same purpose. A Corps of Engineers member proposed that leftover rubble from weir construction be used to create temporary dikes or build artificial islets, and it seems that some of this occurred.[68]

Niagara Diversion Treaty

There was tremendous growth in manufacturing in the Niagara region during and after the Second World War, as well as a tourist explosion – the postwar era was the height of honeymooners.[69] Ontario, the leading manufacturing province, was already experiencing power shortages, and Ontario Hydro predicted that it would have adequate power supplies only until 1952.[70] Between 1945 and 1957, Ontario's population grew by almost 50 percent while the personal income of its citizens doubled to an estimated total of $9.2 billion and the gross value of manufactured goods produced in Ontario surpassed $10.5 billion.[71] There was substantial industry on the Canadian side of Niagara, much of it American-owned with the products sent south of the border. There, the electricity demands of US commercial and defence industries, particularly those along the Niagara River, kept mounting. Niagara Falls was the leading electrochemical and electrometallurgical centre in the United States, boasting a veritable who's who of American heavy industry. By the 1970s, an estimated 700 industrial operations – chemical plants, steel and aluminum mills, oil refineries, and others – were discharging over 250 million gallons of wastewater into the Niagara River every day.[72] As world war gave way to a fragile peace that slid into cold war, the United States stood unchallenged as the world's leading power. Electricity was indispensable not only for the North American military arsenal but also for the competition with communist nations in standards of living and material goods that would prove just as decisive. Maximum power production along the border at Niagara remained a top priority.

But the provisional nature of the Niagara diversion situation was also leading to financial uncertainty. New York bond houses were unwilling to issue securities for new industries that wanted to set up in the Niagara frontier if the electricity they relied on was not guaranteed.[73] The "temporary" power diversions agreed to during the war continued afterward, though on a technically illegal basis.[74] About 88,000–89,000 cubic feet was being siphoned off the Niagara River, leaving roughly 110,000–128,000 cfs of the river's average volume to go over the Falls.[75] This was close to the proportions that the 1929 convention had recommended, provided that these diversions were accompanied by extensive remedial works. Most of these works had not been executed, however, and the appearance of the Horseshoe Falls was suffering.

Both sides did not want to give up the extra power but were wary of seeking formal permission – namely, through Congress – because power development was a politically sensitive topic. That Congress was unlikely to agree to any accord that maintained the existing ratio of diversions in favour of Canada gave officials in Ottawa and Toronto pause. Electricity export contracts for power sent across the border would also need to be cancelled or renegotiated in the event of a new diplomatic accord.[76] American and Canadian diplomats therefore agreed to keep the diversions going after 1945 with mere correspondence exchanges, such as another round of diplomatic notes in 1948 renewing Canada's 4,000 cfs Niagara diversion (whose power was exported to the United States) and a 2,500 cfs DeCew diversion (see Table 3.1).[77]

Table 3.1 Niagara power developments in 1949

Name of plant	Approximate net head (ft)	Installed/ rated capacity (kW)	Maximum water use (cfs)	Average daily allotment (cfs)
United States				
Schoellkopf	215	365,000	23,400	23,700
Adams	135	80,000	8,800	8,800
Total		445,000	32,200	32,500
Canada				
Queenston	294	373,000	17,030	15,200
Ontario	180	138,000	11,100	10,700
Toronto	137	108,000	15,500	15,000
Rankine	135	80,000	10,600	10,600
DeCew	280	149,000	7,600	3,930
Total		848,000	61,830	55,430

Note: Note that DeCew drew an additional 2,500 cfs during non-navigation season, and that the Schoellkopf installed capacity had been as high as 373,400 kilowatts.
Source: Drawn from National Archives and Records Administration II (NARA II), RG 59, Decimal File, 1945–49, box 3306, file 711.4216-N51/9-149: Federal Power Commission, "Possibilities for Redevelopment of Niagara Falls for Power" (Washington, DC: Bureau of Power, Federal Power Commission, September 1949).

Power needs kept growing all the same. The 1941 agreement, meanwhile, languished in Congress. In fact, a Senate Joint Resolution was put forward calling for the elimination of the Niagara articles from the agreement. Considering the stalled status of the St. Lawrence Seaway, not to mention that power could be developed much more quickly at Niagara than on the St. Lawrence because of existing infrastructure, consideration turned to a treaty focused just on Niagara Falls.[78]

Ontario Hydro had been planning for a new Queenston–Chippawa power plant for some time. This new plant was feasible not only because of the prospect of increased Niagara diversions but also because several of the water rental agreements that had been made in the early twentieth century were set to expire: the Canadian Niagara Power Company agreement on May 1, 1949, and the Ontario Power Plant agreement on March 31, 1950. These plants had both been taken over decades earlier by Ontario Hydro, which planned to use their water rights at a new and more efficient hydroelectric station.

To the annoyance of Canadian federal officials, Robert Saunders, the new chairman of Ontario Hydro, took it upon himself to travel to Washington to lobby on his commission's behalf with the Federal Power Commission, PASNY, and the Niagara Hudson Power Company.[79] He repeatedly pressed powerful cabinet minister C.D. Howe and the rest of Prime Minister Louis St. Laurent's government to negotiate a Niagara treaty. In Ottawa, a Niagara treaty was enticing because it could relieve some of the pressure to get immediate movement on the St. Lawrence question. At the end of May 1949, the avuncular prime minister wrote to President Harry Truman, appealing for St. Lawrence action but also asking for a separate treaty on Niagara Falls diversions.[80] Truman replied that his government would be happy to reopen the subject, but only after the necessary Federal Power Commission report was submitted. To help smooth the way, Saunders returned to the United States for another round of negotiations, this time with someone from the Department of External Affairs in tow.[81]

The FPC report, titled "Possibilities for Redevelopment of Niagara Falls for Power," was completed in late September 1949.[82] It was written by Robert de Luccia, chief of the FPC's Bureau of Power, in consultation with PASNY, the Niagara Hudson Power Company, and Ontario Hydro. The FPC report went over, in minute detail, the diversion and power production history and contemporary situation on both sides of Niagara Falls. It sidestepped the delicate issue of whether the power should be developed by public agencies or private interests, and was geared not to offend "nature lovers" concerned about the preservation of the Falls.[83] The report contended that the Ogoki–Long Lac diversions balanced out the power exported from Ontario across the Niagara River, and thus a new treaty should be built on the principle of an equal national distribution of diversion volumes. The report also recommended a new pumped-storage plant, probably on the cliff at Lewiston, New York.[84]

The FPC report outlined a crucial reversal for the persistent power-versus-beauty tension. The Boundary Waters Treaty had set a maximum volume of water that could be abstracted for hydroelectric production, with any water in excess of that volume sent over the Falls to enhance its scenic appeal. The 1929 Niagara Convention incorporated that ratio. But the FPC report, based on the Special International Niagara Board's recommendations, switched that set-up around by establishing minimum volumes for water that contributed to scenic preservation, with anything above that amount available for power generation.[85] Officials from both nations would adopt this approach in the ensuing diplomatic negotiations.

In October 1949, the United States informed Canada that the president now wanted to initiate the drafting of a treaty.[86] Truman was chiefly concerned with the power aspect and, as a New Dealer, preferred that the federal government control hydro power and sell and distribute it at the lowest possible rates, with preference given to public utilities and cooperatives.[87] American diplomats offered the FPC report as a basis for negotiations. The State Department felt it had the upper hand in negotiations because of Ontario's urgent need for power.[88] It also wanted quick action so that an agreement could be presented to the Senate in early 1950. The treaty negotiation process would in fact be concluded in about three months – perhaps because, as one American diplomat said of Canada-US relations, "I don't know if it is that we are so friendly because we are so frank, or so frank because we are so friendly."[89]

Both countries scrutinized the host of important technical and political questions that needed to be cleared up (see Figure 3.4). The placement of hydro-power station intakes, for instance, was vital since placing them above the submerged weir would effectively nullify its benefits.[90] How would the water be split between the two countries: according to water usage or power potential? How would existing export contracts be handled? And what about the Canadian Niagara Power Company, the one remaining privately owned plant on the Ontario side? Would control works be necessary at the head of the Niagara River so that there would be no adverse impacts on Lake Erie levels?

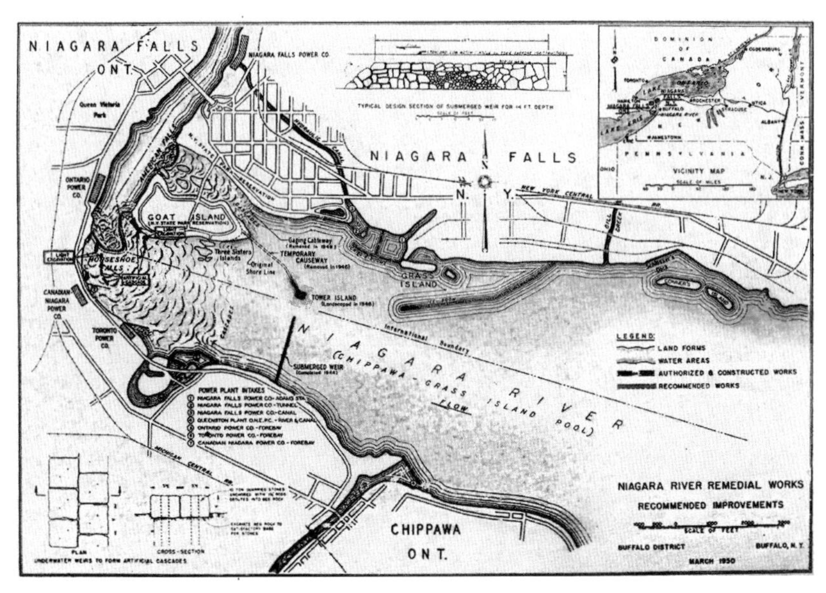

FIGURE 3.4 Recommended improvements – Niagara River remedial works. *Dwight D. Eisenhower Presidential Library and Archives*

Most in the Canadian government, as well as Ontario Hydro officials, thought they should accept the US demand for an equal division of water diversions, though for the time being (i.e., until the export contracts were concluded) Canada should insist on a greater allocation of water. Other officials contended that the water used to produce electricity that was exported should be counted as water diverted by the United States. They reached a compromise. Including a statement about "equal shares" of the water would, some felt, give the Canadian government the appropriate basis for refusing power exports once the United States was fully using its portion of the water volume. To further address this concern, they later arrived at an understanding that Canada would, in conjunction with signing the treaty, send a letter to the United States stating that it would cancel the export contracts when the United States was prepared to accept its share of the new diversion rates. The State Department in turn would not object, and both sides would endeavour to keep this arrangement quiet.[91] Canadian officials did not want the DeCew diversion from the Welland Canal included in the treaty since, like the Chicago Diversion, it predated the Boundary Waters Treaty. But they ultimately decided that it was not worth jeopardizing a wider arrangement to get their way on this issue.[92]

To save some face, Ottawa successfully argued that the treaty should recognize Canada's exclusive right to the water from the diversions into Lake Superior.[93]

The Truman administration had proposed a minimum flow of 100,000 cfs over the Falls during tourist hours, but Canadian officials wondered whether that should be closer to 80,000 cfs, even if just on an experimental basis, so that the scenic aspect shared more of the hardships of low water.[94] This was mostly a smokescreen for Ontario Hydro's worry that, since the Niagara River was running at about 170,000 cfs in 1949 due to low lake levels, if 100,000 cfs was guaranteed to flow during tourist hours, there would not be enough water left over to supply the volumes that Ontario and New York power stations had been diverting since the war.[95] There was debate about how to define these tourist hours in terms of precise start and end times. The US government, for its part, was keeping an open mind on the subject; it was "apprehensive of possible agitation by the scenic beauty supporters, including the Special International Niagara Board."[96]

While the mandarins left most of the scenery specifics to be worked out later by others, they did discuss the potential for coloured lighting to cover up curtailed water volumes. They spoke of the waterfall as if it were a faucet to be turned on and off according to aesthetic whim. "In the evening the Falls are floodlighted and 50,000 cfs may not be enough water to provide an adequate spectacle at that time," wrote the Canadian secretary of state for external affairs. "It may prove necessary to maintain a flow of 100,000 cfs up to midnight in the tourist season," he continued; "on the other hand, it is probably unnecessary to turn on the full flow at sunrise. It may be better to define 'night-time' as the period from midnight to 9:00 a.m. We shall have to discuss this problem with the authorities responsible for lighting the Falls."[97] After further discussions, the "night-time" hours shifted, primarily to align with normal tourist visitation hours. In terms of floodlighting, officials believed that an uninterrupted sheet of water was an even more important consideration than water depth, and settled on 50,000 cfs as the minimum amount that would "adequately reflect the beautiful light display" as well as sufficiently flush ice.[98]

Such high-level discussions revealed the politicization of expertise and uncertainty – and the extent to which imperfect or incomplete knowledge did not get in the way of proceeding with political or engineering plans. Equally fascinating were other exchanges during the treaty talks about the ways the water bodies of the Great Lakes–St. Lawrence basin did, and did not, operate as one connected system. Regarding other diversions throughout the basin, American and Canadian officials debated the length of time it would take water to get from Lake Superior to the Niagara River, and the impact of evaporation on the volume that would ultimately arrive downstream.

With the studies that produced the 1929 Niagara Convention and Protocol still serving as a basis, representatives of both governments met in Washington in December 1949 for several days of exploratory talks to reconcile the draft treaties both sides had prepared. Following these discussions, the two diplomatic services worked on minor alterations to the wording of a draft treaty, while the preamble was more substantially revised. In the December talks, the American side had favoured permitting power diversions above 81,500 cfs only on a seven-year experimental basis to test the scenery impacts of diversions at the highest contemplated levels, but the Canadian side argued for making these permanent from the start, and eventually won out. Ontario representatives contended that International Joint Commission oversight of the remedial works should be optional rather than mandatory, but the Americans held firm, pointing out that scenic preservation advocates needed to be appeased if this treaty was to pass the Senate.[99] It was clear that the US government strongly desired a simple treaty that avoided complicated technical regulations because of the need to get it through the Senate. In light of this, the State Department cautioned the Department of External Affairs that any mention of private companies should be left out of the treaty.[100]

Meanwhile, a Canada-Ontario agreement also had to be signed to govern their respective domestic rights and obligations when it came to implementing a Niagara treaty. The provisions of a federal-provincial agreement had been discussed in a desultory manner over the previous year, and with an international treaty imminent, coming to such an agreement was fairly straightforward. Ontario would receive the water diversions specified by the treaty, and agreed to build and pay for the Canadian share of the remedial works.[101]

On February 27, 1950, the two nations signed the Niagara Diversion Treaty (technically it was titled: Treaty between Canada and the United States of America Concerning the Diversion of the Niagara River). The text of the finished form of the treaty was quite brief, containing ten short articles. The bilateral accord virtually equalized water diversions while restricting the flow of water over Niagara Falls to no less than 100,000 cfs during daylight hours (of what they deemed the tourist season: 8 a.m. to

10 p.m. from April to mid-September, and 8 a.m. to 8 p.m. during the fall), and no less than 50,000 cfs the rest of the time (i.e., during nighttime and at all times from November to the end of March).[102] The remainder of the water could be redirected for hydro power, and would be divided equally between the two countries. Thus, the two countries could collectively take between half and three-quarters of the total flow over the Falls; at a flow rate of approximately 200,000 cfs, only 50,000–100,000 cfs would go over the brink of Niagara Falls.

Canada had given up the more favourable Niagara distribution, 36,000 cfs to 20,000 cfs, which it had enjoyed under the 1909 Boundary Waters Treaty. But this was not the negotiating defeat it might first appear.[103] First, we need to recognize the context: in addition to the inherent power asymmetry between the two nations, the reality that anything other than an equal volume distribution was not going to get through the US Senate, the political impossibility of including the Chicago Diversion, and Ontario's urgent need for greater hydro-power production. Second, there were several linked areas where Canada made out well. It had secured the under-the-table agreement on cross-border power exports. The treaty gave Canada credit for Ogoki–Long Lac – though for only 5,000 cfs, even though these two diversions together have debouched into Lake Superior an average of about 50 percent more.[104] Since New York hydro facilities were not yet in a position to absorb the full American allotment of the new diversion rates, Ontario would end up getting extra water until the early 1960s. Finally, the economic benefits from remaking the Horseshoe Falls were more likely to redound to the benefit of Ontario's tourism industry, which turned out to be the case.

The Boundary Waters Treaty called for any measures affecting boundary water levels to be approved by the International Joint Commission, unless the two federal governments made a direct diplomatic agreement. The latter was the case for the Niagara Diversion Treaty, though in such cases the Boundary Waters Treaty still required that the IJC approve the construction or maintenance of any infrastructure that affected border flows. The 1950 treaty therefore specified the creation of shared remedial works following the Special International Niagara Board's 1929 report, with the designs supervised by the IJC. The remedial works would "distribute the flow so as to produce an unbroken crestline on the falls" and be finished within four years. The 1950 treaty superseded the limitations on diversions in the 1940s diplomatic notes, and amended Article V (terminating the third, fourth, and fifth paragraphs) of the Boundary Waters Treaty. Incidentally, this was the first and only time that this 1909 agreement has been modified.[105] The Niagara Diversion Treaty would run for fifty years, a period primarily designed to allow for amortization of the bonds to finance power redevelopment. After this half-century span expired in 2000, each country could withdraw from the treaty with one year's notice.

But the Niagara Diversion Treaty still faced US Senate ratification, which was far from a certainty given that the Senate had nixed the 1929, 1932, and 1941 US-Canada agreements concerning water resources in the Great Lakes–St. Lawrence basin. Truman referred the Niagara treaty to the Senate for approval in March after some handwringing about tactics.[106] In early May 1950, twin bills were introduced in the Senate and House of Representatives. Learning that Canadian legislative approval might help grease the wheels in the US Senate, the St. Laurent government decided to go ahead with parliamentary ratification.[107] After a few days of consideration in the House of Commons in mid-June, the resolution received rapid and unanimous approval.

No comparable legislation had yet been approved in the US Senate, however, since Congress was preoccupied with developments in Korea and with McCarthyism. Canadian ambassador Hume Wrong made the rounds of US officials, stressing Ontario's drastic power needs and warning that failure of the treaty would "be apt to have an adverse effect upon United States–Canadian relations in the broader sense."[108] Lester Pearson, the secretary of state for external affairs and future Nobel Prize winner, followed up with a phone call to express his "intense anxiety" about the situation – exacerbated by "present world conditions," including inaction on the St. Lawrence Seaway – and underscore the "irritation up here" that would soon result.[109] The Ontario Hydro chairman let it be known that Pearson had privately told him that Canada would have to "take retaliatory action," with Columbia River developments being a "logical example."[110]

In transmitting the Niagara treaty to the Senate, Truman stated his preference that Niagara power be publicly developed and tied to a St. Lawrence project. This raised concerns within the Senate about whether the Niagara treaty was a front for either private or public power, or federal versus state development of the power.[111] But the attraction of increased Niagara power for national defence swayed many. By the end of July, it looked like the treaty was ready to be reported out favourably to the Senate by its Foreign Relations Committee.

However, the Senate Foreign Relations Committee attached a reservation to the 1950 treaty that was purely about a domestic matter. This stipulation reserved to Congress the right to determine the disposition of the water volume that the United States received under the treaty – this was an attempt to avoid opening the can of worms that would accompany a decision on whether Niagara power would be developed by federal, state, or private interests. This so-called Wiley Bill read:

> The United States on its part expressly reserves the right to provide by Act of Congress for redevelopment, for the public use and benefit, of the United States' share of the waters of the Niagara River made available by the provisions of the Treaty, and no project for redevelopment of the United States' share of such waters shall be undertaken until it be specifically authorized by Act of Congress.

This seemed innocuous enough, but in putting off the issue to the future, this clause would turn out to be a political landmine in the United States.

Nevertheless, on August 9 the US Senate unanimously ratified the Niagara Diversion Treaty, including the reservation. After some deliberation, Canadian officials decided that this reservation did not require Parliament to reapprove the treaty. The treaty came into effect on October 10, 1950.

Conclusion

Agreements to change the literal brink of Niagara Falls put Canada and the United States on the metaphorical brink of a new era, albeit one that had been several decades in the making. The First World War dangled the possibilities of what a more lax diversion regime would allow in terms of energy production. Both nations set out during the interwar years to have it both ways: increase hydro-power diversions without sacrificing the aesthetic beauty of the Falls. Between 1929 and 1941, they forged several international agreements, based on the reports of the binational engineering boards they formed to study how to produce both beauty and power from the Falls of Niagara. But the greatest obstacles during the interwar years proved to be political rather than technical, as these international agreements could not get through the US Senate, foundering on the rocks of politics and statecraft.

These political barriers largely faded away when the world plunged into another global conflict and producing extra energy became paramount. During the Second World War, the two nations ignored limits on Niagara water diversions, and made binational changes to the Niagara waterscape to facilitate greater abstractions. And once the electrical genie was out of the bottle, it was too difficult to put back in: neither country wanted to give up the extra production when the war ended. In the next half-decade, the United States and Canada worked their way toward permanently enshrining a new water regime, and a new fallscape, through the 1950 Niagara Diversion Treaty. In that accord, the two nations agreed to increase their diversions over and above the levels that already obtained before 1950; the difference was that a suite of remedial works would offset heightened diversions. Thus, in the logic of planners and engineers, taking more water would actually improve and enhance the scenic appeal of the Falls because it would coincide with a facelift of sorts for Niagara. Of course, repeated incisions risked death by a thousand cuts – or a thousand diversions, as it were.

4
Empowering Niagara: Diversions and Generating Stations

For man is the magician ... he dips his hand into Niagara, and gathering a few drops from its waters, educes a force from Niagara itself, by which he confronts and defies it.

George William Curtis, 1851

THE OBSERVATION DECK at the Robert Moses Niagara Power Plant offers a powerful vista (see Figure 4.1). Named after the famed New York City builder, whose tenure as head of the Power Authority of the State of New York is overshadowed by his other roles, this power station was the largest in the Western world when it was completed in the early 1960s. On the Ontario side stand the two Adam Beck generating stations. The American and Canadian stations face each other across the Niagara Gorge, resembling oversized concrete cash registers locked in an endless staring match. Looking downriver, the horizon line is dominated by the international Queenston–Lewiston Bridge, through whose arch is visible the drop of the Niagara Escarpment, after which the Niagara River broadens out as it runs to Lake Ontario.

Walking back to my car from the observation deck of the Moses station, which is part of the elaborate visitor centre, I look at the forebay, conduits, pumped-storage generating facility, and reservoir that together constitute New York's Niagara Power Project. Then I drive several miles upstream, past the Whirlpool and various sets of rapids, almost to the Rainbow Bridge. I park my car in a lot by the Niagara Gorge Discovery Center, where prior to the 1950s the hydraulic canal would have pooled water in a reservoir supplying the factories and generating stations that made Niagara a household name for industrial development. Not far away are some remains of the original ALCOA aluminum plants; this company

FIGURE 4.1 Contemporary view from observation deck of the Robert Moses Niagara Power Plant.

was at one time the largest user of hydroelectricity in the world. I hike down the gorge to the river level, partially following the bed of the long-abandoned tourist railway called the Great Gorge Route.

Recently, the New York Niagara Parks Commission installed an elevator, obviating the need for a strenuous hike. But the elevator is not in service on this occasion, and I prefer to walk anyway. My destination is the ruins of the Schoellkopf station, which for a time was the largest generating station in the world. The Schoellkopf station was crushed in June 1956 by over 100,000 tons of rock. Now the site is used as a docking and repair site for the *Maid of the Mist* boats that take tourists for a cruise up to the waterfall. The remnants of a major rockfall are obvious even today. The textured stone wall that covers the penstocks, intended to make the generating station blend in better with the rock face, is still there. So are the old tail-races, which propelled the water into the river after spinning turbines. I hop onto the remains of the sill in front of the tailraces, peering into the bowels of the old plant. Turning to look upriver, I see the Rainbow Bridge framing the western edge of the Horseshoe Falls and the many towers in the Fallsview area on the Canadian side (see Figure 4.2).

FIGURE 4.2 View looking upriver from tailraces of the old Schoellkopf generating station

The last time the Schoellkopf station was generating power, the privations of the Great Depression and the Second World War were just starting to fade in the rear-view mirror. People were ready to do many things again, like get married, buy a house in the suburbs, and start a family. This baby boomer formula often meant a stop at Niagara Falls, which in the postwar period was at the peak of its status as a honeymoon destination. Its attractiveness for those with a "Just Married" sign hanging from the rear fins of a ridiculously long sedan was further enhanced when Marilyn Monroe arrived to make the 1953 film *Niagara.* This movie, featuring the actress's famous "long walk," helped propel her to superstardom, and made Niagara more fashionable still.

While Niagara Falls boasted cultural power, power of many different kinds was on offer. In the 1950s, New York State and the Province of Ontario embarked on massive hydroelectric projects that significantly altered the Niagara landscape. Their respective public power utilities, PASNY and Ontario Hydro, were headed by two Bobs – Robert Moses and Robert Saunders, respectively – famous for getting things done. Ratification of the 1950 Niagara Diversion Treaty unleashed a wave of engineers, who flooded across the Niagara peninsula. Government planners had been eagerly waiting for decades, updating and fine-tuning their blueprints. The remaking of Niagara Falls was at hand. Two new power stations, one on each side of the border, would channel most of the Niagara River's water for electricity production, while a range of remedial works would physically change the river and cataract to hide the impact of diverting all this water. The power stations are the focus of this chapter, and the remedial works the subject of the next chapter.

Beck No. 2

Energy scholars have separated energy regimes into "stocks" and "flows," with the latter generally consisting of "organic" energy – such as wood, water, and human/animal muscle power – while stocks (coal, petroleum, uranium) are generally "mineral" energy forms. Unlike carbon sources of energy such as coal and petroleum, which society deploys as non-renewable stocks of fossil fuels, to get hydro power humans harness theoretically renewable water flows, usually rivers, and transform them into electricity. Flowing water produces power that needs to be used on demand and at a scale that justifies the construction and maintenance of the system designed to convert and deliver that power as electricity. Fossil fuels can be removed from their place of origin and then burned and utilized at a desired location. But hydroelectricity typically has to be generated at the site of falling water, and the resulting electricity can be transported only if and where transmission wires make that possible, such as North America's massive electricity grids.[1] Since hydro power involves both water and electricity, it is a hybrid form of energy regime: it has features of both flow and stock, both mineral and organic.[2]

Powered by a variety of energy types, the 1940–70 period witnessed tremendous growth and prosperity in North America. The concomitant rise in income and class equality has cheekily been called the "great compression."[3] However, as economist Robert Gordon shows, this growth was actually not as intense as that which took place during the first four decades of the twentieth century, the most revolutionary period of concentrated expansion in human history. Put another way, while the second half of the twentieth century was the Great Acceleration globally, in America the post–Second World War age was not the period when most of the acceleration took place, but rather the era when it reached its top end speed.[4] These great leaps forward were characterized by inventions that could happen only once – such as electricity, which was an indispensable growth factor, revolutionary in both private homes and factories.[5] By 1940, most

American homes were networked and had electricity, heating, running water and sewers, telephones, and so on. Electricity not only enabled mechanization at home and work but also radically changed life since it upended nature's daily cycle of light and darkness.[6]

At the midpoint of the twentieth century, Canada produced 87 billion kilowatt hours of hydroelectricity (a kilowatt hour, or kWH, is a measure of electrical energy equivalent to the power consumption of 1 kilowatt for one hour) while the United States produced 50 billion kWh. Ontario had developed over 40 percent of its hydroelectric potential, compared with 22 percent for the rest of Canada.[7] Granted, with the exception of Niagara power, Ontario Hydro began using its hydroelectricity facilities for peaking purposes rather than baseload generation, which would come from sources of thermal power such as coal, natural gas, and nuclear.[8] Canada has since been among the top, or at the top, of global per capita users of energy in general and electricity specifically. Of course, the United States has long outpaced all other nations in total energy consumption. The US had been consuming electricity from all sources at much higher rates than Canadians prior to 1939, and its energy production further soared during the war through interconnections and greater generating capacity, and continued to rise tremendously in the decade after the war.

The negotiations that produced the 1950 Niagara treaty speak to the harmonization and integration of the two countries that had taken place during and after the war. When the treaty officially came into effect, Ontario was most concerned with starting work immediately on its power phase – not only because the plans for the new power station were ready to go but because Ontario Hydro craved electricity. The moment the treaty entered into force, it launched into full-scale construction of its new Queenston–Chippawa power station, though the Ontario legislature did not pass a Niagara Development Act authorizing its construction until April 1951. The new power facility would be called the Sir Adam Beck Generating Station No. 2, after the founding Ontario Hydro chairman. The original Queenston–Chippawa Generating Station was renamed the Sir Adam Beck Generating Station No. 1 and remained operational right beside the new development.[9] Beck No. 2 would be generating power by 1954. Signifying the pace of technological change, the new station was almost twice as long as its older but smaller sibling, which had been the largest generating station in the world when it was built three decades earlier.

Moving ahead with Ontario's new power station required negotiations and discussions with a wide range of other entities – aspects of the power development as well as remedial measures required consultation and coordination with the International Joint Commission, the Canadian and American federal governments, and New York State. Moreover, work on Beck No. 2 necessitated extensive meetings with local officials to handle the practical logistics of an enormous construction zone in and near a small city.

Then there were the logistics and negotiations involving a tricky environment. An internal Ontario Hydro report noted that the design and construction of Beck No. 2 presented the engineering division with "many difficult structural problems, some of which could not be accurately solved because of the lack of reliable data."[10] One of these problems was rock creep, shrinkage, and movement.[11] Over the previous half-century, the existing power plants had all experienced some rock shifts on the order of over half an inch.[12] Engineers took measures to address these concerns, and diamond-drilled thousands of holes to test the subterranean conditions. Erosion was another obvious concern close to the waterfall. For example, because of the gnawing effects of erosion, the scenic tunnels below Table Rock that lead behind the Canadian Falls had been receding.[13] They were closed in 1939 because of wartime security concerns, and in 1944 it was discovered that there was only 5.5 feet left between the tunnel and the face of the gorge. Studies estimated that the rock had dissipated about 25 feet in twenty years. The Niagara Parks Commission rightly deemed this condition a threat to public safety and began constructing new tunnels, which were opened in August 1945. Out of concern for the welfare of tourists, the park commissioners also extended ornamental iron and stone walls south of the Horseshoe Falls for about four hundred feet, and much of the bank of the upper river was riprapped with stone.[14]

The new hydroelectric station was placed just south of the existing Beck station, near the Niagara Escarpment. This siting allowed for a maximum drop, or head. When completed, the sixteen-unit powerhouse was capable of generating 1,328,000 kilowatts (note that between the construction of the first and second Beck stations it had become more common to use kilowatts rather than horsepower to rate hydroelectrical output).[15] Because it was roughly five miles downriver from the Falls as the crow flies – and closer to seven miles following the route of the Niagara Parkway – water would have to be conducted a long way to the new station. A cut-and-cover method or open canal would not work for most of the route because of urban buildup. Thus, two parallel tunnels would run for 5.5 miles under the City of Niagara Falls, from the upper Niagara River near Chippawa, to a 2.5-mile open-cut canal leading to the power plant forebay (see Figure 4.3).

FIGURE 4.3 Constructing penstock trenches for Beck No. 2. *Ontario Power Generation*

Models had been used to help determine certain aspects of the tunnels, foreshadowing their use on Niagara remedial works.[16] After leaving the water intakes above the Horseshoe Falls, the tunnels went under the city and then climbed back up to a higher elevation with the aid of inverted siphons and pumps. The tunnels were also built to accommodate a third power development in the future.[17] Ontario Hydro had to get approval for its route and methods from the Niagara City Council as well as surrounding townships. In early 1951, Ontario Hydro shared with the council the location of the tunnels and the need to close some streets for up to two years. It pledged to settle any claims from damage to the streets from heavy construction equipment and traffic.

The new tunnels worked in concert with the extant hydro-power canal built for Beck No. 1.[18] All of these conduits fed into the existing forebay for Beck No. 1, which was enlarged. This required new techniques, including novel uses of air curtain cushion blasting by Canadian Industries Limited.[19] From the forebay, the water fed into penstocks put into shallow trenches cut in the face of the gorge, where the water would drop about 295 feet – significantly more than the drop of the Falls and the head in

FIGURE 4.4 Construction of the Beck pumped-storage generating station.
Ontario Power Generation

other Niagara power stations in Ontario.[20] Rock scaled from the cliff was used to construct a cofferdam in the river so that construction work at the station's base could be done in the dry.

The 1950 treaty placed limitations on how much water could be diverted during daytime tourist hours, precisely when power demands were generally the highest. But the engineers had a plan for maximizing water availability during the day: build a reservoir and pumped-storage plant (see Figure 4.4).[21] The reservoir was surrounded by raised walls made of rock-fill and an impervious clay blanket. Raising the reservoir above the surrounding landscape in this way was more cost-effective than digging deeper, but these raised walls meant that the new reservoir was off limits to recreational activities. The pumped-storage plant was situated on the southern side of the reservoir and worked by pulling up water from the forebay into the reservoir for storage; it later sent water in the reverse direction, through six pump turbines each connected to an electricity generator, creating a maximum of 170,000 kW. The advantage of this set-up was that the reservoir could be filled during nighttime, when power demands were low and it was allowable to reroute higher volumes of water from the river. Then additional power could be generated from the pumped storage during daytime, when power needs were higher (peak demand). The pumped-storage plant would in effect turn the reservoir into an enormous aqua storage battery.

The city council was generally supportive of this new economic driver, but understandably had a number of worries about community impacts. Would the influx of workers bring in hooligans? Who would be responsible for schooling and medical care for the workers' families? Would Ontario Hydro open up a commissary to supply its workers – and in doing so compete with local merchants? Ontario Hydro reassured the council that it would build housing for several thousand workers, and swore that only a low percentage were troublemakers. It would also build a thirty-bed hospital, with potential for expansion.

The three "Hydro Cities" were miniature communities with recreation halls featuring billiards, bowling alleys, church services, movies, cafeterias serving 5,000 meals a day, fire departments, and sanitary sewer services.[22] The power commission purchased twenty-two dwellings and built thirty others.[23] Some houses were moved by special vehicles. Workers could later purchase their rentals at market value, though without credit for rent already paid. However, compared with some of Ontario Hydro's recent and ongoing power project builds (for example, on the Ottawa River and in northern Ontario), the house rental rates in the Niagara region were much higher than workers expected. This disparity was partly due to the proximity to major urban centres. Workers circulated a petition to express their displeasure. Ontario Hydro countered that the current rental rates were already subsidized by about 20 percent.[24] Moreover, from its perspective, undercutting the local rental market would generate "serious antagonism" toward it and the idea of public power ownership.[25]

Ontario Hydro also needed to expropriate a range of properties, though not on a scale close to what it was about to undertake while remaking the St. Lawrence River; in fact, when negotiating with those displaced by the St. Lawrence hydro-power development, Ontario Hydro Chairman Robert Saunders would cite the fair treatment of the Niagara community.[26] Ontario Hydro consulted other agencies that had tackled similar relocations, such as the Tennessee Valley Authority and the US Army Corps of Engineers.[27] One of the larger transactions came in 1954 when Ontario Hydro expropriated property from A.A. Holman in Stamford and Niagara townships, compensating him to the tune of $250,000.[28] Ontario Hydro made a range of grants and payments to Stamford Township, some of which would cover various municipal services, including the costs of extra policemen, garbage disposal, and fire protection.[29] Ontario Hydro's director of property, Harry Hustler, was regarded as a fair man despite his unfortunate surname. Depending on the amount and the issue, the power commission sometimes deemed it easier to just grant some money rather than quibble and use up social capital, particularly if the recipient was deemed deserving. In the end, Ontario Hydro spent several hundred thousand dollars on payments of various types to the surrounding communities.

But there were certainly limits to Ontario Hydro's generosity and, as the public utility frequently claimed, it could not overpay for property because of its moral obligations to the taxpayers. Moreover, it was conceptually restricted by its logic of progress, which tended to frame everyone that stood in the way of its construction plans as selfish obstacles to electrical modernity and the greater good. In response to some claims it denied, Ontario Hydro pointed out that "the overall influx of Commission employees into your Township has already proved to be an asset to the community at large in terms of increased purchasing power and stimulation of local trade and development."[30] Citing a booming mill rate, the Council of the Township of Stamford asked for more assistance from Ontario Hydro, but Saunders balked at some of these requests. Though the power commission would pay for the burial of any workers who might perish, for instance, it could not be responsible for the education taxes of those who opted to live outside the three worker camps.[31] During the course of the Niagara project, three workers needed post-sanatorium care, but Ontario Hydro argued that it was not liable.

At its peak, the construction force was 7,655-strong. All types of trades were represented: cement finishers, electricians, pipefitters, welders, and so on. Ontario Hydro signed a general union agreement, a "master contract," with its various workers and contractors. For instance, seventeen American Federation of Labor (AFL) construction trade unions in Ontario formed the Niagara Development Allied Council AFL, a unique occurrence at that point in Canadian labour history. Each side made major concessions: the craft union abandoned the forty-hour work week for a forty-four-hour week, since the target project completion date was 1954, while Ontario Hydro conceded the union shop and waived the necessity for certification. Aside from a handful of engineers and specialists, thousands of workers were covered by the blanket agreement.[32] Another collective agreement was inked in October 1953, adding fringe benefits concerning sick leave, health insurance plans, savings and insurance plans, and an additional statutory holiday for the Queen's birthday.[33] In addition to construction workers, the project required all types of support staff,

FIGURE 4.5 Aerial view of construction on Beck No. 2 in 1956. *Ontario Power Generation*

such as nurses and mechanics. Coordinating a project of this complexity and magnitude also took a small army of office staff. Outside of certain segments of the ancillary staff, the construction zone was an exclusively male domain. Consequently, the waterfall and river environment were often gendered as female, as something potentially powerful and generative but irrational and out of control – in other words, it just needed to be harnessed and domesticated by male expertise.

The creation of the Beck power plant and allied infrastructure had an enormous footprint and changed the landscape along the Niagara Gorge. The area housing the generating station was leased from the Niagara Parks Commission, but much of the reservoir area was not parks commission property and Ontario Hydro compensated the communities for the lost tax assessment. The excavated rock and earth would have filled enough fifteen-ton dump trucks which, if placed bumper-to-bumper, would have stretched east to west across North America. Enough cement was used to fill a train of boxcars that would have covered the nearly fifty miles to Hamilton; the amount of steel would have been sufficient to build almost five Rainbow Bridges or 18,000 cars.[34] In aerial photos, all the enormous construction vehicles – trucks, cranes, scrapers, shovels – looked like a bustling ant colony (see Figure 4.5). Some construction vehicles fell into the Niagara River and were never recovered.

Excess material from construction was dumped into spoil zones that covered about 640 acres, which included filling in nearby streams and ravines such as Smeaton's Cove.[35] Ontario Hydro was also authorized to dump excavated material in the bed of the Niagara River.[36] This dumping did not escape attention, however. Noting that similar disposal in 1919 had changed the river levels and distribution of flow, the IJC worried not only about whether dumping was legal according to the Boundary Waters Treaty but also whether such dumping would invalidate riverbed surveys that had already been performed.[37] According to internal Ontario Hydro reports, about 750,000 cubic yards of material was deposited on the riverbed and banks on both sides of the Queenston–Lewiston Bridge, presumably because this placement would interfere least with hydro power or remedial works – though that amount could alter the river's flow regime. This dumped material was graded and levelled and the parts above the high-water mark were planted, seeded, and landscaped.[38] These new formations also hindered the activities of six commercial fishermen, with whom settlements were reached.[39] In response to complaints that Ontario Hydro's water intakes had ruined a prime bathing beach near the confluence of the Niagara and Welland Rivers, the commission created a swimming area.[40]

A new parkway along the gorge ran right above the two power stations.[41] Ontario Hydro covered the expense of rerouting other roads. The utility built twenty-five miles of road, including routes to reach the disposal areas that received approximately 30 million tons of excavated rock and earth. Cutting a road down the flank of the gorge to the power plant construction site was particularly challenging. As will be discussed later in this chapter, a number of efforts were made to show off the power both of the generating station and the state, the rationalization and improvement of the waterway serving as a display of the government's legitimacy and vitality. But the state's expertise and projection of infallibility was certainly not absolute, for slight errors or uncertainties could be compounded into much bigger issues. For instance, a problem soon arose with the reservoir dike, stemming from a spot where a telephone pole had been located prior to construction. The hole allowed water to get into the underlying rock, which contained many seams. This little oversight ended up costing $450,000 to fix, as well as an estimated $150,000 in lost revenue.[42]

Nonetheless, we should not overlook the fact that, once completed, the new station and its connected infrastructure did operate as planned. Beck No. 2 was officially opened in August 1954 by the Duchess of Kent. By the end of 1955, the initial twelve generating units were in operation. An additional four units were generating by 1958. Ontario Hydro placed the six units in the pumped-storage generating station into service in 1957. The new Beck plant had a much higher head and used water for electrical generation much more efficiently than the Canadian Niagara and Toronto Power plants, both of which became backup stations. The Rankine generating plant remained in full-time service, still privately owned by American interests. And across the gorge, the Americans were trying to get their own version of Beck No. 2 up and running.

Getting Licensed

Although as of 1954 it had not yet been tasked with building new Niagara power works that could use the American water share under the 1950 Niagara Diversion Treaty, PASNY was optimistic. In anticipation of this megaproject – along with the St. Lawrence power works, the first that PASNY would build – it had recruited famed New York City builder Robert Moses to be the public utility's new chairman on top of his myriad roles in the Big Apple (see Figure 4.6).[43]

This was the era of big dam building in North America. Many hydroelectric projects, especially in the western United States, were touted as multi-purpose because they included features such as navigation or irrigation. But these factors were not part of the equation at Niagara, which was unique in that the scenic and park aspects of the undertaking received equal billing with power production. Moses had relatively little interest or background in hydro power, but he had extensive experience in the types of parkland and parkway developments that were part of these power projects, which he called "conservation in the truest sense."[44] To him, building a massive new generating station was as much a means as an end. And, of course, Moses was the ideal candidate for moving people and buildings out of the way in northern New York, as well as attracting the hundreds of millions of dollars of bond sales to finance the project. But before he could build the new generating station that would come to bear

FIGURE 4.6 Robert Moses (middle) at generating station.
Niagara Falls Public Library, New York

his name, Moses would have to run a political gauntlet that would stretch on for much longer than he initially anticipated.

This was not Moses's first official involvement with planning the Niagara landscape: his long-standing role as chairman of New York's State Parks Council meant that the State Reservation at Niagara fell within his domain. In the 1920s, he had sparred with the commissioners of the Niagara state park about several things, including parkways.[45] An additional motivation for Moses to run PASNY was therefore the chance to make up for not fully getting his way decades earlier. Interestingly, Moses's eight years as the head of PASNY are an underappreciated part of his career, and even Robert Caro's magisterial biography *The Power Broker* accords but a handful of pages to it. The historiography of Moses is still largely defined by Caro's work, which paints a decidedly negative portrait of the man. In recent decades, there has been a palpable revisionist bent arguing that Moses was not the "evil genius" Caro makes him out to be, invoking a "man of his times" defence and pointing to New York City's contemporary success as testament to Moses's many projects.[46] Moses's time with PASNY, however, indicates that Caro's interpretation is closer to the mark.

Moses already had many critics by the 1950s, and not everyone was thrilled about his appointment to lead PASNY.[47] David Cort, writing in *The Nation,* called him a "dictator" and said that "the prospect of Moses, unsupervised, unchecked and unaudited, running wild with Niagara Falls and the great Saint Lawrence River is, to put it quickly, disquieting."[48] Moses's presence and temperament – and even just the anticipation of his legendary ripostes – raised the hackles of some while heartening others who saw him as the right man for the job. To be sure, with his appointment as chairman, PASNY increasingly lost its commitment to public power beyond emphasizing industrial development.[49]

In the fall of 1954, PASNY was already well into preliminary planning so that it would be in a position to quickly submit the necessary plans to the Federal Power Commission if it was tapped as the agency to develop the new Niagara power. PASNY started working with Uhl, Hall and Rich, an experienced engineering firm that attracted staff from agencies such as the Bureau of Reclamation and, importantly, had no "entangling alliances" with private utilities.[50] The firm worked all over North America, including Tennessee Valley Authority dams and the St. Lawrence Seaway and Power Project. Ontario Hydro engineers and workers had been involved in the St. Lawrence undertaking too, as well as previous Niagara developments and many other hydro-power projects throughout Ontario, such as the recently completed Des Joachims power dam on the Ottawa River.

At this point, a half-billion-dollar hydro-power plant, along with a pumped-storage plant, was envisioned just south of the Niagara Escarpment in the Village of Lewiston, New York, immediately north of the city limits of Niagara Falls and directly across the gorge from the new Beck development. To finance it, PASNY had to rely on public revenue bonds.[51] Preparations were already well advanced for a new parkway to run from Niagara Falls to Lake Ontario, which was unsurprising considering Moses's penchant for covering everything with concrete.

Moses's diatribes were equally predictable. He began railing against the private utilities' record at Niagara, which "on our side have bedeviled the Niagara River for seventy-five years. In spite of a record of shameless exploitation, they still have the effrontery to claim that the only question here is that of making power." "The record shows," he continued, "that the private companies have never had any genuine interest in the preservation of these public assets."[52] Moses lauded the history of public development on the Canadian side, though whether he really cared about the public ownership principle, or the public, is quite debatable. After all, the master builder was simultaneously being accused of giving too much St. Lawrence electricity to private interests at too low a price. Nonetheless, he argued that the United States should emulate the forward-looking Canadian side, where people received cheaper electricity and where, he claimed, Niagara aesthetic and economic values complemented each other.

But PASNY was still awaiting the decision of the 84th Congress, which had to enact legislation to delegate authority to the Federal Power Commission to license the Niagara development. A bill, HR 5878, was introduced to that effect, but it was only one of many competing bills, some of which promoted various forms of public development, whereas many others were not so friendly to PASNY or public power.[53] In fact, the House of Representatives passed a bill giving a five-company consortium the rights to develop Niagara power, though it did not get Senate approval. Some favoured private enterprise because of ideological or economic commitments; others did not like the fact that public developments did not have to pay taxes and, according to some, gave preference to groups that also avoided taxes. This so-called preference clause was a sticking point for many. The president of the Rochester Gas and Electric Corporation wrote in the *New York Herald Tribune* that the ability of public utilities to avoid tax was "most unfair and un-American."[54] Moses could not resist the bait, writing back that "the development of the state's power at Niagara by a private corporation would be an outrageous gift to that corporation."[55]

Multi-purpose public developments such as the Tennessee Valley Authority and Bonneville Power Administration were alternatively brought up as good or bad precedents, as were other ongoing hydro-power developments, such as those at the St. Lawrence and Hells Canyon.[56] Hoover Dam offered an example of public/private compromise.[57] Some Republican legislators from the Empire State who favoured private development were accused of trading votes with southern Dixiecrats: public power for school integration.[58] After the expansion of public power under Presidents Franklin D. Roosevelt and Harry Truman, President Dwight D. Eisenhower now sought to curb and contain it. His "partnership policy" involved cooperation between the federal government, states, other subfederal agencies, and private enterprise.[59]

In 1956, the Senate approved a bill calling for development by PASNY, but the House of Representatives adjourned without taking action. Then

Figure 4.7 Collapse of the Schoellkopf generating station caught on camera.
Niagara Falls Public Library, New York

an act of God intervened, levelling both the political playing field and a major generating station. On June 7, 1956, Niagara Mohawk's Schoellkopf generating station, located in the gorge not far downstream from the Falls, collapsed (see Figure 4.7).[60]

Dating back to the nineteenth century, the water infrastructure at the Schoellkopf power complex had been enlarged over the years, along with additional generating stations (see Chapter 2). Water was drawn from the upper Niagara River, feeding a canal and tunnel leading to a forebay at the top of the gorge, from where it dropped into three adjacent but connected stations that collectively generated 334,800 kW, making Schoellkopf at the time the largest Niagara power plant and the largest privately owned generating station in the world. The precise cause of the collapse has never been conclusively established. An earthquake has been mentioned as the culprit, but the likely cause was water seeping from the plant's unlined canal basin into the gorge wall, which was honeycombed by old tailraces. Small rockfalls already had everyone on alert, and after five rockfalls in quick succession around suppertime on June 7, two-thirds of the station dramatically crumbled into the river. Luckily, there was only one fatality. Chris Nelson, who worked in the station at the time, remembers: "All of a sudden I saw a crack open up ... and I made tracks, headed down the floor to the north. I thought maybe I could get the elevator up, but it was too late ... there were rocks falling behind me, and I was running. The water was chasing me down the floor."[61]

The plant's destruction caused a massive power outage in the City of Niagara Falls, which at mid-century had a population of about 100,000. Niagara Mohawk decided to rebuild unit 3-A, which was still standing, though damaged by fire and flood. It purchased 25-cycle power from Ontario Hydro to sell to industrial consumers,[62] but this was nowhere near enough since Schoellkopf had generated about one-half of all electricity in New York State. The Schoellkopf disaster created a wider power vacuum, literally and metaphorically. It would eventually help bring the congressional impasse over the Niagara power licence to a head.

Even after the plant collapse, Moses was planning an end run to circumvent Congress. PASNY had been studying the possibility of asking the courts to rule that the treaty reservation attached by the Senate to the 1950 Niagara Diversion Treaty, which called for Congress to designate the entity to develop Niagara power, was constitutionally invalid. PASNY's application was based on the premise that the reservation illegally attempted to amend or repeal part of the Federal Power Act. It was also related to the basic question in the ongoing Bricker Amendment dispute: could the president or the Senate put something into an international treaty that affected purely domestic laws? In August, PASNY filed an application to the FPC requesting that it be granted authority to develop the new power, as well as two-thirds of what had been generated by private interests at the Niagara frontier, including that of the Niagara Mohawk Power Company.[63]

Within a few months, however, the FPC ruled that it did not have jurisdiction to determine the validity of the treaty reservation and dismissed the PASNY licence application.[64] PASNY immediately petitioned for rehearing of the application; the petition was rejected a month later since there was no new cause.[65] PASNY then took the matter to the United States Court of Appeals for the District of Columbia Circuit.[66] In June 1957, the court handed down a majority ruling that the Senate reservation dealt with a purely domestic concern and was *not* part of the treaty. In

FIGURE 4.8 Water conduits on New York side. *Niagara Falls Public Library, New York*

other words, the FPC could, in fact, issue licences for a Niagara power project. After years of delay and confusion, matters now moved quickly. On August 1, Congress passed bills creating the Niagara Redevelopment Act, which designated PASNY as the responsible agent. By the end of the month, the FPC had issued a conditional licence to New York State to develop Niagara power.[67] The Niagara Redevelopment Act represented a compromise between private and public power. Niagara Mohawk's water

FIGURE 4.9 Planning the water conduits. *Niagara Falls Public Library, New York*

rights went to PASNY, and the pioneering but outdated Adams plant, which had been on standby, was retired. Niagara Mohawk would receive one-quarter of the PASNY station's output. The rest would be distributed between private industries/utilities and neighbouring states, and a "preference clause" reserved some power for municipally and county-owned systems, state departments and agencies, defence agencies, and rural electricity cooperatives.

But with a final power licence still hanging fire, PASNY could proceed only so far with planning for what would be officially titled the Niagara Power Project (which was made up of two principal parts: the Robert Moses Niagara Power Plant and the Lewiston Pump Generating Plant). It was still debating whether to use tunnels, open canals, or covered conduits to bring water from the upper river to the new station (see Figures 4.8 and 4.9). It was also uncertain about the shape, size, and location of the station's reservoir. The existing plans placed the reservoir primarily on the reservation of the Tuscarora Indian Nation. This was potentially problematic, not only because of Tuscarora objections but also because it was difficult for PASNY to obtain bond financing without knowing what the final works, and thus the cost, would be.

Because of the Niagara River's turn to the north after the Falls, the New York water conduits did not need to be as long as Ontario's tunnels, and this geography also made it easier to avoid built-up areas.[68] Conduits built by cut-and-cover, which required removing whatever was on the surface, were envisaged.[69] These conduits cost in the neighborhood of $100 million – half the projected bill for deep tunnels. In addition to cost, PASNY also emphasized the more dangerous work involved in a deep tunnel, pointing out that fifteen people had died during Ontario Hydro's conduit construction (twenty would perish during construction in New York State). But local officials demanded deep tunnels, arguing that cut-and-cover would be too disruptive, and called for the power reservoir to be substantially reduced in size. The Department of the Interior requested assurances that fish and wildlife resources, including the water flows in the lower river, where the generators would expel their water, would not be adversely affected.[70] Multiple hearings before the FPC on the tunnel and reservoir plans were required, throwing PASNY's schedule off-kilter and preventing it from issuing bonds. PASNY issued bid requests for equipment and letters of intent, but until bonds were subscribed, binding contracts could not be signed and construction could not formally start.

PASNY met with industries in the area to ascertain their needs and expectations. If assured affordable power, most planned to expand. PASNY could offer 60 Hertz (Hz) power from its imminent station at a lower cost, or mill rate, than had Niagara Mohawk. But this raised questions about 25- versus 60-cycle power. Within a ten-mile radius, twenty-five industries used 550,000 kW, all but 100,000 of which was 25 Hz.[71] Within fifteen to thirty miles, which included Buffalo, another hundred-plus industries used similar amounts and proportions.

Given the types of products that local industry produced, and the lax standards of the time, the Niagara region had long been a dumpsite for hazardous waste. But at places such as Model City, the problem was not just toxic chemicals. A sparsely populated hamlet east of Lewiston and just below the Niagara Escarpment, Model City owed its origins to William Love's ill-fated canal. Beginning in the Second World War, Model City became a host for radioactive refuse, first from the Manhattan Project and then from other Cold War–era nuclear endeavours.[72] Niagara Falls was not only a receptacle for radioactive waste but a producer of it. At least thirteen local facilities received government nuclear contracts and secretly contributed to nuclear weapons development.[73] But radioactive materials did not always end up in officially designated disposal areas.

At the end of January 1958, the FPC issued a full licence to PASNY. It affirmed cut-and-cover conduits but eliminated much of the open canal that PASNY had planned for the tract leading to the reservoir forebay. PASNY immediately issued bonds, and work started on all phases of the project, with first power anticipated for early 1961. The new Lewiston development would take the full American allotment of Niagara water under the 1950 treaty. Its maximum kilowatt output would be slightly higher than that of North America's largest generating station, the Grand Coulee Dam on the Columbia River. Like the Beck No. 2 station across the river, the new station was chiselled into the gorge wall and would have an enormous footprint requiring far-reaching land acquisition: the powerhouse area was expected to require 167 acres, plus 4,610 acres for the reservoir and another 220 acres to dispose of the resulting rock and dirt spoil.

PASNY had the state legislature pass a bill giving it the right to appropriate, without first condemning, lands needed for the project.[74] The power authority simply had to file a map with the secretary of state and deposit the market value of the land with the state comptroller. PASNY's director of land acquisition was in charge of negotiating purchase agreements, and had the authority to close a settlement as long as it was within 10 percent of the appraised value (above 10 percent required approval of the authority's trustees). Thorne Appraisal Service was hired to assist. Condemnations would result if no agreement could be reached. In cases where an owner resisted condemnation, the power authority would need to file a petition in court, which was time-consuming and involved additional costs. Those who held out for higher prices often succeeded since PASNY was on a tight time schedule and was keen to avoid protracted legal dealings for small properties. For example, one large property owned by a Mrs. Whittell was appraised at $200,000, but PASNY ended up paying $275,000 for it in order to get the transaction closed promptly. On the other hand, many people felt pressured or bullied into giving up their properties quickly.

About one hundred residences would have to be moved, mostly from the "Alphabet Street" area, with the owners given the option of repurchasing their residence.[75] Another 110 homes would be constructed, which was approximately the number that would be needed to house Uhl, Hall, and Rich staff. The ability to offer residences was linked to the ability to attract first-class engineers. A neighbourhood called the Veteran Heights Subdivision was built from scratch, sewers and sidewalks installed, with large lots for up to 168 houses. The power authority hired Hartshorne house movers to transport residences (see Figure 4.10). This company had

FIGURE 4.10 Moving houses. *Niagara Falls Public Library, New York*

relocated many buildings for Moses in New York City and was finishing up its St. Lawrence work for Ontario Hydro.[76] Land was also required for the long-distance transmission lines that would connect the Niagara and St. Lawrence developments and take the resulting electricity toward the mid-Atlantic states and New England. Agreements were inked with the Niagara Falls Bridge Commission to abandon the bridge between Lewiston and Queenston, which crossed the Niagara River immediately below the escarpment face. The commission built a new multi-lane international crossing at the top of the bluff. It provided a sweeping panorama of the new power developments, but was described by a local historian as "a minimalist's concrete fantasy, something only an engineer could find beautiful."[77]

Apologies to the Tuscarora

Robert Moses had adopted a "use value" approach to New York City slum clearance, believing that property acquired for the common good by public institutions like PASNY should not have to pay market value for land. Yet PASNY reached an agreement with Niagara University at the tidy sum of $50,000 an acre, since some of this school's land was needed for the new generating station and reservoir.[78] PASNY also had to acquire a range of privately owned properties, as well as large swaths of land from Niagara Mohawk.[79] For the bulk of the reservoir's footprint, Moses targeted the Tuscarora Reservation. He argued that it was less "developed" than other surrounding areas, meaning that it would be cheaper to obtain than other properties in the vicinity, and he clearly thought taking land from the reservation would be simpler since the area outside the reservation had a higher concentration of houses and a graveyard.

In the long run, PASNY would end up spending so much on legal fees, delays, and other costs that it probably would have been cheaper just to acquire non-reservation land. For Moses, however, quashing opposition seemingly trumped other practical considerations. To him, no type of land use was more valuable than the power and parkland he had in mind, and no one, particularly not 700 Indigenous Peoples using the land for what Moses deemed non-productive purposes, could be allowed to stand in the way. Neither he, nor most of society, could conceive of a use for this property beyond its commodity and economic value; the Tuscarora's cultural connection to their land seemed alien. It became a fight over different conceptions of property and progress.[80]

At some point in the distant past, the Tuscarora had migrated south from the Great Lakes region to what would later become the Carolinas. After conflict with other First Peoples and European settlers, including the Tuscarora War of 1711–13, they returned north, becoming the sixth nation in the Iroquois (Haudenosaunee) Confederacy. Following the Treaty of Canandaigua in 1794, some of the Tuscarora moved to land near present-day Lewiston, New York, owned by the Holland Land Company, and then purchased more in fee simple title – the territory that would be targeted for the power reservoir.[81] Initially, PASNY had planned to take 1,684 acres from the Tuscarora Reservation, a little more than one-third of what was required for the reservoir. The FPC licence contracted the size of the reservoir to about 950 acres, but then PASNY announced that it had to bump up the amount needed from the reservation to 1,250 acres, and then to 1,383 acres. The Tuscarora Indian Nation objected, arguing that it was feasible to redesign the reservoir so that it did not impinge on their land. The group sent formal letters of objection to the president and the secretary of the interior, and began appearing at FPC hearings to which they had conveniently not been invited.

Moses quickly became impatient and combative. He feared the delay would hurt bond sales. The Tuscarora community was well aware of Moses's recent dealings with another Haudenosaunee group astride the St. Lawrence. The PASNY chairman was brusque, arrogant, and condescending, as were many of the delegates he sent to negotiate.[82] The Tuscarora council took legal action, contending that PASNY could not take their land via eminent domain under state legislation or the FPC licence since they were a sovereign Indian nation, part of the Iroquois Confederacy and its treaty rights.[83] Federal consent, through an act of Congress, as well as Tuscarora concurrence, was what they argued it would take to dispossess reservation land. PASNY countered that Tuscarora land in question had been privately purchased, and thus could be taken like any other private property. A state court ruled in favour of the Tuscarora; this was followed by a series of legal appeals that would stretch for two and half years and reach all the way to the highest court in the land.

Complicating matters, there was internal tribal disagreement about how to fight the legal case, i.e., whether to make a deal or to resist. There was a long history of mistreatment by the American state, and the federal government was in the midst of its assimilationist "Termination Policy" and appropriating vast amounts of Indigenous land throughout the country, such as in the Missouri River basin through the Pick-Sloan Plan.[84] While some members called for militant action, the Tuscarora augmented their legal challenges with a campaign of passive resistance. At first, this included impeding PASNY surveyors and pulling up or burning survey markers; later, it also took other forms. Often women and children were at the front lines of protests, since the men were working.[85] PASNY showed up with police bearing riot gear, tear gas, and submachine guns.[86] Physical altercations occurred, and some Tuscarora were arrested. A judge issued a restraining order to stop the surveyors. The Tuscarora's phone lines were tapped, and there were many other forms of intimidation, bullying, and legal legerdemain, such as failing to provide advance notification about court dates. When the restraining order expired in June 1958, surveyors poured back in, and even though the Tuscarora appeal was still pending, a judge allowed PASNY to begin erecting power lines on reservation territory.

Moses tried to balance his many sticks with a few carrots: he offered to build a $250,000 community centre, and upped the amount that the Tuscarora Indian Nation would receive for their collective property to $1.5 million – though that was just a fraction of what PASNY had paid to the nearby university. By 1959, this figure had risen to $3 million but was rejected. PASNY also dangled the potential for a land exchange on the other side of the power reservoir, but when the reservation expressed interest, speculators bought up the land and raised the price beyond what Moses was willing to pay.[87] At court hearings, PASNY went to great lengths to show that Tuscarora land was quite worthless and unmodern: such a small and insignificant group should not be able to stop the kind of progress that would benefit so many. Moses himself wrote that "the bulk of the land is not used for any purpose at all"[88] and tried to turn the surrounding communities against the Tuscarora, implying that if the "intransigent Indians" did not budge, the reservoir would have to move further onto white property.[89] Since Indian land was non-taxable, he pointed out, taking their territory wouldn't adversely impact local coffers, compared with removing other properties from the tax rolls.

The Tuscarora appealed to the media. A series of articles were published in the *New Yorker* by Edmund Wilson and then collected in a 1959 book titled *Apologies to the Iroquois*. The Tuscarora themselves thought the book flawed, though it did accurately divulge the "low tricks" that Moses and PASNY employed.[90] Moses replied with publications of his own bearing titles such as "Niagara Desperately Needs More Power" and "Tuscarora: Fact or Fiction."[91] In these tracts, he made many claims to discredit the Tuscarora, and suggested that it was just a small group of recalcitrants blocking the reasonable majority within the reservation. Public opinion, locally and around the country, was sympathetic toward the plight of the Native Americans. Many others could relate, indirectly or first-hand, to what it felt like to be shunted aside by the Moses machine. Roger Alexander Millar of Lewiston wrote an open letter to the governor in the *Niagara Falls Gazette:*

> What a shameful spectacle it is when the great State of New York, of many million people, of whom you are the Chief Executive, browbeats, and intimidates a pathetic handful of Tuscarora Indians, whose mistake it is to spurn the almighty dollar and to stand up for what they and many, many white people hereabouts believe to be their historic rights. At the instigation of the State Power Authority, four separate arrests of Tuscaroras have been made. In each instance the case has been thrown out of court and the Indian freed. Once a writ of habeas corpus has been required to set [PASNY] back on its heels ... As for me, I say more power to the Indians ... [Moses is] one of the master politicians of our time, even as Hitler.[92]

Legal challenges went back and forth, each side scoring some victories. The Court of Appeals held that Tuscarora land was a "reservation" within the definition of the Federal Power Act. In February 1959, the FPC found that the licence it had granted to PASNY "will interfere and will be inconsistent with the purpose for which such reservation was created" and the tribal community could not be compelled to sell it.[93] This decision was upheld by the United States Court of Appeals for the District of Columbia Circuit, but the FPC appealed the ruling to the Supreme Court, hoping for either a final judicial decision or an out-of-court settlement.

While all this was going on, PASNY continued with construction but prepared contingency plans in case the courts ultimately ruled against it. If that occurred, it planned to build a smaller reservoir not touching Tuscarora land, but with the walls raised by ten feet. This would cost more to build and inflate slightly the electricity cost for future customers.[94] It would also lower the final generating capacity by 100,000 kW. Additionally, PASNY drew up blueprints for a reservoir that would need only 350 acres of Tuscarora territory. In the meantime, work proceeded on just three sides of the reservoir, and was done in such a way that would later permit any of the potential designs.

In March 1960, the Supreme Court issued a 6–3 decision against the Tuscarora. The majority held that it was not a reservation since the title to land was vested in the Tuscarora and not the federal government. The dissenting opinion disagreed with this interpretation of the legal definition of an Indian reservation. One of the dissenting justices, Hugo Black, lamented that "some things are worth more than money and the costs of a new enterprise ... I regret that this Court is to be the governmental agency that breaks faith with this dependent people. Great nations, like great men, should keep their word." The Tuscarora Indian Nation filed a petition for hearing, but it was not successful.

"A Niagara of fictional treacle of molasses has been poured on the Indians," Moses crowed, "a sticky flow finally stopped by the United States Supreme Court. We have never harmed the Indians. We have been more than generous in delaying with them." Now, he condescendingly and paternalistically chided, "we only hope that they will use the money we pay them for a fraction of their land wisely and in the interest of their children."[95]

Since PASNY had made its contingency plans, it was able to build a 60,000 acre-foot reservoir that required less Tuscarora land than originally sought. PASNY took the southwest area of the reservation, about 550 acres (55 acres were for a transmission line) – a bit less than a square mile – for the reservoir that covered about 1,880 acres in total (see Figure 4.11). Technically, it was an easement, and the Tuscarora retained ownership of

FIGURE 4.11 Aerial view of the Niagara Power Project under construction.
Niagara Falls Public Library, New York

its land. The Tuscarora council reluctantly consented, characterizing the payment as a damages settlement.[96] PASNY moved the eleven affected houses, provided new ones to their owners, or gave them a cash settlement. The authority spent a total of $370,091 on Tuscarora relocations, about half of that on residences, which included a range of improvements (e.g., heating, lighting, landscaping). For example, William Farnham took over Kenneth Patterson's former home, albeit in a new location. Charles Johnson had his house moved, and Tracy Johnson took the new house option. PASNY made a blanket payment of $850,000, or $1,100 per acre (far less than the $3 million offered in 1959), and every tribe member, regardless of whether or not their property was directly affected, received $800 as part of the final settlement.[97]

In a placatory move of sorts, the new generating station had initially been named after the Tuscarora. But an editorial in the *Niagara Falls Gazette,* probably planted, suggested that it be named after Moses. This was quickly implemented. With the stroke of a pen, the Tuscarora went from being not only physically but also symbolically displaced. The reservoir left behind many scars, both across the land and within the community.[98] The construction of the project reworked the hydrology of the area, destroyed fish and wildlife habitat, and hurt agricultural production.[99] Obliterating the creeks that provided most of the reservation's freshwater supplies created a water crisis that continues to this day, with wells contaminated by lead and bacteria.[100]

The Tuscarora lost any remaining faith in the US government, contributing to a Red Power movement among not only the Haudenosaunee but Native Americans across the continent. In retrospect, the deck was surely stacked against the Tuscarora, and given Moses's stature and the project's importance, it was inevitable that the colonial legal apparatus would be mobilized against them – even if one court rebuffed PASNY, this was just a temporary delay until other legal avenues could be found. All this was the logical extension of several centuries of settler societies taking Indigenous land for supposedly higher purposes.

Indeed, those in power tut-tutted about the silly attitudes of the Indians: Did they not realize that everyone, including themselves, would be better off if the Niagara Power Project went through? Those planning the project did not understand that for the Haudenosaunee the land was an integral part of their heritage, community, and identity. As one of the many judges who dealt with the Tuscarora issue had suggested, and which Moses repeated in one of his propaganda pieces, "if the Indians are to enjoy equal protection of the laws and all the other benefits extended to each citizen it may well be that they should bear some of the burdens including that of being subjected to having their lands condemned for public purposes, beneficial to the State in which they live."[101] But it is environmental racism and injustice when the same groups have to repeatedly shoulder disproportionately greater burdens while also receiving lesser benefits and lesser legal protection.

Moses Parts the Waters

PASNY could now move full-steam ahead. Out of vindictiveness as well as the need to protect his position, Moses continued to fulminate against private power propaganda. In a glossy PASNY pamphlet intended for public consumption, the chairman noted the many local improvements his authority would make, and the wider benefits that cheaper electricity would bring in terms of attracting industry, creating jobs, and improving the standard of living. All of this would compensate for any "small losses of taxable real estate occasioned by land acquisition by the Power Authority and, in the final analysis, will result in more taxes."[102] But considering that the new power facility would be outside the City of Niagara Falls proper, and that the Adams and Schoellkopf stations it was replacing had been inside the city limits, the city stood to lose substantial tax revenues. Moses therefore advocated for a payment of $2.5 million to the City, Council, and School District of Niagara Falls in lieu of taxes.[103]

While the confrontation with the Tuscarora played out, PASNY was busy building the Niagara Power Project (see Figure 4.12). The major contractors were Merritt-Chapman and Scott.[104] A preeminent American hydraulic construction company, it had built many iconic hydro-power structures in the United States and consulted around the world. Other contractors included Peter Kiewit, Morrison-Knudsen, and Perini-Walsh, all of which joined to form Channel Construction.

All told, 11,700 workers would toil on the project.[105] Ken Glennon, who was employed on the project as an eighteen-year-old, later published an oral history of worker recollections, in both book and video form, which effectively captures many of the social and technical aspects.[106] The workers exuded pride in their role. Many were from nearby, but many followed construction jobs around the country, and pockets of the Niagara region took on the atmosphere of a western boom town. This caused friction with locals. Schools and public services were overburdened. Blue-collar workers were not provided with residences like engineers, and had to find their own accommodations. At some establishments for single workers, they slept in shifts, called "hot bunking" since the bed was still warm from the previous occupant.[107] For those with families, housing became an acute problem – trailer parks sprang up for miles around, including one on the Tuscarora Reservation.

Work continued day and night, summer and winter. Though the latter brought cold and ice, it obviated the dust and mud problems that plagued work during other seasons. As of the end of 1959, almost all excavation, land acquisition, and building relocations had taken place, and concrete pouring began in earnest. By the time all was said and done, over a million cubic yards were poured, enough to build a sidewalk four feet wide and three inches thick around the world. The construction site was a hornet's nest of activity, the gargantuan Euclid trucks and dragline cranes joined by the din of wailing concrete vibrators, roaring sandblasters, clanging steel, and pounding hammers that only served to heighten the controlled chaos.[108] In 1960, with the February 1961 deadline looming, an army of 9,000 workers was at work on all facets of PASNY's Niagara development.

Above the Falls, the Robert Moses State Parkway, as it was christened, would run along the edge of the river. Thirteen million cubic yards of rock and earth was excavated from the trench for the dual tunnels taking water to the Niagara Power Project, and much of it was used to reclaim land along the American shoreline of the Chippawa–Grass Island Pool (see Figure 5.9).[109] The parkway ran above this extended shoreline, which gave drivers an excellent view of the water but cut off non-vehicular access to the water in many locations. It ran right through Prospect Park, allowing motorized tourists to see the Falls without even getting out of their cars, but forcing those approaching on foot to climb cumbersome pedestrian ramps within the Niagara reservation just to get to the cataracts. Motorists could also drive onto Goat Island and leave their vehicles at the new parking lot, which had replaced a forested area in 1951. Spoil dumped by the various agencies in the riverbed further altered the flow patterns of the river, such as adding 8.5 acres at the upstream tip of Goat Island for another parking lot and a helicopter landing pad.[110]

Former landmarks along the New York shoreline, such as Willow Island and the intakes for the hydraulic canal and the Adams plant, were mostly obliterated by the parkway.[111] Unlike many other arterial roads built by Moses, this parkway was not very popular; in fact, parts near the Falls, such as the section running through Prospect Park, were soon removed.[112] New York State also took the opportunity to construct new freeways: the Niagara Expressway and the Niagara Thruway. The latter ran from Buffalo to the Rainbow Bridge, crossing Grand Island on new bridges, while the former ran from the Rainbow Bridge to the new Queenston–Lewiston Bridge, connecting to the Queen Elizabeth Way in Ontario. Railway grade crossings in urban areas were eliminated. Authorities relocated many miles of roads and railway tracks, and some crossed over the reservoir forebay or the top of the new power station.

PASNY built new water and sewer systems for the communities of Lewiston and Niagara Falls, installed playgrounds, and expanded Hyde Park, including a nine-hole addition to the golf course. It funded improvements to the state park at the Falls and Goat Island and other parts of the Niagara reservation, such as adding an observation tower at Prospect Point, while creating the 132-acre Reservoir Park behind the new generating station (though, as on the Canadian side, boating and swimming

FIGURE 4.12 Robert Moses Niagara Power Plant, with pumped-storage generating station and reservoir in upper half of the picture. *Niagara Falls Public Library, New York*

were not allowed in the artificial lake).[113] Ten million yards of excess material, the equivalent of two and a half Central Parks, was deposited at a spoil pile north of the Moses powerhouse, which had to be cleared of buildings. The spoil material reshaped part of the escarpment and covered the historical portage route up "the mountain." This area was later transformed into Art Park to serve as community space for the Village of Lewiston.

By virtue of its Niagara undertakings, PASNY controlled the shoreline and underwater rights on the US side of the river. Moses shot down plans from the Niagara City Council to erect a motel at the ruins of the Schoellkopf station, but did approve many private applications for easements along the water's edge in other places, such as docks for houses and cabins on Grand Island. Two water intakes for the new power plant were placed on the extended shoreline about two miles above the Falls. Huge frames were built to house the gates, which could be lowered to block off the twin conduits that conveyed the water to an open canal leading to the reservoir.[114] These mirrored the intakes Ontario Hydro had erected on the other side of the river. Many of the companies that made Niagara industry widely known were situated just north of the parkway, along Buffalo Avenue. Because of the new hydro station tunnels and parkway, these companies had to change their own water intake and outfall lines. Corporations such as DuPont complained that effluent from upstream companies, such as Hooker, fouled the water quality, yet were unwilling to treat their own outflows.[115] They also lodged complaints about air pollution from surrounding industrial concerns.

Air and water pollution had long been hazards in the Niagara corridor as well as at other connecting channels in the Great Lakes basin. Worries about pollution from one country befouling the other extended back to the signing of the Boundary Waters Treaty. Some of the International Joint Commission's earliest dockets involved cross-border pollution studies in places such as Detroit, though the commission's recommendations were not heeded at the time. In 1946, Canada and the United States sent the IJC another reference on pollution in the Great Lakes connecting channels, with the Niagara River added to the study in 1948.[116] The following year, New York passed the Water Pollution Control Law, along with a control board, to maintain "reasonable" purity standards. The City of Niagara Falls, NY, was also taking steps to change and improve its public water supply, which was drawn from the Niagara River.[117] By then, the river was more or less an open sewer. An environmental study released in 1954, though it promoted the utility of dilution for industrial wastes, noted that "some 20 fish kills have been reported in the Buffalo–Tonawanda–Niagara Falls area during the past 17 years. Cyanides were implicated as the cause of 9 of the kills with phenols a contributing factor in two cases. A lethal dose of chlorine presumably was the cause of two of the kills." The study went on to note that the "Niagara River is an important gathering point for migratory wildfowl and occasions have been reported of large number of birds being swept over the falls as a direct result of oil pollution."[118] It is therefore difficult to believe that these companies truly considered their discharges benign, as they repeatedly and publicly claimed.[119] After all, if the water quality was not good enough for their own industrial intakes, how could it be good enough for living organisms and the local ecosystem?

PASNY had passed several bond series to finance the entire project, and the total came to $720 million.[120] The power authority began marketing its power before the system came online and signed a range of contracts before and after the powerhouse was finished.[121] PASNY was required by its FPC licence to supply 445,000 kW of firm power to Niagara Mohawk, which in turn sold the power to companies previously supplied by the Adams and Schoellkopf plants (the remaining generators in the Schoellkopf station were decommissioned). This list of over thirty companies included well-known American metallurgical and electrochemical businesses: Bethlehem Steel, Hooker Chemical, DuPont, Atlas, the Carborundum Company, General Mills, International Paper Company, Pittsburgh Metallurgical Company, Union Carbide, Dunlop Tire and Rubber, and so on. Another 250,000 kW was expansion power that was sold to industries within thirty miles of the Niagara project. PASNY claimed that low-cost replacement and expansion power in the Niagara area was "responsible for at least 2,500 jobs resulting from expanded activity, an additional 2,500 jobs resulting from the restoration of low power costs to existing industry, and will provide an estimated 2,000 jobs when the program of industrial expansion established on the strength of low cost expansion has been completed."[122]

Over the last half of 1960, construction continued at a frantic pace as PASNY pushed its workers and contractors to meet its deadline. They were motivated by pride but also by financial necessity. First power was indeed achieved by the target date of February 10, 1961, with two units coming into service. Granted, it would take until October 1962 before all twenty-five units were operational, and other parts of the development, recreational facilities, and clean-up would take until 1963. Nevertheless, this pace of construction was a remarkable feat, considering all the obstacles.[123] For PASNY, the ends justified the means: "Soon forgotten will be the inconvenience suffered by the few people discommoded by construction activities."[124]

New York State's entire Niagara project required over 39 million cubic yards of excavation. The completed pumped-storage plant housed twelve reversible turbine-generators with a total capacity of 240,000 kW, while the main station featured thirteen turbines that, connecting with Niagara's past, were fabricated by Westinghouse Electric Company. This allowed for a total generation capacity of 1.95 million kW (2.19 million kW counting the pumped-storage plant). This constituted about a third of the state's electricity, and it sold for about 4.1 mills, which was "comparable to rates in such areas as TVA [Tennessee Valley Authority] and actually is lower than in practically all other areas of the country, with the exception of the Pacific Northwest."[125] PASNY bragged that it was the free world's largest hydroelectric installation and formed, in conjunction with the Beck stations, the largest hydro-power complex in the world. "Niagara's scenic splendor is unspoiled," the *New York Times* enthused; "it is, in fact, enhanced."[126]

Over 1,231 parcels of land had been acquired by PASNY, but at the start of 1963 many cases remained unsettled.[127] At that point, ninety-one claims had been filed in the Court of Claims: thirty-eight cases were pending, with ten awards under appeal.[128] One of these involved the Hewitt family, to whom the Court of Claims awarded $374,000 for about 105 acres of appropriated farmland in Lewiston; however, on hearing that the Hewitts had offered to sell for $67,000 in 1954 to Niagara Mohawk, PASNY appealed this award as well as others. A number of condemnation cases had gone to the Supreme Court, though about one-quarter were still pending a settlement or ruling.[129] Since it had to dispense with leftover property in a relatively depressed real estate market, the power authority ended up unloading twenty-four houses and thirty-five vacant lots for $345,000, about $50,000 less than the appraised value.[130]

PASNY also required many properties for its two 345,000-volt transmission tie lines between Niagara Falls and Rochester, followed by one line that went to Utica, where it joined with power from both PASNY's St. Lawrence development and Niagara Mohawk.[131] From the Utica interconnection, electricity flowed to the populous downstate areas. This built on the cross-border electricity connections recently established between Ontario, New York, and Michigan. In 1965, PASNY and Ontario Hydro signed a memorandum of understanding to formalize procedures for operating and maintaining electricity interconnections between their two systems at Niagara and the St. Lawrence.

This kind of energy, economic, and political coordination was made possible by the previous integration of Canadian and American industrial technologies, standards, and practices.[132] In fact, up to the 1960s, the majority of energy of any kind, including fossil fuels, exported from Canada to the United States consisted of hydroelectricity from Ontario. St. Lawrence and Niagara power had played the leading role in shaping the approach of Ontario and the federal government to electricity exports. Non-firm – i.e., interruptible – power sales dominated the Canada-US electricity trade in the decades after the Second World War.[133] These hydropower megaprojects therefore further entrenched Canadian-American energy relations and paved the way for the development of the transborder electricity grids that proliferated in the Cold War era and tied the two countries together physically and symbolically. The oil crises of the 1970s further increased American imports of Canadian electricity. As of 1975, which also marked the peak of the national American grid, sixty-five interconnections crossed the international border, with a total transfer capability of over 6,000 megawatts, including the Canadian allowance of long-term firm power as part of the Columbia River Treaty.

Hydro Tourism

The Niagara power projects speak to the ways in which national identities are bound to environmental features as well as human artifacts. Niagara Falls is aptly characterized as Canada's front door and America's back door.[134] Though prior to the twentieth century the Falls of Niagara were not seen as predominantly "Canadian" to the same degree as other parts of the basin, such as the St. Lawrence River, Niagara Falls gradually came to resonate with Canadian nationalists for various reasons, including the great cataract's proximity to the national heartland, its location in the Great Lakes–St. Lawrence system, and its sites of resistance to American invasion during the War of 1812. The Great Lakes–St. Lawrence region, home to about one-third of Canada's population, has historically been that nation's political and economic centre and offers one of the primary access routes by which Canadian people and goods intermingle with their southern neighbour. In order to stimulate tourism and national consciousness, the Canadian government increasingly framed Niagara as a "patriotic topography" that Canadians should visit.[135] By comparison, after the nineteenth century, the Great Lakes–St. Lawrence basin did not figure nearly as prominently in the American popular consciousness as other waters such as the Columbia, Colorado, and Mississippi, and freshwater narratives emphasized the scarcity of the arid regions of the country rather than the liquid abundance of the Great Lakes basin. Only about 15 percent of the American population now reside in this borderlands basin that, outside of places such as Chicago, the nation seems to view as a backwoods area or a sort of sacrifice zone.

But in the middle of the century, the construction of the Niagara megaprojects tapped into a unique form of *hydro tourism* based on a heady blend of technological, environmental, state-building, and nationalist appeal.[136] Hydro tourism can be considered a type of industrial tourism, which attracts people to massive engineering projects such as bridges and buildings, but is also a type of nature or landscape tourism. The development of hydro power was acted out on a public stage, representing engineering progress that heralded perceived heritage connections to specific natural features and landscapes while simultaneously reshaping them to provide a heroic future. Tourists wanted to see sublime nature, but also flocked to see it controlled by sublime technology.[137]

The fin-de-siècle Niagara generating stations, particularly the Adams plant, were enormous tourist draws when first opened and had arguably started the power-station-as-tourist-attraction trend, which subsequent efforts such as the Hoover Dam continued. The post-1945 Niagara developments were among the earliest to explicitly incorporate extensive tourist infrastructure and amenities into the finished product, and perhaps the first to make the technology of tourism a principal concern during the construction period. After all, the new Niagara power stations needed to be seen by the general populace in order to fulfill their role as symbols of state power. New York offered displays about its power project in a temporary exhibit building, which over 700,000 people viewed between April 1959 and October 1962. Tours were offered of the construction area and proved to be a big draw, with almost 53,000 people participating in 1,900 tours.

Robert Moses presided over an elaborately staged opening ceremony for his eponymous generating station at nearby Niagara University, designed as a nation-building exercise. The three-day affair featured celebrities and VIPs such as Governor Nelson Rockefeller and pre-recorded encomiums from new president John F. Kennedy as well as three former occupants of the Oval Office: Dwight Eisenhower, Harry Truman, and Herbert Hoover.[138] It was "an outstanding engineering achievement," JFK rhapsodized, "and a happy example of teamwork and cooperation between private and public enterprises"; Ike enthused that "the mighty power of the Niagara has been harnessed for public good, and the beauty of historic Niagara Falls has been preserved for all time."[139] The visual backdrop to the ceremonies was rich in symbolism, replete with bright lights mimicking transmission lines. Composer Ferde Grofé was commissioned to produce an orchestral work, *Niagara Suite,* which debuted at the opening ceremony. PASNY struck 150 commemorative bronze medallions and distributed them to deserving recipients.

PASNY went to great lengths to make sure hydro tourism would continue after the completion of the project. It had installed a visitor centre with an observation deck atop the south end of the new power station, called the Power Vista, which attracted 300,000 people in its first year.[140] The centre featured a working diorama of the main plant and an accurate scale model of the whole project and the surrounding area, along with a seven-by-twenty-foot painting by the noted American artist Thomas Hart Benton of Father Louis Hennepin viewing the Falls in 1678.[141] Specially built observation stations, overlooks, and parking lots allowed visitors to view the Beck power complex in Ontario, before and after construction, and a permanent visitor centre was built right into the new plant, offering a museum, a large model of the works, and tours down to the generating hall. This coincided with other early Cold War efforts to capture tourism: as Niagara became more of a middle-class destination in the 1950s, large sums were invested in tourism developments and amenities, and motels sprang up all over the region catering to vehicular traffic.[142] By the last half of the 1950s, no less than 3 million people annually visited the Canadian side alone, with just a little less attracted to the Niagara reservation in New York.[143] By the middle of the 1960s, both sides were hosting closer to 5 million per year.

The power stations were visited not just by the general public. In 1962 alone, 2,166 official foreign visitors from sixty-seven nations toured PASNY's project, many of them heads of state, dignitaries, or engineering specialists. The generating stations became a sort of international finishing school for engineers. Moreover, many engineers and contractors who took part in building power stations at Niagara later went around the world to lend their expertise on hydraulic projects, both as representatives of governments and as private consultants.[144] Niagara engineers also disseminated their experiences and knowledge through their networks, conferences, and professional journals.[145] Thus, the Beck and Moses plants were designed within the context of a global engineering fraternity and the international spread of engineering techniques and ideologies framed by the Cold War.

Richard P. Tucker avers that "much of the world's dammed rivers reflect Cold War zones of competition, and the concentration of fiscal and industrial resources at many dam sites in remote locations cannot be fully explained outside the framework of Cold War rivalries."[146] The same was true at more central and renowned sites. The Niagara power and remedial developments not only projected the strength, scientific ingenuity, and technological capabilities of capitalist and democratic North American society but were tied to the high standard of living on the continent, itself one of the major weapons in the Cold War. The strategic value of Niagara's power, considering that it was provided to many industries linked to North American defence and armaments, meant that the Cold War was being engaged right there on the US-Canada border. And political leaders were well aware of all this, directly contrasting the Niagara developments with those of the Soviet Union. The Soviets were equally busy building massive hydroelectric installations and rerouting water systems, such as the Volga River and Aral Sea, not to mention helping with Egypt's Aswan High Dam on the Nile River. Many other nations, especially China and India, were damming any and every large river. The hubristic drive to dominate nature was equally apparent in communist, capitalist, and non-aligned nations.

Conclusion

Robert Moses had been brought in to head PASNY while it built the Niagara and St. Lawrence power projects. At the end of 1962, with both projects completed and after eight years as PASNY chairman, Moses resigned from this and his other state posts too. He had made a habit of threatening resignation, and with his popularity on the wane, Governor Rockefeller called Moses's bluff. Two years earlier, Moses had resigned from or lost most of his New York City posts as well. Partly because of the lucrative salary, he took a position as president of the 1964–65 World's Fair.[147] Robert Saunders, who died from injuries suffered in a 1955 plane crash, was able to see the new Beck plant produce its first kilowatts.

There were both important similarities and important differences in the ways that the two public utilities treated those affected by the new power developments. Ontario Hydro made a number of concessions so as not to generate concerns about its public ownership, but also exhibited a siege mentality at times, and its lack of accountability as a quasi-Crown corporation helped perpetuate its self-image as the representative of the public interest.[148] Moses was more prone to running roughshod over those who stood in his way. Both PASNY and Ontario Hydro had an institutional engineering mentality in which their versions of progress overrode other concerns. Most locals were very wary of tangling with either utility, and often felt pressured into settling with them when their property was affected. To be sure, the Tuscarora did successfully stand up to Moses for quite a while, garnering media attention and public support that forced alterations to his plans. Moreover, a number of Moses's schemes for more New York City expressways were successfully blocked around this time – perhaps the Tuscarora's fight emboldened others to resist him. Still, whereas the Tuscarora tended to see themselves as paying the costs without reaping the rewards, the wider non-Indigenous community believed it would benefit economically. The majority of citizens in both nations shared their governments' commitment to that ineluctable talisman of industrial modernity: progress.

The types of power projects built at Niagara in the 1950s were, to be fair, more ecologically benign than most other large-scale hydroelectric developments of the same era. Other hydro-power developments of a comparable size generally required flooding out the upstream river for a long stretch to turn it into a power reservoir, and it was from the creation of such reservoirs that the greatest negative ecological impacts usually resulted: the radical reshaping of a river system, its fluvial geomorphology, and the ecological processes within, including changed water speeds and temperatures; the potent greenhouse gases, notably methane, which such reservoirs might emit over time because of decomposing organic material (trees and foliage) under the water, along with mercury accumulation; and so on. Conversely, riverine habitat was not flooded out for the Moses and Beck power developments since their water was held in long tunnels and canals, forebays, and then built-up reservoirs placed on land.

This is not to suggest that the Moses and Beck plants had no major environmental repercussions – far from it, in fact. The construction phase for the post-1945 stations had many direct environmental impacts. The massive amounts of spoil deposited in the river and shoreline changed flow distribution, water levels, water temperatures, and benthic communities. The diversion of water to the new plants extracted significant volumes of water from the river for almost the entire length of the Niagara gorge – more than five miles of the lower river, including the Maid of the Mist Pool, Whirlpool Rapids, and the Whirlpool – with attendant ecological impacts, until it was returned to the waterway near the Niagara Escarpment. Symbolizing the extent to which natural patterns were overturned, the counter-clockwise rotation of the Whirlpool actually reverses when the flow is too low due to diversions. The turbidity and temperature of the water was changed too, and fish habitats obliterated.[149] As a local historian wrote in the early 1990s, the pollution, power plants, and changed water regimes alienated people from the lower Niagara River, and the unique botanical heritage of the area praised by turn-of-the-century naturalists looked very different.[150] Diverting so much of the water away from the river affected not just water levels but related mist and ice patterns as well, thus altering biological processes and biodiversity at Goat Island and in the Niagara gorge.[151] We need to keep in mind indirect impacts, too. This includes all the industry that set up shop at Niagara Falls, attracted and made possible by its abundance of power. The many toxins and persistent organic pollutants (POPs) that resulted and were deposited in the surrounding counties and townships need to be seen as part of the environmental legacy of hydroelectricity at Niagara Falls. And it isn't just toxins that have persisted: the mindset that humanity can, and should, control nature is another poisonous inheritance.

5
Disguising Niagara: The Horseshoe Falls Waterscape

I look forward to the time when the whole water from Lake Erie will find its way to the lower level of Lake Ontario through machinery ... I do not hope that our children's children will ever see the Niagara cataract.

Lord Kelvin, 1897

THE THREE SISTERS Islands are an often overlooked excursion, a hidden gem, for those visiting Goat Island. A series of small bridges connects these tripartite isles, named for the progeny of a prominent Niagara family, to the western side of Goat Island. The Three Sisters Islands are interspersed by rivulets, small falls, and rapids (see Figure 5.1). At the edge of the outermost sister, Celinda Eliza, the ferocity of the upper Niagara River rapids is on full display. Looking upriver feels like facing a staircase of roaring water. In fact, the drop of some of these rapids is so significant that they might be considered important waterfalls unto themselves if located anywhere else but Niagara.

On a recent October day, beset by hail and sleet, I walked out on the Three Sisters Islands after dusk. My intention was to try to visually represent the drawdown that occurs each night for hydro-power production: a three-foot drop in the wide upper river, but closer to ten feet in the constricted lower gorge. By the time the drawdown begins – at 8 p.m. in the fall, 10 p.m. in the summer – it is already pitch-black. Since I had attempted this before, I knew to bring a camping headlight. I had absconded with some of my children's coloured chalk, and now use it to mark the water level on rocks along the channels that separate these islands, and stand other types of markers in the shallow water along a rock shoreline that gradually slopes into the water. I set up a camera on a tripod, and take a series of time-lapse photographs. After an hour, there isn't much apparent difference, so I walk to the big waterfall to kill some time. I come back

FIGURE 5.1 Low water in a channel between the Three Sisters Islands

later at night, and then again before dawn. It is obvious that the waterline has receded, even though the elements weren't kind to my chalk lines.

This daily recession is the result of the 1950 Niagara Diversion Treaty. At 10 p.m. during the tourist season, the Province of Ontario and the State of New York increase their diversion of the Niagara River's water from about half to three-quarters; they will turn the Falls back "on" (or should I say "up") for tourists early the next morning. Niagara Falls, despite being freshwater and hundreds of miles inland, has its own tide.

After the Niagara Diversion Treaty was ratified in October 1950, the United States and Canada referred the remedial works to the binational International Joint Commission, which established Docket 64 titled "The Preservation and Enhancement of Niagara Falls." "Remedial works" or "control works" are the catch-all names used by Niagara engineers for the various engineering interventions, including weirs, control dams, excavations, rock fills, shoreline removal/stabilization, and so on. The previous chapter explored the power stations built in the 1950s and early 1960s; this chapter focuses on the remedial works, which were designed in the early 1950s, built between 1953 and 1957, and then enhanced between 1957 and 1963.

These remedial works had several primary purposes. First, to increase hydroelectric production by diverting water to tunnels above the Falls, which then ran to downstream hydro stations. Second, to "beautify" the Horseshoe Falls by reshaping the flow of water over its rim and reshaping the crestline itself, in order to retain visual appeal and tourism. Both of these were also intended to reduce erosion. These works were a joint undertaking by Canada and the United States, with work for the former handled by Ontario Hydro and for the latter by the US Army Corps of Engineers.

The remedial works remade the hydrology and landscape of the Niagara River and Falls. The water regime was radically changed, as were the physical topography and bathymetry of the riverbed and crest of the Falls. Yet, the engineering techniques of disguised design allowed the Niagara to continue to look like itself, even as it became a flowing facade. In fact, it was intended to look like an improved version of itself. The waterfall was transformed into a technology, a unique form of hybrid envirotechnical infrastructure blending not only steel and concrete but water and rock, weeds and ice. On the face of it, this appears contradictory, inasmuch as tourism depended on ample amounts of water flowing *over* the Falls, while the power industry required water going *around* the Falls. In the high modernist mindset of state-sponsored planners, technology allowed the experts to have their cake and eat it too: a cataract that was rationalized and improved for hydroelectric production, transformed into infrastructure forming a larger energy network, but that sacrificed none of its aesthetic appeal and thus none of the connected tourist economy.

A Model Waterfall

As historian Martin Reuss has stated in his work on the US Army Corps of Engineers, for river engineering, all objectives are subordinate to control. Even today, however, engineers "do not fully understand the dynamics of river flow, and complex natural systems exceed our ability to mimic them in laboratories. Consequently, engineers must fall back on art, intuition, and close empirical study. River engineering remains in many ways experimental and tentative."[1] These qualities were evident in the engineering of the Niagara environment. What is striking, however, is the extent to which the engineers at the time believed that they could achieve full control, even when their own methods belied this belief. And key to this engineering approach was the use of scale models.

The Niagara Diversion Treaty called for 100,000 cubic feet per second (cfs) of water to go over the waterfall during daylight hours of what was deemed the tourist season (8 a.m. to 10 p.m. EST from April 1 to September 15, and from 8 a.m. to 8 p.m. between September 16 and October 31), and no less than 50,000 cfs the rest of the time (at night during the tourist season and at all times from November 1 to March 31). Anything in excess of the so-called tourist water – the 50,000/100,000 cfs base minimums – could be diverted to the hydro-power plants. Article II of the treaty gave the International Joint Commission responsibility for the design of the remedial works. At an IJC meeting in Detroit, representatives of the US Federal Power Commission, the US Army Corps of Engineers, and Ontario Hydro agreed to establish the International Niagara Falls Engineering Board (INFEB) to help with the design of the remedial works. The IJC also began holding a series of public consultations about Niagara Falls remedial works.

The International Niagara Board of Control and the Special International Niagara Board, both formed in the 1920s, still existed on paper; they both were now effectively dormant, though they weren't dissolved until 1956. A new iteration of the International Niagara Board of Control was created in August 1953 to supervise the remedial works recommended by the IJC and the INFEB. An International Niagara Committee was also formed. This committee was mostly responsible for what happened downstream from the soon-to-be built International Control Structure, while the International Niagara Board of Control oversaw the construction of the control structure and what happened upstream from it, including Lake Erie levels.[2]

The cascades, or rapids, channel leading to the Horseshoe Falls had a width of 3,200 feet at its upstream end, but narrowed to only 1,200 feet at the cataract. However, the lip of the bending waterfall had a length of around 2,600 feet. The water depth in the cascades channel varied from 6 to 12 feet, studded with shoals, ledges, boulders, and islands. A large central shoal divided the flow into two main channels near each shore, which then converged toward the central, and deeper, part of the Horseshoe Falls. At the crest, water was pulled by gravity toward several dips, or divots, which were the result of eroded or fallen rock. These recesses detracted from the appearance of an unbroken crestline and led to a vicious cycle: they exacerbated erosion at the middle of the horseshoe, further deepening it, bringing in even more water, causing even more erosion. Granted, because of increased diversions in the first half of the twentieth century, the IJC reported that the rate of recession in the middle of the Horseshoe Falls had been decreasing: 4.2 feet per year from 1842 to 1905, 3.2 feet per year from 1905 to 1927, and 2.2 feet per year from 1927 to 1950.[3] But increased diversions meant less water at the sides, at times completely unwatering and exposing the riverbed abutting the shoreline. Moreover, the deeper water at the centre of the horseshoe had a green hue, while the thin flows at the flanks (e.g., a foot deep) were white in colour.

Engineers considered physical hydraulic models – representations at a reduced geometric scale – to be indispensable planning tools for deciding on the form and location of the Niagara remedial works (see Figure 5.2). This privileging of models was a relatively new development for North American governments, and epitomized the high modernist penchant for expert planning. It would be the first time that high-precision scale models were used in Canada to plan a large civil engineering project. The US Army Corps of Engineers began employing models only when it opened the Waterways Experiment Station (WES) in 1929 in Vicksburg, Mississippi, and many of its engineers remained skeptical about their benefits.[4] This hydraulic laboratory included very large outdoor models and indoor facilities. The lab and the discussions that preceded its creation represented in many ways a clash of competing visions about proper science that hinged on debates about larger and smaller scale models, and pure and applied research methods.[5] The Corps of Engineers established other hydraulic laboratories and models throughout the nation, and the Bureau of Reclamation and the Tennessee Valley Authority also began employing hydraulic models. WES broadened its program of experimental and applied research following the conclusion of the Second World War. By the midpoint of the century, it had become the largest facility of its type in the world, which reflected the Corps of Engineers' full organizational acceptance of models.

The transnational Niagara engineering boards of the 1920s had also tried models, but at that time the models were deemed not reliable or accurate enough, in large part because the width and speed of the Niagara River made accurate surveys impossible. This was still the prevailing opinion in the lead-up to the Second World War. In the words of the International Joint Commission, however, by the post-1945 era "the hydraulic model [had] gained recognition as a new and valuable tool of engineering design ... It is the modern method of arriving at a solution that will work effectively and satisfy the economic requirements."[6] Models had become so attractive in large part because they purportedly saved significant amounts of money: Ontario Hydro would later brag that its use of Niagara models saved about $5 million.[7] The basis for this accounting was debatable, however. Of course, inflating the potential savings was a way of ensuring that models were adopted – which in turn was a form of occupational

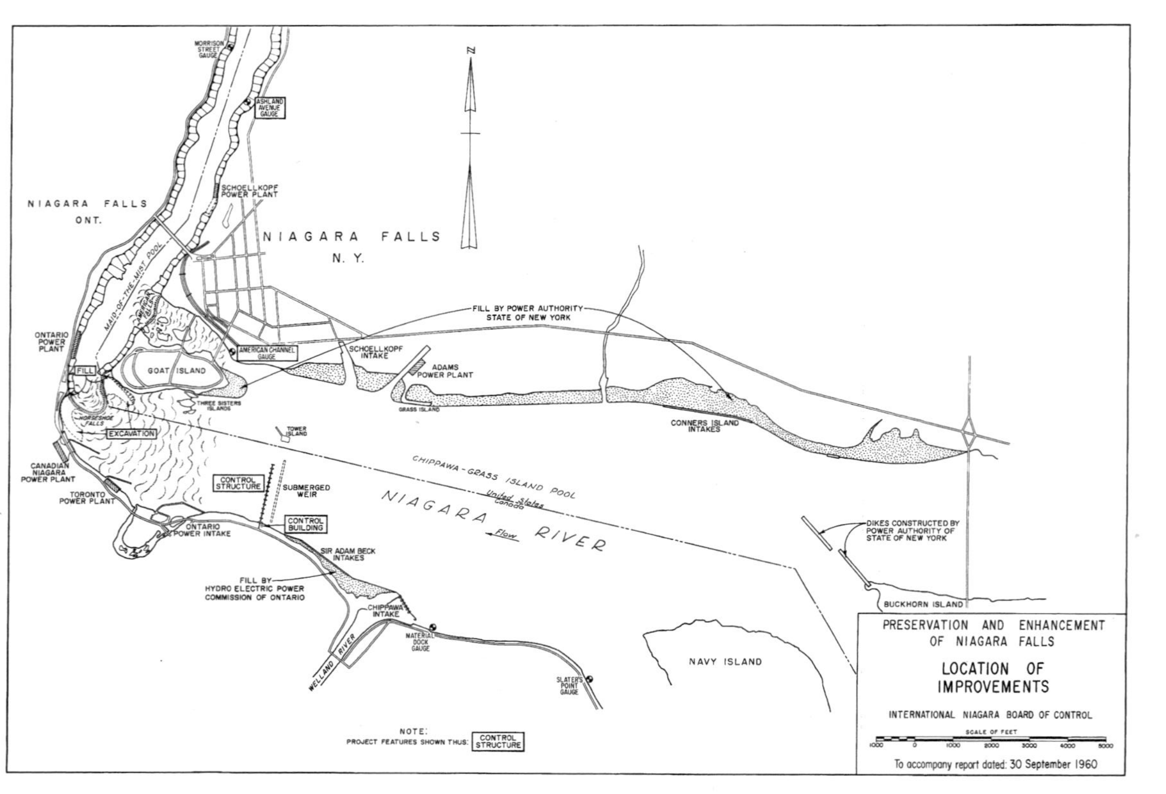

FIGURE 5.2 Location of recommended remedial works. *International Joint Commission*

gatekeeping since the hydraulic engineers were the only group with the requisite expertise.

According to the authors of a study on the history of WES, "perhaps the greatest advances" in physical hydraulic modelling came from the international Niagara and St. Lawrence projects, including unique methods for obtaining field data in hazardous situations; model verification; models at different sizes; and portrayal of complex hydraulic situations with distorted models. However, the authors of this study also indicated that these advances were joined by "some unanticipated setbacks" stemming from the transborder nature of these undertakings, which gave rise to more complicated political situations than the WES had previously experienced.[8]

Ontario Hydro had been busy preparing models for the Niagara remedial works design before the 1950 treaty was even ratified. At early IJC meetings, much of the discussion centred on the use of models, and it was decided to use two models, one in each country. Ontario Hydro built one in a warehouse in Islington (now part of Toronto), and the Corps of Engineers built another at its Waterways Experiment Station in Vicksburg (see Figures 5.3 and 5.4). The Corps argued that it should construct the "official" model, which Ontario Hydro resisted as a matter of prestige. Indeed, politics and pride were motivations for using two models, because each country wanted a piece of the action and aspired to show that its experts and methods were reliable. The officially stated reason for deploying two models was that it increased accuracy due to the "uniqueness and complexity" of the situation, since each model would be built at a different scale and feature different stretches of the river basin, thus allowing for unique insights. Furthermore, both countries also fabricated a range of additional models to portray specific features such as dams and sluiceways.

The American model ran from the brink of the Falls to Lake Erie, covering the entire upper Niagara River. The Ontario Hydro model covered the brink of the Falls to Navy Island, depicting five miles of the river from the tip of Grand Island to the Rainbow Bridge, which allowed for inclusion of the Falls at the largest practicable scale. Encased within concrete walls, it was 95 feet long, 37 feet wide, and approximately 4 feet high. A custom, electrically driven gantry crane, capable of lifting a ton, allowed changes to any part of the model. To combat frost action, the model was built on a reinforced concrete slab with supports running to hardpan. Plywood and sand were used to create templates over which concrete was

FIGURE 5.3 Model at Vicksburg, Mississippi. *International Joint Commission*

FIGURE 5.4 Model at Islington, Ontario. *International Joint Commission*

poured. Water was pumped to a constant head tank from a reservoir below the model, and released by pipes equipped with controlling and measuring devices, and additional gauges and metres were installed elsewhere on the model.

An Ontario Hydro press release enthused that its model bore "an almost breath-taking resemblance to Niagara Falls and the Upper Niagara River."[9] Built on a distorted scale, which is usually the case when modelling long river stretches, of 250 feet to 1 foot horizontally and 50 feet to 1 foot vertically, "the distortion is necessary to provide sufficient depth in the model to ensure a flow pattern similar to the actual river and also enable depths to be measured accurately." Moreover, the press release continued, the model "has a most realistic appearance, and the water plunging down the upper rapids and over the two falls duplicates the actual phenomenon. Approximately 91 percent of the water passes over the Horseshoe Falls in the models and 9 percent over the American Falls as it does at Niagara."[10]

The 7,000-square-foot Vicksburg model included the upper reaches of the Niagara River, and since it represented a greater geographic distance it was on a smaller scale: 360:1 for the horizontal, and 60:1 for the vertical. Given the mildness of Mississippi winters, this model could be built outside in the ground.[11] Wire screens were fitted to the bed of each model in order to simulate riverbed roughness. Small metal strips were embedded upright to reproduce turbidity in the cascades area of the model. The engineers cautioned that, though these took away from the appearance of the model, this was "an unavoidable limitation," for "the model is an instrument, primarily, for measuring distribution of flow, not for direct comparison of natural beauty. If the models show a certain flow over the Falls, the best way to judge the corresponding appearance is to inspect photographs showing the same flow, or nearly the same, in the prototype."[12]

Before the models could be built, experts had to undertake various hydrographic surveys of the "prototype" – the field-size situation, in this case the actual river and waterfalls – in order to get baseline information from which they could construct a model. In the stretch of the river from Lake Erie to near the head of the upper rapids, conventional survey methods and echo sounding were sufficient. But finding accurate baseline information for the cataract and the upstream cascades was a challenge: they had to find out how to map and measure a hostile environment – the streambed topography – that was difficult to access because of the fast and turbulent waters. They came up with engineering advances to take soundings of the riverbed and water velocities. One of these was a lead weight suspended by piano wire from a helicopter, with a "rigged-up" counterweight balance in the helicopter cab to take up the wire slack. In areas where the helicopter could not navigate, "a system was devised for using three Kytoons [helium-filled balloons shaped like cigars] lashed together from which was suspended a similar sounding weight and appurtenances all of which were operated from ground crews on shore."[13]

By spring 1951, the draft models were complete. The next step was to test them against their prototype to assess their accuracy, a process called "model verification." Accomplishing this required "careful adjustment of channel roughness until an accurate and detailed reproduction of all observed hydraulic phenomena of the prototype river is obtained."[14] This involved taking water surface elevation measurements in the model and comparing them with gauge readings from the Niagara River. But here they experienced problems. There was an "apparent dissimilarity of flow conditions over the crest of the Horseshoe Falls in the American and Canadian models."[15] The offending variables had to be isolated. They conjectured that their aerial photographs "were probably inaccurate due to the time lag between the stereoscopic pairs"[16] while water contour maps were "suspect" because of "the probability of false parallax" resulting from the high-velocity flow.[17] The engineers resolved to take further surveys to mediate the differences. The Corps re-examined the bed levels using the Kytoon method, and proposed, instead of aerial photographs, a new method for measuring water surface elevations in the cascades that used a searchlight; Ontario Hydro and the Corps carried out "Operation Searchlight" in late November 1961, taking some 6,000 water surface observations.[18]

Both the Canadians and Americans produced new contour maps and prototypes, and rebuilt their models according to the revised specifications. As part of this rebuild, the Ontario Hydro model was changed to a cellular form so that individual pieces could be quickly removed or replaced in order to simulate excavations and other remedial works. Then it was time to retest. Verification of the models was "a time-consuming trial and error process." The adjustment process was repeated until the models showed "satisfactory" agreement with the prototype, and with each other.[19] In other words, engineers used the "cut and try" method until the models matched the prototype, which was based on the river, with the point of the model to go back and rework that river. Engineers generate general rules – called "building standards" – to be used as parameters, and then through trial and error the uniqueness of any given situation, or river, is incorporated and plans and models revised accordingly.[20]

On April 21, 1952, the revised models were declared satisfactory. With the models verified, the engineers could use them to trial the different combinations that they were considering at different flow levels. For distributing the flow over the Horseshoe Falls, the main options were weirs, dams, excavations, and fills. Tests indicated that without remedial works the flows over the Falls would be reduced below what was considered necessary "to maintain the existing spectacle." Other tests showed that it would take upward of half a day to move the flow rate from 50,000 to 100,000 cfs. That pace was unacceptable, because in the summer it meant that Ontario would take so long to get to the 100,000 cfs diversion level that it would almost immediately need to start cutting back to 50,000. This would result in considerable power and revenue wastage.[21]

Before the Second World War, engineers had envisioned a series of submerged weirs as well as various configurations of control dams. But the models now indicated that these would not be flexible enough to meet the treaty's flow parameters. A control structure with gates, what some might term a dam, which could maintain the pool level for power diversions while also distributing the flow to the cataracts in a satisfactory way, emerged as the backbone of a new water control regime, along with fills and excavations to reconfigure the riverbed and lip of the cataract. The 1949 Federal Power Commission report on Niagara development had placed this control structure near the outlet of Lake Erie at Buffalo, some eighteen miles upstream. But engineers now believed that the optimal location was right above the Falls (see the location of the control dam in Figure 5.2). In addition to a control structure, the remedial works would include reclaiming the flanks of the Falls so that no unsightly trickles would be obvious and a greater volume of water could be channelled to the shortened crest of the Falls. These fringes would be reclaimed and landscaped to serve as viewing areas.

Studies showed that the Horseshoe Falls flow distribution was not affected nearly as much by the level of the Chippawa–Grass Island Pool as the American Falls. Instead, the Horseshoe Falls crest distribution was chiefly governed by the volume of flow from the rapids, or cascades, between the Chippawa–Grass Island Pool and the brink of the cataract. Without a control dam, Chippawa–Grass Island Pool levels would drop up to four feet.[22] PASNY and Ontario Hydro both wanted higher levels in the pool in order to replenish their hydro plant intakes; if levels were too high, however, then there was the possibility of causing high levels on Lake Erie, and the governments might be liable for any resulting damage.[23] Moreover, the engineers had to account for the seasonal growth of weeds, which could displace a not insignificant amount of water, such as raising water levels half a foot.

This speaks to the hybridity inherent in the design. Organic elements were accommodated and factored into their mechanical creation, incorporated as part of the dynamic infrastructure: i.e., the new envirotechnical system engineers were creating would not work properly *without* normal weed growth and ice formation.[24] Designing remedial works and power station intakes to withstand or deal with ice, particularly frazil ice, was a paramount concern. Formative studies of the impact of ice formation had been done in the Niagara River over the previous century, and many of the engineers from Ontario Hydro and the US Army Corps of Engineers were on the leading edge of ice studies. There was still "a dearth of factual data" about ice, however.[25]

Whereas a "controlling criterion" in the studies underpinning the 1929 Niagara Convention and Protocol was a satisfactory depth of water in the high-flow section of the Horseshoe Falls in order to create a satisfactory colour effect, now complete coverage of the flanks was equally important. But what was the acceptable minimum depth and distribution? Several aspects had changed, after all, since 1929: there was the submerged weir, and additional diversions had been authorized in the 1940s. The engineers "agreed that an attempt would be made to establish a scenically satisfactory flow per foot over the flanks in advance of the tests in order to present a target at which the tests would aim."[26] Thus, they agreed to the following guidance criteria:

> Remedial works should distribute the flow so as to present an unbroken crestline at the Horseshoe Falls and to give, during the daylight hours of the tourist season, a flow of from ten to twelve cubic feet per second per foot of crest near the Canadian end of the Horseshoe and from six to eight cubic feet per second per foot near the Goat Island end.[27]

On the model, engineers simulated dikes and barrages in the cascades above the horseshoe to deflect water to the edges – eleven different model designs in all – and then they tried with fills and excavations. The latter were ultimately recommended since they were more aesthetically pleasing, no more expensive, and less susceptible to ice damage. But the precise placement and size of fills and cuts needed further tests: for example, lesser amounts of excavation at the flanks could produce an unbroken crestline on the model with the river flow at 100,000 cfs, but not at 50,000 cfs.

While a control structure had emerged as the preferred method in place of weirs, what should its dimensions and features be? The engineers went to the model to test the different options: A dam across the entire river, or only part of it? Or another dam at the head of the river by Buffalo? Should a control dam have sluices or movable gates so that water flows to certain sides of the river could be better controlled, and to deal with ice? The models were used to progressively eliminate the less suitable solutions. The control structure would be most useful on the Canadian side of the river, since the riverbed slanted in that direction. The proposal for a short dam from the American shore to join a longer structure from the Canadian shore, with a gap in between, was rejected because the costs would be out of proportion to the resulting benefits.[28] Settling on a control dam extending from the western shore, experiments indicated that it needed to be a minimum of 1,550 feet long and feature thirteen gates/sluices. There was a good chance a longer structure would be necessary once an American plant (namely, the Moses station) was built and its intake locations were known.

By late 1952, the engineers were ready to demonstrate their recommendations to the IJC. A number of officials came to see the Islington model in action. It had become the favourite model for binational meetings because of its proximity to the actual Falls. Even if some observers could not properly interpret the models because of the distorted scale, the engineers were confident in their own ability to do so, and, moving forward, the models served as the basis for Niagara remedial works.[29]

Preserve and Enhance

In spring 1953, the IJC approved the plans for the remedial works, at a price tag of about $17.5 million split between the two countries, via its "Report to the Governments of the United States of America and Canada on Remedial Works Necessary to Preserve and Enhance the Scenic Beauty of the Niagara Falls and River."[30] This 354-page report had lengthy appendices, including photos of the various models and remedial work options, graphs and tables of data, and blueprints. The objectives remained basically the same as in the interwar years: to ensure the appearance of an unbroken and satisfactory crestline while allowing for the diversion of water for power production. But the emphasis seemed to be on doing the bare minimum that would maintain a "sufficient" and "satisfactory" flow and inhibit erosion, rather than create an exceptional or outstanding spectacle.

According to the IJC report, remedial works needed to ensure a dependable and satisfactory flow over the waterfalls, including an unbroken crestline, without harming levels in the Chippawa–Grass Island Pool, but with the ability to promptly step up and down the permissible power diversions. To achieve these goals, the IJC called for three key types of interventions. The first was the control structure stretching out from the Canadian shore, parallel to the submerged weir but downstream from it. The second involved excavations in the horseshoe cascades near the shoreline, and crest fills on that flank about one hundred feet long. The third consisted of the same procedure on the Goat Island flank of the Horseshoe Falls, except with a crest fill of about 300 feet. Each country would generally construct the works within its boundaries, giving the United States the Goat Island work and Canada the other flank and the control structure, though there would be some overlap. Both countries approved the commission's recommendations in July 1953. Construction could now begin. Prime Minister Louis St. Laurent formally tagged Ontario with the responsibility, and the province in turn designated Ontario Hydro to fulfill this role. Congressional appropriations equipped the Army Corps of Engineers to handle the American share.

Given that segments of the population were reportedly worried that the new developments would damage the Falls, the International Niagara Board of Control's first activities included publicity measures and dissemination of information to assuage public concerns and stave off rumours. According to a former public relations official who worked for Ontario Hydro chairman Robert Saunders, the utility was very wary of the public response to decisions that affected the Falls.[31] Ultimately, Ontario Hydro and PASNY spent about $40,000 each on public information films, signs, guides, and brochures, primarily intended to glorify their undertakings and shape the narrative. They also opened the hydraulic models for public viewing.[32] The IJC ran its own public relations program, including exhibits and films, in order to "minimize unguided speculation which might be the basis for unfounded rumors leading to unfavorable newspaper publicity."[33] All of these entities regularly distributed press releases.

Planning for the form, type, and location of the control structure continued. An accelerated construction schedule was adopted since the 1950 treaty stated that construction should take a maximum of four years.[34] Engineers had to take into account a ledge of rock where the upper rapids commence above the Falls, which serves as a natural weir regulating the

FIGURE 5.5 International Control Structure under construction. *Ontario Power Generation*

Chippawa–Grass Island Pool. The submerged weir built in the 1940s was a secondary control on pool levels. Power diversions taken from above the submerged weir (i.e., by the Moses and both Beck stations) would affect Chippawa–Grass Island Pool levels, while any diversions from below the submerged weir (the "cascade plants": Ontario Power, Toronto Power, and Canadian Niagara) would affect only flow over the Falls. The control structure was initially envisioned 200 to 250 feet downstream from the existing submerged weir, but model studies indicated that it should be moved another 50 feet (see Figure 5.5). The International Niagara Board of Control directed that the levels of the Chippawa–Grass Island Pool should be the same as those that would have occurred before the remedial works and hydroelectric expansion. Unsurprisingly, this was not possible to consistently achieve in all conditions, and studies showed that Lake Erie levels would be lowered three to five centimetres.

Cofferdams were required to still the water so parts of the control structure could be built "in the dry" below the waterline. Cofferdams are the behind-the-scenes heroes of the hydraulic engineering world – they are indispensable but get none of the glory. Inherently transitory and

Figure 5.6 Niagara riverbed dewatered by cofferdam. *Ontario Power Generation*

utilitarian, they are the first thing that gets built, but are intended to be dismantled as soon as their purpose has been fulfilled. For the Niagara control structure cofferdam, steel frame works were faced with steel sheet piling, forming open cribs into which concrete blocks were dumped. This was executed in a series of six stages, the structure progressing out into the river in a sort of "leapfrog operation." Frequently the fill for cofferdams gets washed away, as happened on other cofferdams used at Niagara, so even more is layered on until it sticks, and the area is enclosed and then dewatered (see Figure 5.6). Submerged weirs, wing dams, and deflecting dams are all a step up from the cofferdam, for they are meant to be permanent, if not as prominent as dams that hold back a waterway's entire flow. While engineers and architects do not spend much time, if any, plotting out the aesthetics of a cofferdam, the appearance of grandiose power stations, on the other hand, received detailed treatment.[35] To illustrate, the IJC was involved in extensive discussions about the technical requirements and architectural design of the International Control Structure. Planners wanted to minimize its intrusiveness on the vista without compromising its ability to perform its hydraulic duties. This meant keeping its profile as low as possible – or "graceful" as an Ontario Hydro pamphlet subsequently described it (see Figure 2.1).

By the end of 1953, various political and economic delays and engineering uncertainty about model results had put the remedial works behind schedule. This became a bilateral point of irritation. Other water-related issues festered as well. Canada formally protested a proposed increase to the Chicago Diversion, while the US House of Representatives Foreign Affairs Committee demanded that the United States cancel the 1940s notes authorizing the Ogoki–Long Lac diversions. More significantly, in 1953, the attempt to build an all-Canadian seaway was coming to a head; by 1954, Ottawa had reluctantly backed down and agreed to a joint waterway, to be built in conjunction with a joint New York–Ontario power project. Remaking the St. Lawrence for hydroelectricity was an even bigger job for PASNY and Ontario Hydro, as some 40,000 acres of land and infrastructure – including a number of communities – would be submerged. The St. Lawrence and Niagara megaprojects had reciprocal influences on each other, since many engineers, workers, and bureaucrats worked on both.

It became apparent that Ontario Hydro planned to begin diversions to its new Beck development in 1954, before the remedial works were completed. An American member of the IJC objected, arguing that remedial works needed to come first since the spirit of the treaty was to protect and increase the beauty of Niagara Falls. Starting the new diversion regime before the remedial works were in place would leave the Falls looking pathetic. The State Department followed up with correspondence to that effect.[36] The Canadian government, and the Canadian section of the IJC, held that this "beauty-first" interpretation of the treaty was flawed, even if that sequencing had been the intent of the treaty negotiators. Nothing prevented Canada from diverting water, they averred, before the remedial works were complete. Ontario Hydro contended that its model studies showed that the cofferdams would effectively double as temporary control works for river levels.[37] Ontario Hydro gave repeated assurances that no damage would result. The United States sent a diplomatic note stating that no diversions for power purposes should take place until both governments were assured there would be no scenic impairment. But it reluctantly let the matter slide, assuming that the IJC's International Niagara Board of Control, which was created to oversee construction works and subsequent water levels and diversions, would prevent any untoward problems.[38] Levels also happened to be high throughout the Great Lakes–St. Lawrence basin. The United States and Canada conducted separate talks about how

FIGURE 5.7 Dewatered Canadian flank of the Horseshoe Falls (note the shape of the US flank and contrast with Figure 5.8). *Ontario Power Generation*

to deal with this problem, and American concerns were alleviated by Ontario's trimming of its Ogoki–Long Lac diversions.

While it is difficult to generalize about the views of an entire government apparatus and its citizens, the United States cared more about protecting the Falls than did Canadians and their officials. On the American side, there was considerable public pressure from civic groups for scenic works, and the Special International Niagara Board seemed genuinely concerned about the scenic quality of the Falls. Though there were clearly vocal public elements in both countries opposed to manipulation of the waterfall, they tended to be louder in the United States. Nonetheless, most governmental elements in both countries exhibited a technological and environmental hubris about their ability to configure Niagara Falls.

Confirming his recurring fears that scenic beauty was much less important to Ontario Hydro, American official Ray Vallance complained about the state of the remedial works after a tour in mid-November 1953. Vallance was well known among Canadian officials for his slogan "No diversion before substantial completion."[39] Noting that the United States had pressed hard to get congressional appropriations quickly approved

FIGURE 5.8 Creation of Terrapin Point on the US flank of the Horseshoe Falls. *Ontario Power Generation*

for the US share, Vallance said he now found thousands of workers toiling around the clock at the power station, but only three or four employees engaged on the remedial works. The chair of the Canadian Section of the IJC, A.G.L. McNaughton, reacted with derision, pointing out that it was the Americans who had caused delay at every turn.[40] Construction progress reports appeased Vallance for the time being, but American worries would resurface the following year.

In early 1954, Ontario Hydro awarded contracts for several major control dam features. The US Army Corps of Engineers began soliciting tenders for the required work on the Goat Island flank. For technical reasons, the engineers had decided that fills and excavations on the two edges of the Horseshoe Falls should take place separately, with the Goat Island side addressed first. The control board approved plans for the Goat Island flank, and a Buffalo construction firm commenced work at the end of April (see Figures 5.7 and 5.8).

On the Goat Island side, about 24,000 cubic yards was excavated, much of it to create a channel running from the deeper middle section of the cascades toward the eastern part of the Horseshoe Falls crest. The point of this was to stop so much water from concentrating in the middle of the horseshoe. The improvement of the Goat Island flank involved a stone-faced retaining wall enclosing a triangular area "with a base about 400 feet long on Goat Island, a side of about 300 feet along the brink of the Falls, and the other about 500 feet along the new deep channel" that was "filled with excavated material and landscaped to provide additional viewing space."[41] Initially, more square footage was going to be reclaimed, but there were worries about using unstable rock as a public platform. After dewatering this section, engineers found that the upstream area needed further adjustments: the riverbed elevations produced by the surveys and soundings, and thus reproduced on the models, were higher by several feet. After model tests to determine the solution, an additional excavation of 4,000 yards, an overrun of 15 percent, was performed, taking the form of a succession of high and low areas about one foot above and below the specified grades in order to approximate the turbulence of the natural riverbed. All these fills appreciably increased the size of Goat Island.

Engineers and park officials debated about how to finish the new landform, christened Terrapin Point, at the western fringe of Goat Island. Some advocated strewing it with boulders to look as "natural" as possible. But this approach lost out to arguments that visitors probably wouldn't prefer the resulting "trickles" of water – a solid curtain was preferable – and walls and reclaimed land were employed to broaden Goat Island.[42] This area had previously consisted of rocks and boulders that supported the walkway to the nineteenth-century Terrapin Tower overlooking the waterfall. About seven to eight feet of debris and soil was dumped there and graded and landscaped to harmonize with Goat Island.

Additionally, at the American Falls, a projecting ledge of rock at the crest of the Bridal Veil Falls, between Goat and Luna Islands, was blasted off. Furthermore, as noted in the previous chapter, fill was also placed along the length of the American shore for the length of the Chippawa–Grass Island Pool, extending the mainland into the river (see Figure 5.9). This also occurred, albeit to a lesser extent, on the Canadian shoreline of the pool. The many channel and shoreline fills, along with bridge piers and other interventions, cumulatively constrict and change flows in the river.[43]

The cofferdam for work on the Canadian flank went in during May 1955, then 70,000 cubic yards of rock was removed from the river bottom, mostly adjoining the Ontario shore. As on the American side, a channel was etched from the deep part of the cascades leading to the sill of the Horseshoe Falls, spreading more water to the western edge of the cataract. A stone-faced concrete wall was supposed to be erected along the brink

FIGURE 5.9 Reclaiming shoreline of the Niagara River in front of the Adams plant and intake canal. *Niagara Falls Public Library, New York*

of the waterfall about one hundred feet from the old shore line, but model tests indicated that fifty-five feet would be sufficient (and would not de-water the portal of the Canadian scenic tunnels below). This area was filled with excavated material and landscaped. The new observation area on the Canadian side was on cap rock from the remnants of Table Rock, and thus remained under tensional stress. To address this, workers removed more of the overhang, injected grout to prevent moisture ingress, and built a retaining wall.[44] Requests to dispose of the excavated rock in the Maid of the Mist Pool were rejected, however, by the International Niagara Falls Engineering Board (see Figures 5.10 and 5.11).[45]

Almost 100,000 cubic yards of rock, about twice the volume of the US Capitol Rotunda in Washington, had been excavated from both sides of the Horseshoe Falls. The bed of the riverbank leading up to the Falls, and the lip of the waterfall itself, had been reconfigured. Islands and shoals were added or subtracted. Deepened channels entrained water to spread out over the rim and avoid concentrating in certain spots, while retaining walls held it in. This apportioned water in the desired way, but near the

FIGURE 5.10 Terrapin Point flank of the Horseshoe Falls in the twenty-first century

FIGURE 5.11 Modifications to the Horseshoe Falls. *Image by Jason Glatz (Western Michigan University Libraries) and Daniel Macfarlane*

brink it became more like a canal with hard edges than a river with sloping shorelines. The depth of the water in the middle section of the horseshoe was reduced from about ten feet to an average of two to three feet, with the water that had formerly been in the middle sent to the sides, where it had a similar depth. The point was, as it had been since the 1920s, to create the much-sought-after curtain of water and "impression of volume": a consistent and dispersed volume and appearance that wasn't visually interrupted by rock and incidental flows at the extremities. According to both quantitative and qualitative measurements, there would not be enough water after the power diversions to create an adequately impressive flow over the Horseshoe Falls, so the solution had been to change the size of the Falls. Thus, engineers had shrunk the crest of the waterfall by reclaiming its edges by 355 feet (see Figure 5.11).

The International Niagara Board of Control approved both reclothed flanks. But the Falls sent signals that it didn't like to be corralled. While the Horseshoe Falls was under construction, in summer 1954 there was a large rock slide at Prospect Point on the American Falls (see next chapter). Some blamed the rockfall on the extra water shunted over to the American Falls as a result of the ongoing renovations, but the board of control concluded that this had no bearing on the matter. Another slide then occurred just north of the Horseshoe Falls on the Canadian edge of the gorge.

Remedial Operations

The 1950 treaty had dictated that the remedial works be finished within four years, i.e., July 23, 1957. The fills and excavations were completed in 1955, while the control structure became operational just under the wire. Thus, the remedial works package was finished on time and in an extremely challenging environment that involved numerous crosscutting national and subnational jurisdictions.[46]

On September 28, 1957, a small ceremony was held at the control structure to dedicate the International Niagara Control Works.[47] Since officials wanted to minimize the remedial aspect of remaking Niagara Falls, in contrast to the inaugurations for the power stations the ceremony involved only a handful of representatives from both countries and the IJC. A number of laudatory speeches were given, with the predictable platitudes, and an aluminum plaque installed.[48] Wilber Brucker, the US secretary of the army, proclaimed: "What has been accomplished here at Niagara Falls is one more impressive example of the wonders man has wrought through the application of his engineering skill and genius in altering the physical world to fit the pattern of his expanding requirements." Brucker referenced the "evil purposes" of the Soviets and claimed that the US-Canada relationship "furnished a shining example for all mankind" and that this Niagara project reflected "the genius of Canadians and Americans for the effective conduct of mutual enterprise." "As man has triumphed over his physical environment," Brucker continued, "so must the peoples of the world triumph over the moral environment of falsehood, fear ... envy, hate, and arrogance."[49]

Alvin Hamilton, the Canadian minister of natural resources, summarized the bilateral accomplishments:

> The American side has been enhanced with a greater flow of water, the flanks of the Horseshoe Falls have been re-clothed and the ugly scars removed, an unbroken crestline has been achieved, and the mist cloud, which so often hid much of the beauty of the Horseshoe Falls, has been lessened, new vantage points for sightseers have been provided and landscaped and the central concentration of flow which formed the threat to the future of the Falls has been tapped and dispersed. In the face of increased diversions of water for power there has resulted for posterity a scenic spectacle more captivating than that which we have known in the past.

The minister closed by remarking: "Surely this is conservation at its best."[50]

PASNY was close to completing its new Robert Moses plant and intakes. However, model tests suggested that once the power entities diverted all the legally allowable water from the Niagara River, the Chippawa–Grass Island Pool could not be maintained at the authorized levels during non-tourist hours. As planners had suspected, it quickly became apparent that "the control structure as built is of insufficient length to hold the level prescribed by the Board of Control under all conditions," and Lake Erie levels might be affected. The control dam would either have to be lengthened or power diversions would have to be curtailed; the latter was not an option PASNY or Ontario Hydro wanted to entertain.[51]

But there was concern that the results were skewed by weed growth as well as by the models themselves, which had to be recalibrated after all the recent changes to the fluvial geomorphology of the river. Running additional flow experiments in both the models and the actual river, engineers ultimately determined that the models were not the problem in this particular case.

Adding more gates onto the end of the control structure would require formally going to the IJC for approval. The Niagara engineering boards decided that, with the Moses plant coming into operation, they would observe the operation of the existing remedial works.[52] During this time, they would stick to the legal pool limits "as near as may be."[53] The two utilities claimed that lengthening the control structure would enable them to better control ice above the dam, but their studies also showed that a dam extension would potentially create new problems with passing ice. Using paraffin blocks to simulate ice in the models, they advised the installation of training walls upstream and downstream from the dam to compensate.

Ontario Hydro and PASNY agreed in early 1961 to submit a supplementary reference to the International Joint Commission asking for a five-sluice extension to the control structure in order to make greater use of their allotted water. As part of this reference, they also requested looser limits on diversions to see whether greater flow reductions than those specified in the 1950 treaty would negatively alter the scenic appeal. Specifically, they proposed a test period to see whether a summer daytime flow of 70,000 or 80,000 cfs, rather than 100,000 cfs, sufficed. Ontario Hydro projected that this would save millions of dollars, and any savings for the public power utilities would be passed on to the people. In a bit of circular reasoning, the engineers argued that "a number of responsible opinions [believe] that a Falls' flow of less than 100,000 cfs would provide a very satisfactory spectacle."[54] Furthermore, according to the Canadian representative on the Niagara board, at times faulty gauges had resulted in 20,000 cfs less than the minimum required quantity going over the Falls (though he called these only "minor"). Since no one had complained, he calculated that the proposed reduction would not be noticed either. The two utilities also proposed a 10 percent volume tolerance for Falls flow discrepancies, with changes to the method of calculating the average. Briefing the two federal governments on their plans, the power authorities further rationalized that lessening the flow might actually be advantageous since it would further diminish spray and erosion. The idea of a dam where Lake Erie flows into the Niagara River, or a dam in the lower Niagara River, resurfaced, but did not get much traction.

A Canadian Department of External Affairs memorandum astutely pointed out that these proposed works would do nothing for the beauty of the Falls, but had been designed entirely to ensure that the power entities received all the water they were entitled to divert.[55] Nonetheless, both governments forwarded the reference to the IJC, which, having been privy to the ongoing discussions about the extension, quickly approved it, stating that these enhancements to the remedial works were essentially the logical conclusion of the earlier works. The commission recommended an extension of five sluices to the control structure, the additional training walls, and the removal of the top of the old weir. It had the hallmarks of a bait and switch: get the treaty passed with tight rules on diversions, but then relax them, or change how they were calculated.

The IJC labelled this work urgent and held only one public meeting, but negative feedback came via other forums. This opposition was part of the nascent environmental movement starting to sweep the continent, but stemmed as much, if not more, from the unique historical affinity for Niagara Falls as the archetype of natural splendour. Prior to the 1950s, technocrats planning to alter the beauty-versus-power quotient at Niagara conducted little systematic analysis of public opinion, and more or less assumed that their opinions would, and should, be shared by the general masses. The fact that tourists kept showing up in larger numbers affirmed this view. Starting about the midpoint of the twentieth century, however, officials put more effort into discerning public opinion about changes to the Falls. As will be discussed in the next chapter, in the 1960s planners formally surveyed the potential emotions that tourists might experience when encountering various configurations of Niagara Falls, and then tried to create models that would produce these qualitative reactions.

The chambers of commerce in both cities astride the Falls, as well as city and township councils, grew skeptical about the new proposal. The Electrical, Radio and Machine Workers of America passed a resolution calling for no further diversions for any reason, and condemned the control dam extension. Locals stated in newspaper interviews that the existing remedial works had already been detrimental to the scenic beauty, with one stating that the envisioned additional remedial works would only result in "a completely man-made and artificial cataract." Moreover, the interviewee continued, "it is a far stretch of the imagination to call the works remedial: that is, enhancements of the natural assets of the Niagara. They quite obviously are for the diversion of water for electric-generating development in the Lewiston-Queenston area, and it is completely false to say anything to the contrary."[56]

It is quite possible that some of this dissatisfaction stemmed from people seeing the Falls at the reduced 50,000 cfs stage for the first time, which may have been exacerbated by extremely low water levels around the Great Lakes basin. A newspaper suggested, tongue-in-cheek, that visitors to

Niagara might get the impression that it was a tidal waterway because of the daily changes to the water level.[57] The mayor of Niagara Falls, New York, complained about the water dividing into two deep channels, the lower water levels reportedly causing the whirlpool downstream from the Falls to reverse, and adverse impacts on birds and fish.[58] Another complaint was that, when water discharge was at 50,000 cfs, the upper rapids resembled a rock garden because the water was so low.[59] That is, when flows were at 50,000 cfs, shoals and boulders in the cascades above the Falls were exposed, and foliage and birds appeared on some of the bare rock, turning them into provisional islands. The water levels dropped enough that two of the Three Sisters Islands became one – this would have removed the basis for their name, but rock excavation soon separated the affected islets.

Because power diversions lowered the level of the Maid of the Mist Pool below the Falls, they effectively made Niagara Falls taller as well. Formal studies indicated that the pool's level had decreased by some 20 feet, depending on the time of day, and thus the height of the Horseshoe Falls had increased to about 180 feet. Incidentally, because of the lowered water levels, an 18-foot channel was blasted in the rock bed of the Maid of the Mist Pool so that tour boats would have enough draft to reach their docks.[60]

The IJC conditionally approved the additional control dam extensions in August 1961. Dissatisfaction was even greater in the United States than in Canada, however, and many civic groups complained to political leaders, particularly about proposals to increase daytime diversions during the tourist season. Faced with this popular outcry, in 1962 New York governor Nelson Rockefeller compelled PASNY chairman Robert Moses to withdraw the request by refusing to fill two vacancies on the PASNY board of trustees.[61] Ontario, also feeling domestic pressure against the diversions, concurred and the application was amended to remove the diversion aspect. Nevertheless, the proposed remedial works, at a cost of $7.45 million split between the power entities, including the extension to the control structure and training walls as well as the removal of a shoal at Tower Island, went ahead and were completed by 1963.

That winter, unusually high ice levels caused significant problems for the intakes at the Beck and Moses power stations. Ice also caused about $3 million in shoreline damage. The winter of high ice coincided with several years of below average flows, and in 1964 power generation was reduced by 20 percent at the Moses plant.[62] Both PASNY and Ontario Hydro received permission to dredge and excavate high spots, shoals, and islands on which ice hung up, adding up to 161,000 cubic yards of spoil, and they bought shallow-draft icebreakers to ply the upper river.[63] But it was apparent that something better was needed to keep ice from blocking the intakes. Authorities developed plans for an ice boom – a floating structure designed to prevent ice passage – to be set up at the head of the Niagara River. The IJC approved the boom on a trial basis in 1965, and soon made it permanent since the boom, along with other ice mitigation infrastructure, reduced ice problems to an acceptable level.

As Near as May Be

The fact that the Niagara River is an international border significantly influenced the engineering process, for not only were different federal government agencies and departments involved but also those from the provincial and state levels, as well as bilateral forums such as the IJC. Some forms of knowledge were circulating and migratory, and others stopped at the border, which gave rise to both epistemological conflicts and compromises among experts. Sometimes this was purely the result of self-interest, hydraulic nationalism, and power politics, but it also resulted from differing national, regional, and institutional conceptualizations of the nature of Niagara. Assumptions and knowledge claims that might be self-evident for certain groups had to be explained and justified to other groups, which meant that experts had to confront areas of uncertainty. At the same time, because of their shared occupational backgrounds and training, engineers from different nations could collaborate and come up with a shared way to deal with the uncertainty – i.e., as near as may be – and in other ways these liminal situations served as crucibles that forged and synthesized knowledge.[64] In the 1950s, admitting uncertainty in public was unthinkable, so engineers had to exhibit confidence; by the later 1960s, this facade was beginning to crumble and engineering reports began hinting at the disclosure of doubt.

The phrase "as near as may be" was repeatedly used as part of the Niagara engineering and modelling process, and it is an apt descriptor for the engineering *modus operandi*. The phrase was a proxy for uncertainty, with "satisfactory" or "sufficient" agreement constituting "reliable," or at least reliable enough. While central aspects of this massive project were the product of technological momentum or path dependencies, the use of "as near as may be" was also a purposeful means of incorporating some flexibility into their plans by allowing for contingencies or wiggle room. In other words, engineering uncertainty was beneficial to a certain extent since it required them to leave room for adaptation, thus preventing some negative path dependencies, unfavourable technological lock-in, and likely additional expenses.

While their approach to the models helped avoid certain mistakes, the binational engineering approach of the involved governmental entities, funnelled through the IJC, was nonetheless rife with miscalculations, assumptions, and compromises. Political and partisan preferences were sometimes more influential than accurate technical evidence, for there were many instances of Canadian engineers complaining about American tests and models, and vice versa. The US Army Corps of Engineers and Ontario Hydro sometimes came up with conflicting results, and contended that the other's model was wrong.

Models go even further than maps or images, and at different scales, allowing for a more multi-dimensional representation of a site.[65] Like maps or blueprints, however, models inevitably require abstraction, simplification, and obfuscation. Consequently, model findings were often flawed. This could be the result of incorrect knowledge, such as faulty gauge or survey data about the river prototype on which they were based. The engineers believed that when they had sufficient knowledge of conditions – the types of rock, the composition of soils, the water flows and velocities – they would be able to control the ecosystem. But there were a range of problems that often stemmed from lack of knowledge about actual conditions. Engineers worked hard to solve these as part of correcting the models, but occasionally these errors were not apparent until work began in the field. The hazards of working in the river resulted in some incorrect baselines, which in turn could hamper the "model verification" since that required comparing model performance against known conditions, which was not always possible. The model and the prototype might show matching water surface elevations, but such readings could be obtained from very different combinations of fluvial geomorphological conditions, thus not necessarily revealing the actual shape of the riverbed.

Other aspects of a model's design and material properties could also be a problem. It had been necessary to distort the scales of the models because of the upper Niagara River's width and relatively shallow depth. This inherently skewed nature of the models led to errors. As an internal engineering report concluded as late as 1959, after years of experience with the models, "on the basis of Niagara tests which [Ontario] Hydro itself has run on models of its own design, Hydro Engineers have been led to conclude that the models being used by the Board, which were designed to a distorted scale, produce results which are quite different from those run on true scale models."[66] The types of materials used to mimic nature, while innovative, sometimes were not accurate enough – for example, simulating the turbidity of the river with metal strips. When extrapolated onto a larger scale, any seemingly slight model mistakes could have large and significant real-world ramifications.

The following exchange at an IJC meeting, which involved Ontario Hydro chief engineer Otto Holden and one of the lead engineers, J.B. Bryce, further reveals the experts' approach to models and uncertainty:

Otto Holden: When the distortion gets higher there is a level above the dam that you would not get if it was a natural scale. Then, when you adjust your model you have to take out some friction and resistance. When you take that out of the model, the water takes a different path and goes around to the end of the dam. I don't think you can build a model of a full dam at natural scale. But the effective distortion is something approximating half a foot.

J.B. Bryce: We are not too sure what it is.

Holden: Something like four-tenths.

Bryce: An appreciable amount.

Holden: Quite a substantial amount. I think possibly we might consider, now that we have the prototype there, while we have low flows in the river, we might run a test right on the prototype, and get one point and see how that one point compares with the model test.[67]

Other experts also expressed concerns about the models. A Corps of Engineers colonel "raised the question as to treatment of the exposed, boulder strewn riverbed along Goat Island and stated that he did not believe models were sensitive enough to locate the shore line accurately."[68] The failure to account for the numerous islands directly above the Falls, many of which now appeared because of lower water levels, also created problems "since they were too small and numerous for accurate representation in the models," though engineers swept this concern aside: "However, it was pointed out that the islands are in the area near the crest of the Falls, where the models indicated no excavation necessary."[69] The intentional suppression of certain influences, such as these islands, can create discrepancies that hydraulic engineers now call "scale effect." But this compartmentalization was part of the way they treated problems or complexity for which they had no solution: exclude or conceptually bracket such issues as irrelevant, unknowable, or acts of God.

Models also could not adequately represent the changes to the sensory experience that would result from altering the waterfall. At public forums, perceptive members of the public "emphasized that the scale models would not show the turbulence of the Upper and Lower Rapids, the rainbows in the mist, the ice bridge, nor the impressive rumbling of the falling water."[70] Would its waters continue to run its trademark green, for example, if changes to erosion rates altered the amount of dissolved minerals and fine rock particles? The echoing roar of the Falls of Niagara without question contributes to the overall ambiance of standing at the edge of the cataract, but on a model there was no way to factor in or account for the ways that abstracting much of the river's volume altered the sonic experience.[71] Had the sound of the waterfall changed perceptibly? How could a model predict mist and spray patterns or whether they would still form the beloved rainbows?

The engineers had to walk a fine line between knowing enough about the river to create a model but without being overwhelmed by too much knowledge of specific and local conditions (e.g., every little island). The experts undoubtedly knew a great deal about the river's properties – from a quantifiable, scientific perspective, they almost certainly knew more about the river and waterfall than anyone else. But experts often knew Niagara on an experiential level, too. Many moved to Niagara Falls for extended stretches and spent their recreation time at and on the river. Through their first-hand experiences, they came to know it in a personal way, and it was here they could try to partially account for the ways that changing the waterfall would change the sensory experience for visitors. The engineers were highly skilled at compensating for model shortcomings and surmising, intuiting, or guessing how model performance might translate in the real world; granted, these translations did not always work.

Going inside the black box of hydraulic modelling reveals both the benefits and limitations of the circular modification processes (nature to models to nature) employed to match models to observed data (i.e., the prototype). Based on the various survey methods, Niagara models were modified to create and test desired effects. Models helped create approximations that marked out the workable parameters, and then the actual waterway environment was engineered to reflect the model. Model projections did not always work perfectly when extrapolated to the real world, but some extra finagling could often make things right. To outsiders it may look like guessing, but the "cut and try" method is the way that hydraulic engineers work.[72] But in changing the water flow so significantly, they invalidated most of the experiential knowledge that they or anyone else previously had, leaving only their hydraulic engineering method of knowing. Once the engineers used observed data to design the model, and then used the model as the basis for remaking the river, the original prototype became invalid. In a way, the models became the true record of what the waterfall had been like before its reinvention. Ultimately, the river was intended to look and behave like the model.

The attempt to hide what was being changed at the Horseshoe Falls was a form of "naturalizing industrialization."[73] While the downstream power stations were surely designed to impress the awestruck observer, the remedial works were largely meant to be hidden. Like an umpire or referee, the less they were noticed, the better a job they did. There was as much smoke and mirrors at the cataracts as at the *Ripley's Believe It or Not!* museum, which opened on Clifton Hill just as the extension to the control dam was completed. Niagara was quite different from other contemporary water control projects that obviously imposed themselves on their landscapes, for at Niagara this imposition was mostly hidden. The remedial works infrastructure was the product of disguised design – technologies that are meant to go unseen by mimicking the natural appearance of the waterfall to create a simulated wildness.

But Niagara Falls wasn't just acted on by technology; the waterfall itself became a dynamic technology, a unique form of hybrid envirotechnical infrastructure involving not only steel and concrete but water and rock, weeds and ice.[74] At the waterfall, various forms of networks and infrastructures with a transnational and continental reach intersect at a specific location, such as the long-distance transmission of electricity. As the continental grid and interconnections proliferated, it became possible for someone, say on the Atlantic seaboard, to experience Niagara Falls in a way by turning on a light switch. By the same token, overextending these networks made them more brittle and vulnerable, externalizing and expanding risk just as it did the benefits. This vulnerability was exemplified by the Northeast power blackout of November 1965. This blackout was caused by a protective relay on a transmission line from the Sir Adam Beck Generating Station No. 2 that travelled – or, more accurately, failed to travel – the new transmission lines in New York State built to take Niagara power to denser population areas.[75]

Such technological failures are instructive since stability, fixity, and predictability are defining aspects of infrastructure.[76] An uncontrolled Niagara flew in the face of these qualities, so instead of a messy, receding waterfall, an idealized version was frozen in place to the greatest extent possible: after the remedial works, the annual erosion of the Horseshoe Falls was ostensibly reduced to about one foot. Niagara, after all, had to be protected from itself, before erosion destroyed it, or at least moved it from its current location and the tourist and industrial infrastructure that were premised on a waterfall with a fixed address. The tourist amenities at Table Rock and Terrapin Point, for example, would be wasted if the brink of the Horseshoe Falls migrated upstream.

In his book on the Columbia River, Richard White points out how nature, human labour, energy, and technology are intertwined into an "organic machine." White argues that Columbia planners were mimicking, not conquering, nature. While White is surely correct that they mimicked nature, in the case of Niagara they did so precisely so that they could conquer it. Engineers and labourers working on the Niagara River actually exhibited an intimate engagement with that specific place, even though they intended their expertise and model techniques to be exportable and applicable elsewhere.[77] Many technocrats acquired a respect for Niagara, even a personal affection for it, albeit as a combatant does a worthy foe, and along the way developed first-hand knowledge of specific parts of the river. But motivation is crucial here: the point of this knowledge was to remake a specific landmark, not knowledge for its own sake, for survival, or for the benefit of that environment. Granted, those who remade Niagara did believe that their task was for the betterment of society. Engineers and workers knew the Niagara River through their labour, but this work was filtered through the lens of industrial capitalism, facilitated by intrusive large technologies, which framed the river as a commodity and saw it primarily in terms of cubic yards or feet per second. While there was some respect for the sacredness of Niagara, acquired by personal experiences with the waterway, this was overwhelmed by a desire to dominate and alter the waterfall, with little recognition of limits or unintended consequences.

The plans for Niagara Falls were undergirded by hubris – a faith that technology could remake nature without any negative repercussions. What is most surprising is that those on the front lines, the engineers, knew that this was never wholly true, but they kept acting as though they could achieve complete control. In other words, they knew enough to be aware of what they didn't know; nevertheless, they proceeded as if they *could* know everything through their engineering techniques. This is emblematic of high modernism, which is articulated by James C. Scott in his influential 1998 book *Seeing Like a State:* high modernist plans are predicated on a synoptic view informed by bureaucratic and technocratic expertise, at the expense of local knowledge and without recognition of the limitations of a top-down approach. They involve large-scale prescriptive attempts to make "legible" social and natural environments through simplification and standardization in order to control them and prescribe utilitarian plans for their betterment.[78]

However, Scott contends that fully high modernist projects – what the author calls "ultra" high modernism – can occur only in authoritarian states, and are bound to fail because of their intrinsic contradictions.[79] According to Scott, truly high modernist projects cannot succeed in liberal political economies because of three barriers: private sphere of activity, private sector of the economy (i.e., free market), and effective democratic institutions. Considering that remaking Niagara Falls exhibits the salient scalar features of high modernism, but was successfully achieved in a liberal democratic setting and was predicated on certain types of place-based knowledge, how might we calibrate the high modernist concept for Niagara Falls? The engineering process was defined by negotiation between the prototype and the model, between the technology and the waterfall, between the human and the non-human. The diplomacy and transnational technopolitics that forged the 1950 treaty and then worked out the form of the remedial works were also negotiated. Thus, the US-Canada development of megaprojects on border waters in the mid-twentieth century can be termed *negotiated* high modernism: the various levels of the involved states had to repeatedly adapt, mediate, and legitimize themselves and their high modernist vision to both the specific natural environments and the societies they aimed to control.[80]

Though the synoptic high modernist gaze and its reductionist elements have many flaws, it is also easy to unduly valorize and romanticize local knowledge.[81] There are many benefits to the types of expertise that can come only from the synoptic perspective.[82] Both local and elite expertise has its failings and its upsides; the key is to be aware of what they privilege and obscure, and find a balance. Ultimately, from the perspective of the experts, the models and the remedial works were a clear success. The Niagara models were a pivotal step in the North American evolution of hydraulic modelling practice, and in the understanding of the fluvial geomorphology of large riverine systems and waterfalls. There were a number of cases where the models facilitated important decisions, such as the selection of a control dam over submerged weirs, and did so more cheaply than would have otherwise been the case. For all that, the experiences at Niagara Falls highlight some of the drawbacks of overconfidence in models and technology, and later engineers, at least in North America, started to imbibe the necessity of some humility when confronting powerful water systems.

Conclusion

Perhaps it was the new water regime that saved Roger Woodward's life. On July 9, 1960, the seven-year-old was out on the upper Niagara River with his sister Deanne and a family friend, James Honeycutt, in a twelve-foot aluminum boat. After a rock knocked out their engine, the boat was dragged toward the Horseshoe Falls, and capsized. Onlookers managed to pull Deanne out of the water just above the waterfall. But little Roger, wearing a life jacket, and James, who was not, went over the brink. While the latter perished, as do most who plunge over the cataract, Roger miraculously survived.[83] While it is pure speculation, he might not have been so lucky if the Falls had been running at 200,000 cfs rather than 100,000 cfs.

The 1950 Niagara Diversion Treaty authorized this altered water regime, as well as the remedial works that would compensate for this loss of flow. The remedial works were designed to produce continuity along the rim, an uninterrupted "curtain of water" with a sufficient "impression of volume" and an appropriate shade of water colour. They were also intended to reduce spray and mist, which the parks commissioners were sure discouraged visitors.[84] The official history of Canadian Niagara Parks claims that the spray problems were reduced, at least at the scenic tunnels.[85] While it was debatable whether the mist and spray had actually dissipated near the Falls, and whether erosion had really been curtailed to the extent claimed, the waterfall was taller and cloaked by a liquid veil. Parts of both reclaimed flanks of the Horseshoe Falls were landscaped, walled or railed, and turned into prime vantage points. The public flocked to these viewing stations. This all spoke to the commodification of the Niagara experience: nature was simplified and ordered, sanitized and tamed, and packaged for convenient consumption.

By filling in the flank at Terrapin Point and thus reducing the crestline, the brink of the Horseshoe Falls was almost pushed back entirely into Canadian territory – if that had been the case, the United States would have technically given up its share of the great cataract. Indeed, on some official topographical maps, including maps produced by the United States Geological Survey, the international border appears to cross Terrapin Point, putting the western tip of Goat Island in Canada. However, closer investigation indicates that the border is actually crossing the rock at the base of the island – i.e., at the foot of the Horseshoe Falls at the lower river level – that juts out further to the northwest than Terrapin Point.[86] Thus, the United States did retain a very small portion of the lip of the Horseshoe Falls.

Niagara Falls still involved a huge volume of plummeting water. Granted, it could now be turned on and off – or at least up and down – like a tap, or like workers punching in and out of their shifts. Time itself was further reconfigured with the creation of the pumped-storage plants that hoarded the water according to the needs of the generating stations, rather than nature's clock. The manipulation of Niagara Falls raises interesting questions about the supposed divide between natural and artificial, between human and non-human. Niagara Falls was turned into a model environment, where the river and waterfall were reduced to a prototype, a simplified schematic of numbers and measurements, especially after the waterway was redesigned to reflect the model.

6
Preserving Niagara: The American Falls Campaign

Although the mass of waters fall vertically upon the rocks, there has formed, notwithstanding, by the strong impulse of the current and its great volume, a considerable talus.

Pouchot, 1750s

NIAGARA FALLS, New York, is not a city bereft of charms. That said, it could be a poster child for the rust belt: abandoned blocks, boarded-up houses, and brownfields. The population of about 50,000 is less than half of what it was in the 1950s, and no less than 200 companies have left the region since the 1960s.[1] The combination of deindustrialization, ill-fated urban renewal in the 1960s and 1970s, and flight to the suburbs has left swaths of downtown Niagara Falls in a sorry state.[2] As I walked down Old Falls Street toward the waterfall on an autumn afternoon, I couldn't help but notice that this historic thoroughfare is not nearly as grandiose as it appears in old photos. Falls Street, before it acquired the "Old" modifier, was the gateway to the Niagara reservation, leading tourists to the waterfall from the handsome train station. This station has been gone for over half a century, like many other well-known buildings and establishments. Most people drive here. The central tourist district resembles nothing so much as a giant parking lot, interspersed with shops hawking cheap T-shirts; assorted hotels and restaurants; museums offering wax and horror; and other tourist traps. A few decades ago, boosters tried to convince the Mall of America to set up shop here rather than in Minnesota. Instead, Old Falls Street is now blocked at Third Street, a few blocks due east of Prospect Point and the American Falls, by a monstrosity of a casino that mars the skyline above the waterfall with garish neon lights (see Figure 6.1).

But the biggest gamble isn't at the casino. If you head east from downtown along Buffalo Avenue – also known as "chemical alley" – it is clear

FIGURE 6.1 View of Old Falls Street looking away from the Falls, with the casino blocking the road, in October 2018

that many companies are still in operation. Their synthetic products continue to find their way into the surrounding ecosystem, joining more than a century's worth of chemicals that were constantly dumped or dispersed into the air, water, and ground. Though there are now only a few officially active Superfund sites in this community (i.e., on the National Priorities List), the area is a checkerboard of toxic hotspots, home to the most famous of Superfund sites.[3] Driving further east along Buffalo Avenue, in about ten minutes you will arrive at a postwar suburb called Love Canal. The site of the infamous landfill, where tons of chemicals were dumped over the course of several decades, is easy enough to find just using the mapping features that come standard on any smartphone. In the parlance of regulators, Love Canal has been "delisted," meaning that remediation work there is complete. But the chemical cocktail is still there, buried under a grassy field covered with monitors, enclosed by a chain-link fence, and surrounded by some residential houses and a playground. Delisted isn't necessarily synonymous with safe.

The Love Canal catastrophe followed closely after another local "crisis," one that concerned the state of the American Falls. This chapter explores the genesis, evolution, and resolution of the American Falls preservation campaign between 1965 and 1975. At issue was whether the talus – a slope formed by an accumulation of rock debris – at the base of the American Falls should be removed, and whether natural erosion processes should be circumscribed.[4] However, the talus concerns were also driven by local anxiety about the declining economy and condition of Niagara Falls, New York. The campaign came to encompass the additional questions of public safety related to instability of the rock around Niagara Falls and the impact that the surrounding environment, particularly the industrial and urban setting, had on appreciation of the waterfall. Underlying this approach was the belief that subjective sensory responses to "beauty" could be measured, quantified, and reproduced through engineering techniques and then transplanted to the real-world setting.[5]

Shifting sensibilities about Niagara in the twentieth century stemmed partly from cultural changes, along with a growing appreciation for the technological sublime, as well as the commercialization and vulgarization of the Niagara experience to such an extent that it had become a cliché. But an equally important reason for changing perceptions about its sublimity or lack thereof lies in the fact that Niagara Falls was objectively less impressive in terms of the volume of water falling over it. The 1950s remedial works did mask the impact of the diminished water flows, which improved the appearance compared with when there were few limits (or none at all) on diversions and no remedial compensations. Nonetheless, a tourist at Niagara Falls in the 1960s was seeing, at best, half as much water go over a shrunken waterfall as would a visitor a century earlier. As a result, the failure of the post-1945 tourist to find the Falls sublime may have stemmed not only from societal context or personal experiences but also from physical alterations that had rendered the Falls less impressive by noticeably diminishing their size and volume. Changing cultural attitudes toward Niagara Falls cannot be understood apart from physical changes to the cataract.

Situated within the context of Niagara's modern history, shutting off the American Falls was very normal and was initially uncontroversial. Indeed, it was a logical extension of the transborder efforts to renovate Niagara Falls in the previous decades. The 1950s engineering interventions had resulted in a reconstituted Horseshoe Falls and water regime. However, that was only the most radical step in a long history of taking water and reshaping rock that stretched back a century. Little wonder that romantic and religious conceptions of Niagara Falls as the epitome of the sublime had begun to fade in the latter half of the nineteenth century, precisely when industrial uses of the Falls really accelerated, and even after adjacent parkland was created. The classical sublime is based on a sense of awe derived from a combination of beauty, love, danger, and terror, but for many of those viewing a diminished waterfall, fear had been eroded like the waterfall itself, and replaced with sympathy, even pity. The American Falls were frequently spoken of as an anthropomorphized and vulnerable waterscape that was under siege.[6]

Genesis of the Campaign

Because of water action undermining rock, Niagara Falls is constantly wearing its way upriver. Studies suggested that the Horseshoe Falls, which channelled the vast majority of the Niagara River's total flow, had receded some seven miles in 12,000 years, and 865 feet since 1764, though not at a uniform rate. The American Falls receded at a considerably slower rate since a much smaller volume of water (around 10,000 cfs since the 1940s) poured over it.[7] Niagara is made up of numerous layers of rock strata. Because of these exposed rock layers and the cataract's dynamic erosion patterns, the Niagara area has long attracted the attention of geologists, amateur and professional alike. Consequently, Niagara became a sort of testing ground for various theories about geological processes – and about the very age of the earth itself, which only added to Niagara's mystique.[8] At the top of the waterfall, the hard Lockport Dolomite resists erosion fairly well. Underneath is a layer of Rochester Shale, however, and the friable shale exposed on the face of the Falls is more apt to erode from the effects of water, spray, and seasonal freeze and thaw cycles. This undermines the resistant rock cap until it falls, sometimes in small chunks, sometimes in large and dramatic collapses. Whereas the powerful flow of the Horseshoe Falls breaks up and disperses most rock that collects at its base, the comparatively lower volume of the American Falls is not enough to scour away all the talus, particularly the more durable dolomite (see Figure 6.2).

Two major rockslides had occurred at the American Falls in 1931 and 1954. A total of about 75,000 tons of rock fell from the crest in January 1931, leaving it much more jagged. According to the Special International Niagara Board report: "This caused a sudden recession in the crest of these Falls, probably greater in extent than any single break which has ever occurred during the whole of its comparatively short geological history, and certainly greater than any that had been observed by the white man or recorded by his surveys and photographs." The rockfall was "of such magnitude as to cause a trembling of the cliff along the Canadian side of the river, with the rattling of windows and of china in near-by buildings. The rumble was heard distinctly above the roar of the Falls themselves and a

FIGURE 6.2 Talus at the base of the American Falls. *Niagara Falls Public Library, New York*

column of spray was seen to rise high in the air." The break occurred at the centre of the American Falls, just below Robinson Island, and "extended for a distance of 280 feet along the face of the Falls and produced a maximum indentation in the crestline amounting to 70 feet." The symmetry of this part of the crestline, which previously had been almost a straight line, "now more clearly resembles the arc of a circle, and the total length of the crest has been increased by about 25 feet." The falling rock significantly raised the height of the talus, reducing "the unbroken curtain [of falling water] up to about one-half of its original height." Moreover, projecting ledges at various levels were left behind in this divot, and the water now fell there in a series of steps rather than one larger drop.[9]

On July 28, 1954, a patrolman spotted a crack at the northeast flank of the American Falls. Curious onlookers gathered, and just before dinner a distinct rumbling sound filled the air: 185,000 tons of rock fell from the viewing area abutting the cataract, taking with it most of Prospect Point. This rockfall, the largest in recorded history, further added to the estimated 280,000 cubic yards of talus and altered the waterfall's crest and flank, obscuring the famed "Indian Head" rock formation. Another 900 tons of loose rock was blasted away a few weeks later. In December of the same year, an additional 15,000 tons fell from an adjacent location, and worries about such collapses led the Niagara reservation officials to destroy the remains of the cavern that formed the Cave of the Winds.[10]

The lowering of water levels by fifteen to twenty-five feet in the Maid of the Mist Pool at the base of the Falls, the result of the hydro-power diversions permitted under the 1950 Niagara Diversion Treaty, further increased the height of the rock slope. As a result, in the span of one generation, the talus had reduced the sheer drop of the American Falls by about half. Experts predicted that more rockfalls, completing the transformation of the cataract into a cascade, were inevitable without human intervention.[11]

The campaign to forestall the destruction of the American Falls commenced in January 1965 with Cliff Spieler, the Sunday editor of the *Niagara Falls Gazette*. Early in the month, he pitched to his publisher, Herman E. Moecker, a series of feature articles on the imminent "death" of the American Falls and the need for remedial works. Before any announcement was made to the public, Spieler and Moecker went to work behind the scenes to secure political support for their nascent preservation campaign, including from President Lyndon Johnson. Legislation for Niagara preservation was drawn up at the state level, where New York politicians such as Senator Robert F. Kennedy and Governor Nelson Rockefeller were "tripping over one another to line up in support of the campaign."[12]

On January 31, the campaign went public. The *Niagara Falls Gazette* ran a multi-page feature with the title "American Falls: Death Watch at a Cataract." Featuring background history, explanations of geological processes, and ample maps and diagrams, the point was to show that the American Falls was "destroying itself" by erosion; it could "cease to exist as a waterfall" and this "could happen anytime." Even if it did not crumble into a series of cascades, its slow decay was undermining the grandeur of the cataract, since the talus removed the dramatic plunge of water. Evoking the pity, rather than awe, which some now felt at the waterfall, the newspaper opined that it was "like watching an incurably ill loved one" or, in more gendered language, "like looking at a beautiful woman, her face now showing unmistakeable signs of age." Yet, the writer added, the talus was also possibly preventing erosion or providing structural support. What should be done?

This was a choreographed effort by the *Niagara Falls Gazette* to ignite a wider campaign to preserve and enhance the American Falls. Adopting language similar to that directed toward the Horseshoe Falls in the previous decade, the paper demanded the "preservation and enhancement" of the smaller cataract. It should be dewatered so that engineers could explore options to remedy the situation, which included removing some or all of the talus, various remedial works (such as reinforcing the face of the American Falls, fixing joints and cracks in the rock, installing anchors), and reapportioning the Niagara River's flow to curtail erosion. This manifesto had the backing of expert engineers and geologists. State and federal authorities, and potentially Canada under the auspices of the International Joint Commission, should be involved. Although the costs of such measures to preserve the American Falls were unknown without further investigation, the paper asked: "How can you put a price tag on the world's most-visited natural wonder?"

Because of Spieler and Moecker's advance networking, within days the Niagara Falls City Council had already adopted a resolution urging state and federal agencies to undertake the suggested study. Politicians, government agencies, representatives of the local tourist industry, and the power and industrial companies in the area responded immediately, as did newspapers across New York State and across the country. Headlines were dramatically ominous, playing on the readers' emotional connections to the waterfall: "Fall Menaced," "Save Niagara," "Disaster Threatens." Some newspaper commentators were agnostic or noncommittal about an engineered solution, while a few objected outright to modifying the waterfall and preferred to let nature take its course; most, however, stressed that a dewatering and engineering investigation should proceed. The American Falls at Niagara needed to be preserved at any cost, with cooperation from all levels of government, since it was a national treasure that benefited all. Letters to the editor and other available means of gauging public opinion indicated that the public strongly concurred.[13]

Before the end of the summer of 1965, Rockefeller had green-lighted funds for the state to join in an engineering survey of the American Falls, and a few months later Congress gave the US Army Corps of Engineers approval to play the lead role in a study of measures to preserve and enhance the scenic beauty of the American Falls.[14]

But the sudden alarm about the state of the American Falls was, at least in part, a manufactured crisis. President Johnson had announced in January 1965 the Beautiful America initiative (spearheaded by his wife, Lady Bird Johnson), and Niagara civic leaders saw the American Falls preservation campaign as an opportunity to attract federal funding to resuscitate not only the American waterfall but also the wider community abutting the famous falls.[15] Some perceptive members of the media picked up on this. "Nobody can say that Niagara Falls, NY doesn't have an alert bunch of aldermen and businessmen," the *Chicago Tribune* wryly stated; "as soon as they heard Mr. Johnson tell of the Beautiful America which is to come with the Great Society, they realized what this might mean for Niagara Falls." Like a few other newspapers, the *Tribune* voiced its suspicion that Niagara Falls officials just wanted "Uncle Sam to foot the bill" for local efforts, which were chiefly designed to compete with Canadian tourism.[16] As the *Milwaukee Journal* put it, with tongue firmly in cheek, "all Uncle Sam is being asked to do is improve on nature, to the tune of many millions, so that a community won't be deprived of its prime tourist attraction."[17]

The main instigators of the preservation campaign did indeed have an ulterior motive – to obtain state, federal, and international funds for local improvements beyond just the waterfall. Niagara Falls, New York, was more reliant on industrial development than its Canadian cousin, and some of the impacts associated with the rust belt were already apparent.[18] Moreover, the remaking of the Horseshoe Falls in the 1950s, along with new amenities, had increased the Canadian share of Niagara tourism. National pride was indubitably a factor as well. The editor of Rochester's *Democrat and Chronicle,* for example, claimed that a failure to improve the appearance of the American Falls "would be a blow to this nation's prestige, since the magnificence of the Canadian cataract, in contrast, has been guaranteed for future generations."[19] Indeed, this strategy was already bearing fruit since the Corps of Engineers, other levels of government, and the IJC were eager to take up the cause.

Dewatering the American Falls

Based on exploratory drilling near Prospect Point, the US Army Corps of Engineers' Buffalo District recommended a dewatering of the waterfall. In November 1966, the district reduced the flow over the American Falls for half a day – to a "curtain-like trickle" – in order to perform surveys, take aerial photographs and soundings, and carry out visual inspections of the upper riverbed, falls, talus, and Maid of the Mist Pool. These investigations were supplemented by a public hearing in Niagara Falls: while some thought the natural state was the most desirable, the consensus was that an unbroken curtain of water from crest to base (free of talus) was most attractive. The Corps of Engineers concluded in the spring of 1967 that the removal of talus and other remedial works would be expensive but technically feasible; whether or not such an undertaking was desirable was chiefly an aesthetic question.[20]

What happened to the American Falls plainly impacted the overall experience and setting of Niagara Falls, including the view from across the border. In other words, Canada had a vested interest in all of this. Moreover, since dewatering the American Falls required a cofferdam that would change the level of a border river, Canadian concurrence through the International Joint Commission was required. In March 1967, the American and Canadian governments sent a formal reference to the IJC, under Article IX of the Boundary Waters Treaty, requesting that the commission investigate the aesthetic aspects of the American Falls and recommend: "1) what measures are feasible and desirable a) to effect the removal of the talus which has collected at the base of the American Falls; and b) to retard or prevent future erosion; 2) other measures which may be desirable or necessary to preserve or enhance the beauty of the American Falls." This reference formally became IJC Docket 86, Preservation and Enhancement of the American Falls, Niagara River. The IJC established a four-member American Falls International Board (AFIB), composed of an engineering representative and a landscape architect from each country, to incorporate both technical and aesthetic perspectives. This distinguished board also set up a Working Committee, on which representatives from the state and provincial park commissions bordering the Falls were included. The Corps of Engineers made its studies available to the commission, and the two entities agreed to cooperate moving forward.

The IJC scheduled public hearings for late October. There, the two power utilities that handled most of the hydro power, PASNY and Ontario Hydro, suggested that their generating stations absorb the excess water resulting from a dewatering. Otherwise, the water that would have constituted the American Falls would instead go over the Horseshoe Falls and increase its susceptibility to erosion and rockfalls. But the real reason for the utilities' concern was that it would be wasting potential power. Other groups represented at the hearing were almost unanimously in favour of proceeding with the dewatering of the American Falls. The proposed dewatering and diversions, however, might leave less water flowing over the Falls than was legally required by the 1950 treaty, which complicated matters. Downstream interests, as well as members of local governments and chambers of commerce, complained that this would curtail levels in the lower river and further lessen the scenic appeal, in addition to what had already resulted from the water regime of the 1950 treaty. Some voiced concern that lower water volumes would aggravate existing pollution problems because of the river's reduced dilution capacity. Underlying many of these complaints was anxiety that this "temporary" decrease of water was just the thin edge of a wedge leading to a permanent increase in water diversions.[21] There was also debate about the effect of water rates on erosion and talus: some argued that less water flow meant potentially less debris available for accumulation, while others argued that higher flows would chew up the talus.

On the IJC's recommendation, the two countries agreed in March 1968 to authorize construction of a cofferdam to enable the drying of the American Falls. When the cofferdam was in place, 92,000 cfs and 41,000 cfs would go over Niagara Falls during tourist and non-tourist hours, respectively, rather than the 100,000 cfs and 50,000 cfs required by the 1950 treaty, with the power entities utilizing the difference. The United States and Canada would share the costs of studying the American Falls, which Ontario Hydro and PASNY agreed to supplement with financial contributions, material, and labour in exchange for receiving the extra volume of water. However, the required US Senate approval of this bilateral agreement was not immediately forthcoming. Without higher appropriations from the US Congress, the Corps of Engineers could not move ahead with the cofferdam.[22]

The dewatering was consequently delayed until 1969 (see Figure 6.3). In June that year, the Albert Elia Building Company, contracting with the Corps of Engineers, built a cofferdam from Goat Island to the American mainland to cut off the channel leading to the American Falls.[23] This channel was a dangerous series of rapids, part of which had received the evocative moniker "hell's half acre" because of its churning turbulence. The cofferdam would remain in place until late November. Anxious to get out in front of public opinion, the IJC and other cooperating agencies had planned an aggressive program to inform the populace about the dewatering process. General Roy T. Gorge of the Corps of Engineers North Central Division stated:

> Because of the intensity of the world-wide public interest in Niagara Falls, the remedial study offers the Corps of Engineers a unique opportunity to demonstrate its interest in preserving natural beauty and in responding to the interests of the general public. From a public relations standpoint, such favorable press coverage can enhance the reputation of the Corps of Engineers and can be expected to offset criticism the Corps has suffered on some other projects.[24]

FIGURE 6.3 Dewatered American Falls in 1969. *Niagara Falls Public Library, New York*

Both countries installed a number of outdoor information displays near the Falls. Media from around the world showed up to get footage. Much of the area was fenced off, and the Niagara parks agency built a walkway so that tourists could access the dry streambed, where tomato plants and poplar trees now sprouted. On the advice of a professor hired to conduct an ecological impact study, the Corps of Engineers instituted a watering program for trees on the small islands that dotted the channel. People found many coins and other items of interest in the dry rock bed and talus, including the remains of probable suicides.

The dry American Falls was akin to a sedated patient undergoing a biopsy. Water and sandblasters cleaned rock. Sprinklers were installed to keep the shale layer underneath the Lockport Dolomite rock moist, since it deteriorated more quickly if exposed to wind and sun. Engineers drilled holes in the riverbed several hundred feet in depth so that tests could be conducted to determine the composition of the rock layers, including permeability and strength, and so core samples could be obtained for laboratory analysis. A cylindrical flash camera was lowered into the bore hole and pictures were taken every three-quarters of an inch.[25] Trace dye was injected into rock fractures to see where it emerged. As a safety precaution, sensors were installed to detect imminent rockslides, such as extensometers that used air bubbles to measure any minute rock movement. Piezometers measured hydrostatic pressure on rock joints. Tiltmeters detected deep-seated shear in the rock. Pins and monuments were employed as markers to visually measure any shifts. To conduct tests and scale rocks from the face of the Falls, cranes lowered workers and geologists in suspended cages. Various means of improving drainage were effected. The surface of the talus and underlying bedrock was surveyed, mapped, and photographed in detail.[26]

There had been hopes that the dewatering might stimulate an increase in the more than 10 million tourists who visited the site annually. At first, record-setting crowds did arrive to experience this "once in a lifetime opportunity" – one July weekend, for example, saw 90,000 visitors to the Niagara reservation. Nevertheless, overall tourist visits actually dropped in 1969, and tourists who did come did not stay as long or spend as much, compared with previous years.[27] Some speculated that inflation and exchange rates cut into sightseer spending, but there was no doubt that fewer people came and that the dewatering was the prime reason. Many tourist operators blamed the Greater Niagara Chamber of Commerce for not effectively countering a widespread myth that the Horseshoe Falls was also dry.

Broadening the Reference

In the meantime, the American Falls International Board had become concerned about another issue: the wider Niagara experience beyond just the waterfall. This had not been part of the 1967 American Falls reference. Borrowing a phrase from the 1872 collection *Picturesque America,* the board likened Niagara Falls to a diamond set in lead. The AFIB contended that the waterfall's surroundings should be more like a beautiful gallery built to house great works of art, since the wider scene enhanced or detracted from the object of beauty, or what the AFIB called the culturally shaped "climate of appreciation."[28] That is, the waterfall could not be enjoyed in isolation from everything around it. And, as one government memorandum put it, "the awesome beauty of Niagara Falls is enclosed on either side by two cities of quite awesome ugliness."[29]

Such sentiments bore a strong resemblance to the "free Niagara" movement that resulted in the creation of the Niagara reservation in the 1880s. Rather than mills and factories immediately abutting the upper river and gorge, however, the problem was now a range of large tourist, residential, commercial, and industrial structures. New observation towers cluttered the Fallsview ridge, and another had just been built into the gorge at Prospect Point, joining the numerous power stations at the water's edge above and below the cataract.[30] Power lines radiated out in every direction from the power stations. Advances in building technology had led to taller industrial smokestacks, commercial developments, and residential high rises outside of the areas controlled by park officials, with the result that "the protective park-belt is not effective in providing a visual screen. Developments outside the park boundaries threaten to impose themselves on the view."[31] Moreover, there were numerous applications in the works for other new buildings, such as a convention centre and shopping/office complex on the New York side.

While the skyline immediately at the Falls had been the initial issue that caught the AFIB's attention, the issues soon grew to encompass a variety of considerations across the Niagara frontier: unsightly buildings in the wider area; the snarled traffic situation; various motels, tourists traps, and commercial areas in need of urban renewal; a general honky-tonk atmosphere; industrial blight; and other natural and landscape aspects. Air and water pollution were high on this list. The State of New York had recently passed the Pure Waters Program, which provided funds for sewage treatment plants to control river pollution. The federal government was on the verge of enacting what would become known as the Clean Water Act of 1972, and the studies and negotiations that would produce the 1972 Great Lakes Water Quality Agreement were already well advanced.[32] Nonetheless, municipal and industrial waste threatened the heart of the Niagara experience: the Falls often assumed an unnatural greenish hue, full of sewage effluent, and at times several feet of foam from detergents and chemicals formed at the base of the waterfall and sprayed onto *Maid of the Mist* excursions. The title of a study published a decade later – *The Ravaged River* – said it all, identifying the Niagara as "one of the most chemically contaminated bodies of water in America" and pointing to the thoroughly inadequate standards and permitting system for synthetic organic substances in New York State.[33]

It would take harmonized governmental actions at various levels to address such multi-faceted problems that stretched along and away from the river. In September 1968, the AFIB recommended that the IJC extend the 1967 reference on the beauty of the American Falls to include a special international study on the total aesthetic and environmental setting of the "Fallscape" – a term the AFIB itself employed – which would be "unified around the principle that the Falls are the center of an environmental organism" that included the surrounding rivers, canals, lakes, power systems, urban areas, transportation systems, and countryside. This study would be broken into stages stretched over a year and a half that would identify problems and then prescribe precise solutions, with a budget of $250,000. A joint US-Canada technical team, including both engineers and landscape architects, would undertake the project in conjunction with New York State and the Province of Ontario since many of the relevant issues were not solely under federal jurisdiction. The IJC in turn proposed this to the two federal governments.

To be sure, the proposed study stemmed not only from concern for the Falls themselves but also, as officials admitted, concern for the revenues generated by tourism. Research suggested that many regarded Niagara as merely a stopover on a longer trip, as few visitors stayed more than one or two nights. Honeymooners were on the downswing. A study of the US side speculated that the decline could be linked to "the adverse environmental image the Niagara Falls region has acquired over the past few years ... increasing numbers of visitors seem repulsed by the commercialism of the downtown areas and the prevalence of heavy industry and power installations along the scenic routes of the region."[34]

A debate ensued about whether a wider study could take place under the existing American Falls reference, and it was decided that a separate reference would need to be made by the United States and Canada. The proposal also led to considerable debate within the Canadian government about who should play the lead role in formulating such a reference, as various departments were reluctant to take it on. An ad hoc committee composed of several Canadian federal agencies was eventually formed. Jurisdictional and financial details also needed to be ironed out with Ontario. But the Ontario government was noncommittal, fretting about who would pick up the tab for such a study, and claiming that Niagara had already been "studied to death."[35] Ontario further argued that a plan recently commissioned by the Niagara Parks Commission (and carried out by Richard Strong Associates Limited), constituted their contribution – in spite of the fact that this study focused almost exclusively on traffic access within the territory of the Niagara Parks Commission and struck a tone of competing, rather than cooperating, with New York State when it came to Niagara tourism.

American officials were more enthusiastic about the need to consider the wider impact of the Falls aesthetic. They offered up $150,000 toward the cost of the international study (with $50,000 coming from New York State). The Erie and Niagara Counties Regional Planning Board had used a grant from the US Housing and Urban Development Agency to hire consultants tasked with identifying existing environmental problems and remedies, with input from industry and various levels of government. However, since this would be limited in scope to the American side, it would complicate arrangements for an international study.[36]

Negotiations between the Canadian federal government and the Ontario government proceeded in a desultory manner for some time. With the matter stalled, the Department of External Affairs asked the US State Department to compose an *aide-mémoire,* which it was hoped would put pressure on Ontario to stop dragging its feet. The aide-mémoire arrived in April 1970, stating that the Americans wanted to participate in a shared study and requesting Canadian cooperation. The United States followed up with a diplomatic message stressing the importance it placed on the matter. Ontario seemed more amenable to cooperative study of the Niagara gorge, but remained unwilling to extend it any wider.[37]

To speed things along, the American Falls International Board in November 1970 authored a report titled *Intrusions on Views of Niagara Falls.* It specifically identified as problematic the proposed Canadian construction of an elevated monorail, a fourteen-storey hotel, and a twenty-four-storey condominium, as well as a proposed convention site on the New York side. The report elaborated on the board's previous calls for an international consultative body that would establish guidelines for protection of the area, since local interests "have an understandable preference for immediate cash benefits, rather than long-term measures to preserve and enhance the beauties and amenities of the area."[38] Indeed, it appears that narrower financial concerns were ultimately at the heart of Ontario's reluctance to undertake an international study. Chiefly, Ontario officials believed that an IJC study would likely benefit the American region of the Niagara area more than its Canadian counterpart, and steal away tourism and economic development from Ontario.[39] Canadian officials lamented Ontario's "small-minded smugness about whose side is better than the other."[40] But the Canadian federal government still squabbled internally, which ongoing departmental reorganization only exacerbated.[41] Though an American "Advisory Committee for the International Environmental Study of the Niagara River" had begun meeting, it too soon ran into intra-jurisdictional political problems.

Interim Opinions

Engineers spent several years analyzing and processing the data from the 1969 dewatering. Insufficient congressional appropriations continued to hamper the study. Nonetheless, even while the American Falls were still dry, it had become apparent that the primary cause of rock slides was not the face of the cataracts but joints in the riverbed upstream from the precipice. The water that seeped into these vertical fractures undermined the Rochester Shale, creating thin columns that decreased the stability of the waterfall. The experts eventually determined that this could be best combated by a mass stabilization effort: they created tunnels within the rock to provide internal drainage, tied the rock together with tendons and vertical anchors that connected the tunnels and riverbed, and installed concrete facing.[42]

In addition to the physical evidence, the AFIB wrestled with the subjective question of what constituted beauty when it came to the American Falls. As the board fully admitted, aesthetic assessments and emotional responses to Niagara's splendour varied. It determined that the key elements of the cataract's appeal included "the volume of water, the sculptural form of the talus and bedrock, the surface level and water's edge of the pool. In understanding the American Falls as an immense water-sculpture, these are the controllable elements of the design. The beauty and drama of the Falls depend on the interplay and the relative proportions of these elements." The AFIB surmised that increasing the flow rate might amplify the appeal of the American Falls. A demonstration was arranged in which the flow was decreased and increased by diverting water from the Horseshoe Falls. The logic behind inflating the flow rate by borrowing from the larger waterfall was that even just a few thousand cubic feet per second of water would proportionally be a very substantial addition to the American Falls but would scarcely be missed by the Horseshoe Falls. Indeed, the augmented flow over the American Falls "greatly enriched the appearance, deepening the green water plunging over the crest and adding considerably to the turbulence of the white water pouring over the talus rocks."[43] Making this increase permanent, however, would have required more excavation and another permanent control structure at the upstream end of Goat Island.

The potential physical alterations to Niagara Falls were intended to engender particular emotional responses: a sense of reverent awe, or at least one of being sufficiently impressed. What is striking is the ways in which engineers sought to quantitatively measure and assess what was essentially qualitative – that is, how people emotionally responded to the sight, sound, and feeling of Niagara. In turn, the engineers used this information to physically reshape the waterfall so that tourists would continue to experience the requisite emotions. Granted, the affective response to Niagara Falls was filtered through various lenses, including nationality, but there was still enough common ground about what constituted the grandeur of the spectacle that it could ostensibly be captured in a laboratory model.[44]

One of the other measures that the board considered was a dam downstream from the Falls. Such a structure had first been proposed in the early twentieth century, when several different individuals forwarded what some labelled "Niagara Falls Junior." The dam idea resurfaced periodically, including rumours in the early 1960s that such a structure was being considered as a means of raising Great Lakes water levels and developing additional hydro power.[45] While a downstream dam could not do much for water levels in the lakes above the Niagara River, it could raise the levels of the Maid of the Mist Pool back to what it had been before the water diversions of the 1950s. After all, even if a good deal of talus was removed from the American Falls, the remaining rock might form a series of irregular steps or shelves that would prevent a sheer water drop. But a raised Maid of the Mist Pool could mitigate this by drowning out those rock shelves. Additionally, an international dam placed about four to five miles below the Falls would smooth out the Niagara rapids and provide a one-hundred-foot head of water for hydro-power production.

The Canadian government contracted with Henry Girdlestone Acres's Niagara Falls–based international engineering and consulting firm. Before he passed away in 1945, Acres had been a high-ranking Ontario Hydro engineer; for example, he was the chief hydraulic engineer for the first Queenston–Chippawa power development (Sir Adam Beck No. 1).[46] H.G. Acres Limited produced a "Feasibility Study of a Structure to Control Levels of the Maid-of-the-Mist Pool." This study found that a control dam would not detract from the allure of the Niagara Whirlpool Rapids, nor would it radically increase the probability of ice damage to adjacent

FIGURE 6.4 Model of the American Falls. *International Joint Commission*

structures (such as the power station at the foot of the Horseshoe Falls, *Maid of the Mist* docks, and bridge abutments). It nonetheless recommended a comprehensive ice survey and geotechnical investigations. Four options for different locations and types of structures were presented. The report advocated placing the structure about 450 feet downstream from the Whirlpool Rapids. The envisioned structure would take the form of a rock-filled weir, with either gates or tunnels to serve as bypasses when necessary. Estimated costs ranged from $8.7 million to $18.5 million, depending on the bypass structures chosen.[47]

High-precision scale models had been a fundamental part of designing the Niagara remedial works in the 1950s, and remained the "most practicable tool" for visually determining the effects of removing talus, raising the pool level, and increasing the flow over the American Falls. A scale model of the American Falls was built at Ontario Hydro's Islington facility, at a scale of 1:50.[48] All relevant elements of the surrounding ecosystem were simulated. Dry ice was used to mimic blowing spray, and compressed air replicated water turbulence. Bushes, trees, and bridges were added to enhance the model's realism, and even moss and stains were painted on. The model was constructed with interchangeable talus and various parts so that the model operators could experiment with and photograph the different combinations and conditions that engineers were proposing for the real thing.

The two most pleasing talus arrangements were 1) removing virtually all the talus, and 2) the "Temple of Osiris" configuration (so-called because of its formal and solemn appearance), which would remove the talus from the centre but not the sides, making the central part of the waterfall appear higher and narrower. Based on the model demonstrations, authorities determined that the American Falls would be "more dramatic, majestic and beautiful" if much of the rock was removed, the surface of the Maid of the Mist Pool was raised, and the amount of water topping the American Falls was increased.[49]

But there were limitations: "The behavior of water in its opacity and aeration" could not be reproduced at a smaller scale. Like the hydraulic Niagara models built in the 1950s, these American Falls models were sometimes flawed, and political and nationalist concerns coloured interpretations of the results. However, unlike the 1950s models, which engineers

American Falls, June 1971 – Flow about 8,000 c.f.s.

American Falls, June 1971 – Flow about 15,000 c.f.s.

FIGURE 6.5 Comparison of flow levels on the American Falls. *International Joint Commission*

used to determine technical hydraulic options, in the American Falls case the purpose was only to simulate for landscape architects the visual appearance of the various configurations of the talus, flows, and pool levels (see Figures 6.4 and 6.5).

Based on the models and other studies, the AFIB's interim report, "Preservation and Enhancement of the American Falls at Niagara," was submitted to the IJC in December 1971 (later followed by lengthy appendices). The report summarized the findings of the investigations up to that point, and determined what aspects the AFIB should focus on for the remainder of its study. The board outlined the practical means of increasing the flow over the American Falls (at a cost of $6 million), raising the Maid of the Mist Pool ($8.6 million), and removing practically all the talus ($10 million). The report admitted that all this was feasible and would clearly heighten the waterfall and its dramatic effect.

Nonetheless, in the board's estimation, the desirability of such measures was "debatable." The board questioned whether any of the alternatives would "raise the beauty of the Falls in a truly remarkable way, to an extent that would be fully understood and appreciated by the public." Additionally, there was evidence that the talus actually inhibited more rapid geological deterioration by stabilizing the face of the American Falls. The Falls were so impressive and astonishing, the report postulated, precisely because they were "live, active, and violent" – to control erosion patterns would threaten to strip it of its animating character. Ultimately, the AFIB concluded that "it is better to allow the process of natural change to continue uninterrupted rather than to give permanence to a particular condition and appearance."[50]

Lacking a transborder agreement regarding the total aesthetic setting, all the interim report could do was recommend in passing that "protection and enhancement of the whole environment of the Falls should receive as much emphasis as the measures for preserving and enhancing the beauty of the Falls themselves." But without the participation of the different governments, an international study on the wider setting could not go anywhere.

Though the push to include the wider Niagara setting within the purview of the 1967 American Falls reference failed to gain traction, the reference was broadened in another way. Geological investigations during the dewatered phase had revealed several sectors of instability at the American Falls. Cracks were discovered on the surface of Prospect Point and Luna Island, which, along with the Cave of the Winds walk, was immediately closed to the public. Terrapin Point had problems of its own. Most of these areas had previously been reclaimed, and water undermined these zones that used to be part of the waterfall. The IJC's investigation of measures necessary to preserve or enhance the beauty of the American Falls was therefore expanded in 1970 to look at the extent to which the rock astride the American Falls and the Goat Island flank of the Horseshoe Falls was endangered by the possibility of erosion and other structural problems, and, if so, what measures were feasible and desirable to protect these areas while eliminating any hazard to persons and property. In short, public safety joined scenic beauty as the prime considerations of the reference.

In the context of the momentum to preserve the American Falls and the modern history of manipulating Niagara Falls, the interim report's emphasis on letting nature take its course was surprising. Yet, throughout the report the board's stance that it would be better to allow the process of natural change to proceed uninterrupted was clearly conflicted, even contradictory. To illustrate, in case an intervention route was chosen by the IJC, the board also outlined a range of alternatives for different combinations of physical alterations to the American Falls. In its conclusions, it stated its preference for alternative number 3, the option that, at a cost of $27–29 million and staggered over time, would remove a substantial amount of the talus, increase the flow over the American Falls during selected hours of the day, and raise the Maid of the Mist Pool. The interim report's attention to public safety measures likewise seemed inconsistent with the notion of letting nature take its course: the board proposed that sensors to detect rock movement could be installed, and that the flanks of the Falls could be stabilized to protect the public from further rock slides. This the authors justified by declaring that it was for convenient enjoyment of the spectacle; as the report put it, "when one is considering how to protect the public against the violence of nature, the safest animal is a dead animal and the safest waterfall is an emasculated one." Setting aside the gendered and anthropomorphized implications of such a statement, it was as if the board was saying that it would *probably* be best to leave the American Falls alone, but if that was not possible, they had thoughts about how it should be remade.

Final Report

The IJC approved the AFIB's interim report as the basis for moving forward. The various means that the IJC used to gauge public opinion indicated that attitudes had changed considerably since the start of the reference in 1965. In fact, the various public feedback mechanisms indicated that many people thought that pollution abatement in the Niagara River was a higher priority than remaking the American Falls – a prescient concern, and one that reflected the era's growing concerns about public health issues.

Since both the board and the larger commission felt that the hearings to date had not provided enough public feedback and confirmation, particularly considering that a hands-off tack had emerged, they decided to issue to the public 220,000 copies of a booklet titled *The American Falls: Yesterday Today Tomorrow.* Enclosed ballots asked respondents to voice their opinion on several options: remove the talus, increase the flow over the Falls, restore the Maid of the Mist Pool, or make no physical changes. "How would you like your American Falls?" one newspaper quipped in response to this survey: "On the rocks? Straight up? With a little more water?" For each option, a pictorial representation of the American Falls model under the respective conditions was provided. About three-quarters of the 40,000 people who responded favoured some change. However, the IJC worried internally that the results were skewed because "no physical change" had not been presented as a choice throughout the booklet, apparently because it did not need an explanation, even though it was an option on the mail-in ballot.[51]

The IJC also convened a seminar composed of fifteen distinguished environmental planners and landscape architects from both countries. The seminar participants (nine American and six Canadian) met in June 1973 and reached a remarkable unanimity of opinion about the American Falls: to leave them alone and remove no talus. They opposed raising the pool level or increasing the flow of water over the American Falls, and argued that "irreversible damage" could result from heavy-handed intervention. Moreover, in the eyes of these consultants, the deterioration of the larger scene around the Falls was a matter of much more serious concern than the appearance of the Falls themselves. This heartened the AFIB.

The expert seminar reflected a countercurrent that had been gathering momentum. At the start of the American Falls reference, few voices had been raised against alteration of the Falls, within the IJC or wider society. Exceptions included the general manager of Ontario's Niagara Parks Commission, Maxim T. Gray, who had indicated in 1967 that he preferred to see the cataract untouched; a reporter identified this as "the first and only indication of opposition to a preservation program" for the American Falls.[52] Granted, Gray's stance may have stemmed more from worries about the impact on Canadian tourism. Nonetheless, by the early 1970s environmental groups such as the Committee of a Thousand were vociferously opposing remedial works at the IJC's public hearings. At the March 1972 forum, after the AFIB's interim report, the chairman of the Niagara Parks Commission reported that his organization preferred that the process of natural change continue uninterrupted. Others pointed out at the same hearings that the talus actually added to the winter grandeur of the gorge. The curator of the local Schoellkopf Geological Museum, meanwhile, characterized the potential renovations as "a joke."[53]

The very fact that the seminar was not primarily stocked with engineers suggested a changing ethos within the IJC. Moreover, a member of the AFIB who was also an employee of the Army Corps of Engineers was quoted as saying:

> I've always thought we should just leave the Falls alone. The public has been brainwashed into thinking there's a problem at the American Falls ... It's this kind of misinformation that led to the board's study in the first place – ignorance of the fact that this is a natural geological process here and not some kind of accident or mistake that we should correct.[54]

Moreover, the AFIB admitted that "recent emphasis on environmental values has raised questions about changing natural conditions even for demonstrated and measurable social benefit. When changes are made for a calculated economic and social benefit, measures should be taken to prevent excessive damage to the natural environment."[55] Such statements indicated a greater willingness to acknowledge scientific uncertainty, in contrast to the hubris that had characterized earlier engagements with this landscape.

Within the environmental movement, resistance to engineering nature joined pollution concerns. As Matthew Wisnioski argues, the purposes of engineering within the American engineering profession shifted between 1964 and 1974 – which lines up almost precisely with the period of the American Falls reference – with technology taking on "ambiguous and ultimately sinister connotations in American thought and culture."[56] Within landscape architect circles, moreover, there was a growing movement to be more ecologically sensitive and "design with nature," exemplified by a 1969 book of the same name.[57] Furthermore, many people were manifestly concerned about technology as it pertained to chemical production and nuclear weapons, and by the mid-1970s there was an apparent apprehension, influenced by the likes of Murray Bookchin and Barry Commoner, about large-scale technological solutions that altered nature purely for aesthetic reasons.

This shift in the engineering mindset was firmly linked to the broader North American environmental movement of the 1960s and 1970s. Was the resistance to changing the American Falls just part of the normal trajectory of the environmental movement? In one sense, the answer is in the affirmative: changing mindsets about the environment drove the altered approaches to modifying the American Falls. Yet this story is different in important ways. For starters, this was different because it involved Niagara Falls, one of the continent's natural icons. People *feel* differently about Niagara Falls, and proposals about Niagara receive national and international attention. What is done to a nation's cherished environmental symbols says a great deal about a society's views on nature.[58]

In the 1950s and 1960s, Americans looked long and hard at remaking other charismatic waterscapes for hydroelectricity: damming the Grand Canyon; the Hells Canyon Dam on the Snake River in Idaho; an Echo Park Dam in Colorado's Dinosaur National Monument; and, in New York State, a pumped-storage hydroelectric plant in Storm King Mountain on the Hudson River. In all these cases, as well as other potential water control projects, such as the Tocks Island Dam on the Middle Delaware River or the Meramec River in Missouri, public opposition halted or significantly modified development. In addition to Niagara's status as one of the most revered American natural icons, the American Falls case is also unique because it was the engineers who promoted a hands-off approach – it was one thing coming from landscape architects, environmental groups, and the affluent and upper-middle classes, but a changing ethos is readily apparent when even those whose profession it is to manipulate these environments become wary of doing so.

The campaign to preserve and enhance the American Falls reflects the shift from the rise of modern environmentalism to its mainstreaming.[59] The momentum to remake the American Falls began in 1965, just as modern environmentalism was gaining steam. It followed on the heels of *Silent Spring* and other concerns, the passage of the Wilderness Act and the Wild and Scenic Rivers Act combined with other events – some of them, like the purported death of Lake Erie because of eutrophication and the repeated fires on the Cuyahoga River, were upstream of the Niagara River – and were followed by the first "green" decade, marked by the 1970 Earth Day, the creation of the National Environmental Policy Act and the

Environmental Protection Agency, and the passage of water legislation such as the Clean Water Act.

Even if the critical mass of opinion was leaning away from confronting the talus and erosion, it was a different story for another key component of public environmental sensibility in the 1970s: public safety. Rockfalls had become the primary public safety concern at Niagara. With IJC approval, a cofferdam was built in November 1971 to shut off the flow to Bridal Veil Falls, and remedial work began on the Luna Island area the following year. In addition to clearing away rock and drilling drainage holes, authorities installed warning sensors to detect rock movement. Bolts and tendon cables were put into the rock to increase stability, in addition to other ways of enhancing slope stability protection, such as a concrete parapet at the front of the island. The Cave of the Winds received a facelift too. Along with Luna Island, it was reopened to the public in 1973. Similar work was completed at Prospect Point.

The different forums for public opinion confirmed the AFIB's revised direction. The board forwarded its final report to the IJC at the end of June 1974. This report did not depart much in substance or tone from the interim report. The AFIB explored numerous alternatives and devised a rating system that compared the various options according to different criteria. It found that measures to remove the talus and inhibit further erosion were feasible. Recession at the crest and flanks, for instance, could be prevented and arrested by a program of mass stabilization that would cost about $26 million.[60]

However, the report ultimately came out against procedures that changed the appearance of the American Falls, although the board was open to alterations that would improve public safety. While talus removal would provide a water plunge that was more complete and dramatic, this should not be done at that time since it would be costly and it was "not clear that the removal of the talus would enhance the beauty of the Falls to an extent that would be fully appreciated by the public." The AFIB suggested that this position might be revisited in the future, such as after another massive rockfall or once a study of the overall environment had occurred. Nor should the major measures to stop erosion be undertaken, as that risked making "the Falls static and unnatural, like an artificial waterfall in garden or park, however grand the scale." The board contended that the "guiding policy should be to accept the process of change as a dynamic part of the natural condition of the Falls, and that the process of erosion and recession should not be interrupted."

Nonetheless, the AFIB's final report carried forward the interim report's conflicted stance about the extent to which nature should be allowed to run its course. For instance, in contrast to the above-mentioned position on talus, at one point in the report the board wrote that it "may lean more toward a policy of 'enhancement' ... than a policy of preservation" since others "may conclude that removal of some or all of the talus would enhance the Falls." Moreover, the board left open-ended the questions of raising the Maid of the Mist Pool and increasing the flow over the American Falls, but seemed to implicitly favour them. Granted, for the AFIB, altering water levels was understood as an attempt to return to the natural conditions that existed before large-scale diversions from Niagara. When it came to unstable areas that threatened the safety of tourists, a tension between intervening and abstaining was apparent: the board advocated structural supports, rock stabilization, and safety warning systems, but also suggested that railing realignment and closure of certain viewing areas might be preferable.

The AFIB summarized these findings in a pamphlet distributed to the public. Newspaper editorials and letters to the editor tended to affirm the recommendations, including those in the *Niagara Falls Gazette,* which had started the whole furor back in 1965. Another round of public hearings was held, in March 1975, to receive feedback on the AFIB's report. Most submissions supported leaving the talus at the base of the American Falls, and witnesses almost unanimously agreed that the Falls should not be artificially stabilized and that natural processes should be allowed to continue uninterrupted.[61] At the same time, all witnesses who addressed the public safety issue concurred that steps should be taken.

Based on the AFIB's efforts, the IJC issued its final report on the preservation and enhancement of the American Falls to the US and Canadian governments in 1975. The study had originally been expected to wrap up by 1971, and its costs had also ballooned significantly, up to $2.7 million for each nation. Although the IJC's report was more truncated and matter-of-fact, it echoed the AFIB's principal findings. It concluded that, while it was technically feasible to remove the talus and control the erosion, it was not desirable to use artificial means to do so, at least not at the present time. Though the IJC thought the public should accept that some risk was inevitable, it recommended remedial works for the unstable flanks of the American Falls and the Goat Island flank of the Horseshoe Falls. The commission also stressed the need for an environmental study of the total setting.[62]

The US and Canadian governments generally accepted the IJC's recommendations. There was one key exception, however: the broad environmental study. Despite prodding and pleading, the two nations could not be cajoled into extending their study of the American Falls into a broader and synchronized analysis of the Niagara experience. Without government support, a transborder program to address the wider fallscape could not proceed. As a result, until it was disbanded in 1979, all the AFIB could do was continue to stress that "protection and enhancement of the whole environment of the Falls should receive as much emphasis as the measures for preserving and enhancing the beauty of the Falls themselves."[63]

The AFIB's concern for the wider environmental setting around the waterfall campaign was soon validated by the infamous Love Canal tragedy. The Hooker Company and the City of Niagara Falls had for several decades dumped chemical waste into a ditch, the remnant of William Love's aborted canal. At the time, this was fairly standard practice and few gave it a second thought. Hooker sold the property in 1953 to the City of Niagara's school board for $1, advising them of what was lurking underneath and absolving themselves of responsibility. Soon the city began building new homes and a school. Fast forward to the late 1970s, and the working-class neighbourhood of Love Canal was beset by chronic illnesses and birth defects, which some local people suspected was caused by exposure to the buried waste. Complaints initially met with government apathy and dismissiveness, however, and it took several years, which saw standoffs and other pressure tactics, before all the affected residents were evacuated. Love Canal was only the tip of the iceberg. It became a symbol for the potential public health and environmental catastrophes represented by undisclosed disposal sites all over the Niagara area and around the country. Love Canal spurred a nascent grassroots environmental justice movement, and was a catalyst for the creation of Superfund in 1980, formally known as the Comprehensive Environmental Response, Compensation, and Liability Act (CERCLA), to address toxic contamination.

Even though the final report drew to a close the official American Falls campaign, Terrapin Point still needed fixing. The unstable overhang there had been fenced off in 1972 and stayed closed throughout the 1970s. The AFIB, in keeping with its conclusion that the natural processes of erosion should not be interrupted, recommended that the Terrapin Point viewing area be moved back away from the unsound outer section, with little to no rock excavation, rock bolting, or drainage measures involved. The IJC suggested measures that included a cantilevered platform to hold up the offending overhang, at a total price tag of $2.8 million. Other engineers considered letting the Horseshoe Falls reclaim Terrapin Point, spurred on by the same newspaper that a decade earlier had urged them to remake the American Falls: "Try to work with the river, not against it" since, in the end, "the river gets its way."[64] New York State park officials announced their intention to blast between 15,000 and 32,000 tons of rock from Terrapin Point. The IJC strenuously objected and matters were at an impasse for several years while park officials explored the feasibility of blasting. Action finally took place in 1983. With the public at a safe distance and the *Today Show* carrying the event live on television, over 9,000 pounds of explosives was set off, removing a 10,000-ton rock mass. Bolts, pins, and cables were installed to anchor the remaining rock, among other structural and aesthetic improvements, and slide sensors were installed. A clean rock wall had replaced a jagged rock profile, to the chagrin of the IJC and the US Army Corps of Engineers, which deemed the new look "harsh and artificial" (see Figure 5.10). Nevertheless, after being closed for over a decade, Terrapin Point was reopened to the public in the fall of 1983.[65]

Conclusion

Since the end of the American Falls campaign, no effort has been made to significantly re-engineer Niagara Falls, outside of safety measures such as at Terrapin Point. But this does not mean that heavy-handed intervention has been completely ruled out. The State Reservation at Niagara, now known as the Niagara Falls State Park, is still today the centrepiece of the American side. The park received a $44 million overhaul in 2003, and in 2012 a $40 million makeover was announced, which included work on Luna Island, the Three Sisters Islands, Stedman's Bluff, Terrapin Point, Prospect Point, and the Cave of the Winds plaza.[66] If the money can be secured, park officials are also planning for a 1969 redux: they will dewater the American Falls in order to fix or replace the old bridges to Goat Island. Whether this is even attempted – and, if so, whether it results in an increase in tourist attendance – remains to be seen.[67]

For its part, the City of Niagara Falls, Ontario, has doubled down on tourism since the 1970s, ignoring most of the concerns about the surrounding atmosphere articulated during the American Falls campaign. The incline railway for tourists installed immediately behind Table Rock House has been expanded several times.[68] At the Horseshoe Falls, the foundation and parapet across from the Victoria Park Restaurant were replaced, and an observation deck and pathways were landscaped at the brink. Clifton Hill became an even bigger jumble of humanity, noise, and neon signs, including a casino in a former shopping mall. An enormous new casino complex was also built in the Fallsview area, which in the span of a few decades went from a mostly empty space on the bluff above the Horseshoe Falls to a maze of towers, hotels, and other tourist amenities. These have offered some employment opportunities in the wake of deindustrialization, as many companies and factories on the Ontario side of the Niagara peninsula have, since the 1990s, slashed their operations, shut down, or left.[69]

The interventionist approach to the American Falls "crisis" was driven at first by local interests that extended beyond just the state of the Falls, but the movement to save the American Falls tapped into a wider sentiment involving concern about the cataract for its own sake and as a symbol of natural and national grandeur. This concern morphed over time, as preservation of the Falls came to be understood as leaving it alone rather than intervening. The IJC's decision not to significantly alter the American Falls coincided with the environmental movement of the 1960s and 1970s, as well as changing ideas about the role of engineers and technology in society. In the case of the American Falls, it appears that this philosophical evolution was driven primarily by elites, particularly the American Falls International Board and the International Joint Commission. The decision not to alter nature represented a new direction for the IJC, but a receptive public was also necessary, for the new environmental ethos gave engineers and governments sufficient licence to change their approach to dealing with Niagara. In 1965, only a few voices argued that the American Falls should be left alone; a decade later, it was the majority view.

While American preservationists had always sought to protect "wilderness" and special environmental locales, they often shared some of the conservationist urge to engineer or tinker with nature to make it better for human use. In this respect, prior to the 1960s one of the main preservationist versus conservationist divides was not whether nature should be changed but what the end goals of such manipulation should be: for consumption as natural resources, or for consumption through personal experience for the betterment of the human condition? The meaning of preservation shifted between 1965 and 1975. Many became wary of interfering with nature as the attitudes of the middle third of the century gave way to the realization that there were unintended repercussions and limits to control. This also reflected shifting attitudes about sublimity. By the

1970s, the sublimity of Niagara was returning to a sense of experiencing the setting in all its messy and imperfect glory, as a waterfall that still carried some semblance of its former sense of wildness and danger, even if past sensations of pure sublimity could not be fully recaptured.

Many historians have rightfully pointed out the many ways in which the environmental movement was driven by quotidian concerns, but environmental icons like Niagara are equally important, for they serve not only as barometers of national sentiments about nature but also as drivers of these attitudes. The American Falls was thus very important symbolically, not only because it provided a cross-border contrast with Canada but also because Niagara Falls stood as a national icon, a synecdoche for how Americans perceived the state of their environments.

Nevertheless, the decision not to radically change the American Falls was also calculated, reflecting more realistic cost-benefit engineering analysis compared with prior decades of pork barrel politics, as well as the partial incorporation of ecological values and amenities into this analysis. Cost, concerns about public safety, and uncertainty about whether stopping the natural processes of erosion and talus accumulation would actually make a perceptible aesthetic difference to the average tourist also deflected attention away from intervention. It was not worth spending all the time and money that would be necessary to remove the talus and control erosion if these measures would not be noticed by most people, as studies suggested. The local New York tourist industry, which had jumpstarted the whole campaign thinking that it would benefit from a refurbishment of the American Falls, now thought the opposite: namely, that closing attractions for several more years to reclothe the waterfall would result in a significant loss in tourist revenue. Moreover, for most observers and officials, addressing the potential safety issues discovered during the course of the American Falls study took precedence over improving the beauty of Niagara Falls – tourists might want to experience a sense of danger but not the actual reality. Even though the full range of potential manipulations under consideration for the American Falls were not carried out, the changes executed after 1965 did make Niagara an even more elaborate blend of the natural and artificial, with rock sutured in place by bolts, cables, anchors, and concrete.

CONCLUSION

Fabricating Niagara

We must have new combinations of language to describe the Falls of Niagara.

Thomas Moore, 1804

On one of our many family trips to Niagara Falls, we arrived in the late afternoon. Since it was almost December, darkness soon descended. Leaving our kids with their grandparents, my wife Jen and I walked down to the waterfall, which was lit up in resplendent colours: vivid blues, bold yellows, dazzling purples, striking pinks. Lights were first projected onto the waterfall in 1865, and nightly illumination began in 1925. The lights were upgraded in the 1950s and 1970s, and most recently in late 2016, when the existing xenon gas lamps were replaced by 1,400 LED lights at a cost of about $4 million. This new installation bathed the Falls in "more than twice the previous lighting levels, programmable lighting, and a wider spectrum of colour."[1]

It is apparent to any visitor that the light show is an artificial contrivance, designed to enhance the visual appeal of Niagara Falls. But how many know that the very waterfall on which the colours are cast could be considered fake? Over the last century, Niagara Falls was fabricated in both senses of the word – something that is built and constructed, but also an intentional deception. Beginning in the late nineteenth century, the United States and Canada (and New York and Ontario) began to wring as much energy from the waterfall as was technologically possible. Over the first half of the twentieth century, technocrats and politicians repeatedly sought to increase the amount of water diverted for hydro power, while simultaneously reducing erosion and "beautifying" the cataract. This process peaked with the bilateral 1950 Niagara Diversion Treaty. Under the terms of this treaty, between half and three-quarters of the river's volume is diverted through huge tunnels, leaving the remainder – the so-called tourist water – to tumble over a designer waterfall.

Niagara Falls is dynamic, perpetually transforming and moving itself. But an uncontrolled and constantly eroding Niagara was inimical to the tourist and industrial infrastructure dependent on a waterfall with a fixed address. Instead of a messy, receding waterfall, during the early Cold War a flowing facade was frozen in place to the greatest extent possible. Achieving the desired visual effect with just a fraction of the river's flow required shrinking the crestline of the Horseshoe Falls and strategically manipulating the riverbed directly above the Falls. In the minds of those designing the changes, this was preserving Niagara by protecting it from itself, before erosion disfigured it or moved it too far upriver. Technocrats concealed their manipulation and industrialization of Niagara's waterscape through a process of disguised design, which relied heavily on scale models. Other major hydroelectric projects of the high modernist era dammed and obliterated the rivers they remade. But Niagara was an exception. As Niagara Falls was turned into a tap, it was altered to make it look more like itself, or at least a sanitized version of its past self. Nevertheless, planners hoped that the casual observer or tourist would be none the wiser that they were gazing at a simulacrum of Niagara Falls. The engineers believed they could quantitatively assess and model qualitative aspects, such as emotional responses to different versions of a waterfall, as they sought to empower nature so that it could be even more productive. Experts had faith in their own judgment and acumen, but eventually began using various ways of ascertaining what people – in the aggregate as "tourists" – expected to experience at Niagara, and then used those parameters to help refashion the waterfall. Eventually, the preservation of Niagara morphed from intervention to abstention, driven by concerns about cost, public safety, and the limited benefits that would result from removing talus and arresting erosion at the American Falls.

The development of Niagara-as-technology involved a range of large-scale infrastructures: the power stations, the remedial works, the electrical grid. Indeed, the waterfall itself was enrolled as a type of infrastructure, making it a clear example of a hybrid envirotechnology. Scholars have posited a range of different ways to help conceptualize the blending of human and non-human nature: organic machine, cyborg, hybrid, or an "in-between Niagara" as Ginger Strand phrases it.[2] To invoke Leo Marx, at the remade Niagara Falls it is not so much that the machine is in the garden but that the machine and the garden have become one and the same. Nature and technology have become so intertwined and fused at

Niagara Falls that it is impossible to fully disentangle them.[3] Put differently, the boundary between the natural and the artificial, or between ecological and technological networks, is illusory.[4] It is a mechanical waterfall, fine-tuned and regimented like a clock. Maybe it is appropriate to appreciate the Niagara fallscape in the same way we appreciate built architecture.[5]

The facelift given to Niagara Falls in the mid-twentieth century was definitely not the first time that landscapes had been redesigned on a large scale to mask human intervention. An illustrative comparison might be another celebrated feature of the Empire State: Central Park is an artificial creation that many, probably most, people think is a natural remnant. Or consider Venice, where the dynamic processes of the lagoon have been more or less fixed in place.[6] Even the "natural" Niagara reservation on the New York side of the Falls – which, like Central Park, was designed by Frederick Law Olmsted and Calvert Vaux – had mostly been reclaimed from industry in the late nineteenth century. Nonetheless, the new Niagara Falls was perhaps the most extreme example to that point of remaking a monumental landmark on such a scale and doing it in such a way as to intentionally fool the general public.

As Richard White wisely writes about the damming and hydroelectric process that took place on another major North American transborder river in the same era, "planning is an exercise of power, and in a modern state much real power is suffused with boredom. The agents of planning are usually boring; the planning process is boring; the implementation of plans is always boring." In democratic polities, White added, "boredom works for bureaucracies and corporations as smell works for a skunk. It keeps danger away. Power does not have to be exercised behind the scenes. It can be open. The audience is asleep. The modern world is forged amidst our inattention."[7] While the physical aspects of remaking Niagara Falls were mostly done out in the open, and much of the planning was boring, in this case the entire audience was not asleep, at least not all the time. After all, this was a charismatic waterfall that many people travelled long distances to visit. Fearing that the public might not be fooled, or impressed, governments put considerable effort into controlling the narrative and quashing rumours. For many of those who lived in the surrounding area, inattention or shifting baseline syndrome obscured the alterations to the waterfall. More attentive locals surely noticed at least some of the changes. But given how much of the local economy depended on tourism or industry linked to the waterfall, they either approved, did not really care, or did not want to bite the hand that fed them.

Disguised Design

Precisely how much have the two nations manipulated Niagara Falls since the 1880s? This is a tricky question to answer. Many interventions, especially before the mid-1920s, went unreported, and the cumulative impact of many small and hard-to-notice changes added up over time. And Niagara Falls is of course always changing on its own. I can safely say that the crest of the Horseshoe Falls has been shrunk between 600 and 900 feet since the late nineteenth century, with most of this contraction taking place at the flanks. However, this decrease has been partially offset by erosion expanding the southern bend of the waterfall, which lengthens the lip.[8] The volume of water going over Niagara Falls is 50 to 75 percent less than it used to be, depending on the time of day and the season, and the depth of the water at the crest has been appreciably lessened. We shouldn't forget that power diversions effectively grew the height of the waterfall by drawing down the river level beneath the Falls – by some 20 feet in the case of the Horseshoe Falls.

Remedial works and diversions have only slowed, not stopped, Niagara's erosion. Surprisingly, very few geomorphological and geological studies of erosion at Niagara Falls have been conducted in the last half-century, and this subject remains characterized by a great deal of scientific uncertainty.[9] Official measurements of recession and the crest length of the Horseshoe Falls are based on incomplete baseline information, particularly prior to the middle of the twentieth century.[10] Measuring Niagara's dimensions was historically very difficult, and remains problematic even with modern digital technologies. Recall that, according to the International Joint Commission, the middle of the Horseshoe Falls receded an average of 4.2 feet per year from 1842 to 1905, 3.2 feet annually from 1905 to 1927, and 2.2 feet per year in the quarter-century leading up to the 1950 Niagara Diversion Treaty.[11] Besides the fact that these were best guesses, and different agencies and engineers came up with conflicting numbers, averages can obscure how much recession takes place in any given year. The IJC's International Niagara Board of Control estimates that the current annual erosion rate is about one-third of a foot, although this seems to be the average from combining the rates of both the Horseshoe and American Falls (the current estimate for the latter is only three to four inches of movement every ten years). Furthermore, those IJC numbers are taken from a study by Hayakawa and Matsukara that admits that the basis for this rate is uncertain.[12]

Major rock collapses do seem to have been prevented by the diminished and more equally dispersed volume of water flowing over the lip of the recontoured waterfall as well as the drainage tunnels and structural reinforcements that were specifically meant to curb rock slides. Deterioration at the "feet" and sides of the Horseshoe Falls has noticeably decreased as a result of anthropogenic interventions. However, erosion continues in the middle of the concave Horseshoe Falls. Using Geographic Information System (GIS) software to analyze aerial images, my colleague Jason Glatz and I found that the peak of the Horseshoe Falls eroded a total of about 75 feet between 1962 and 2017, for an average of 1.3 feet per year (compare with past rates of erosion in Figure 0.4 in the Introduction). Much of this 75-foot indentation, however, came from a recent series of small rockfalls from the crest in 2009, 2012, and 2013.[13]

Regardless of whether erosion rates since the 1960s at Niagara's largest cataract are one or two feet annually, that range is undoubtedly less than the average recession in the second half of the nineteenth century, before water diversions were significantly stepped up. But it isn't far off the average erosion rates in the first half of the twentieth century, and because of the recent series of rockfalls from the horseshoe the erosion rate has actually accelerated over the last decade. Because of migration, the Horseshoe Falls has not always resembled the curve of the eponymous equine footwear in the past and may not in the future either (see Figure C.1).[14] Furthermore, the aforementioned Hayakawa and Matsukara study suggests that returning the full flow of the river to the waterfall would increase levels in the lower river by fifteen feet, which would have many positive ecological effects but would likely only increase cataract erosion by a foot per year. The much-reduced volume of water flowing over the cataract suggests that the fallen rock that piles up in the plunge pool at the base of the Horseshoe Falls is not as effectively scoured away. Could the Horseshoe Falls over time begin to look more like the American Falls, with talus crowding the bottom and the sheer drop slowly turning into more of a cascade as the rim of the Falls moves its way back? By staving off more frequent and smaller rockfalls, will anthropogenic efforts to halt the erosion of Niagara Falls actually lead to much larger rockfalls in the future? What other unintended consequences could result? When it comes to erosion and the preservation of the Falls, how beneficial have all the remedial works and diversions really been?

Many variables make guessing about Niagara's future difficult. Isostatic rebound, the continual but slow raising of land that had been depressed

FIGURE C.1 Aerial view of Niagara Falls from upstream in 2018

by glaciers, continues to work its magic. Climate change complicates predictions for future water flows and thus erosion and mist patterns, as well as ice formation. Lately, mist and spray from the Horseshoe Falls has apparently become a problem again. The Niagara Parks Commission claimed that the number of excessively misty days went from twenty-nine in 1996 to sixty-eight in 2003, and hired an engineering firm to study the problem using a scale model in a wind tunnel. This study pointed to the high-rise hotels on the Canadian side – nine went up in the previous decade – claiming they altered the airflow. Conversely, a volcanologist at the State University of New York at Buffalo attributed the mist to climatic temperature change. The matter has not been definitively resolved, but it is hard not to think that changing climate has something to do with it.[15] Either way, Niagara Falls is probably not the place to visit if you want to stay dry.

Similarly, the environmental impact of the Niagara River ice boom at Buffalo has been a subject of debate. While helpful for producing hydroelectricity, changing the river's ice regime is not beneficial for a number of natural riverine processes.[16] There have been complaints that the boom causes large ice fields to congeal, hurting fish populations and increasing shoreline erosion, as well as altering the local Buffalo microclimate by prolonging the cold and ice in the spring.[17] Complaints about the latter were frequent enough for an official study to be undertaken, though it did not find evidence that ice fields altered the climate.

Considering that water in its various guises – liquid, spray, mist, ice – undermines Niagara's built environment, the remedial works, power stations, and related infrastructure would break down without constant maintenance and frequent upgrades.[18] Repairs needed to be made to address structural defects from wear and tear in two spans of the International Control Structure near the Falls, for example. In 1993, the IJC's International Niagara Board of Control issued a new directive, replacing what had been in place since 1973, for the method of regulating the Niagara control works and the levels of the Chippawa–Grass Island Pool.[19] The IJC continued to give some thought to another control dam for the Niagara River between Buffalo and Fort Erie, but this does not appear to be a serious possibility.

The Power Authority of the State of New York, now known as the New York Power Authority, attempted to increase the intake capacity of the Robert Moses Niagara Power Plant, but was blocked by opposition on both sides of the border.[20] Upgrades have allowed for output capacity to be increased without additional water. The Moses plant was successfully relicensed in 2007, though not without some controversy.[21] It underwent a $300 million fifteen-year refurbishment that concluded in 2006, and is in the midst of a $460 million modernization of the pumped-storage generating plant. Its thirteen generators now have an installed capacity of 2,675 megawatts, making it the largest hydro-power station in New York and the third-largest in the United States. Yet none of the resulting electricity goes to residential customers in the surrounding county, and many aver that NYPA takes much more from the region than it gives back.[22]

In 1974, the Hydro-Electric Power Commission of Ontario adopted its long-standing nickname, Ontario Hydro, as its official moniker. As soon as it had finished Beck No. 2, Ontario Hydro was already considering the construction of an enlarged power development, though the most ambitious plans have not been pursued. In the 1990s, it started looking at a higher-capacity tunnel to utilize the extra water rights from the other Niagara power stations: the Toronto Power Generating Station had been retired at the end of 1973, the former Ontario Power Company generating station at the base of the Horseshoe Falls was decommissioned at the end of 1999, and the private Rankine generating station was taken out of active service in 2005. Right before the turn of the millennium, Ontario Hydro was reorganized into five parts, with Ontario Power Generation (OPG) inheriting the hydro-power generation responsibilities. OPG, a Crown corporation, went ahead with the tunnel: "Big Becky," the largest hard-rock tunnel boring machine in the world, carved a new route 6.4 miles long. When it opened in 2013, the tunnel expanded the diversion capability by roughly 25 percent, enabling the Beck complex to generate 2,081 megawatts.[23] Given the recently enhanced efficiencies and capacities of the main generating stations, and the fact that the 1950 treaty could be renegotiated or cancelled as of 2000, there have been calls to revisit the 100,000/50,000 cfs treaty apportionment so that more hydroelectricity can be generated.[24]

The new OPG Niagara tunnel can withdraw about 82,000 cfs, with the Beck generating complex (Generating Stations No. 1 and 2 and the pumped-storage generating plant) capable of handling 64,400 cfs.[25] New York State's Niagara Power Project can divert a maximum of 109,000 cfs. As a result, the combined water diversion infrastructure on both sides of the border could, in all probability, turn Niagara Falls effectively dry, aside from some small rivulets, under average river-flow conditions.[26]

Generating hydro power is surely better environmentally than burning coal and other fossil fuels. In recent years, however, a number of studies, as well as the film *DamNation,* have pointed out that hydroelectricity is not nearly as environmentally benign as many claim. Perhaps it should no longer even be labelled as a "green" or environmentally friendly form of energy production. Scientists have shown that hydro-power reservoirs slowly emit methane, a greenhouse gas much more potent than carbon dioxide, from the decomposing plants and trees submerged under the water.[27] But other experts contend that hydro-power installations could be used to reduce greenhouse gas emissions.[28] Either way, it is likely that the reservoirs created for the Niagara generating stations never emitted much methane because they did not flood out an existing river system. Indeed, the Niagara River was not fully dammed, which in some ways makes it closer to a run-of-the-river hydro-power installation than to megadams that block the entire river, stopping the movement of aquatic species and silt. In its scale, however, the Niagara power regime is akin to megadams and as such has caused a plethora of negative environmental impacts, many of which stem from removing so much water from the river for so many miles.

A free-flowing river has been defined as one whose fluvial connectivity is largely unaffected, meaning that there is an unobstructed exchange of water, material, species, and energy within the river system and with surrounding landscapes. The four measures of connectivity are: 1) longitudinal (river channel); 2) lateral (floodplains); 3) vertical (groundwater and atmosphere); and 4) temporal (natural intermittent change).[29] On a spectrum of waterway types, with free-flowing rivers untouched by humans on the one extreme and manipulated waterways converted into buried concrete sewers at the other extreme, Niagara Falls would be somewhere near the middle. But since it suffers from a lack of connectivity, particularly longitudinal and temporal, it is further along the manipulated side.[30] Generating hydroelectricity removes the river's water and energy from the ecosystem, where it would otherwise perform valuable natural functions: scouring and erosion, transport of sediment, habitat supporting fish and aquatic life, and so on. Reconfiguring a river for hydro power changes the types of ecological benefits a river can provide for non-human nature, as well as the "ecosystem services" – to use a well-meaning but potentially flawed term – that a river can provide for humanity.[31]

A recent extensive scientific study suggests that about half of the pre-1945 flora on Goat Island and other nearby islands has been extirpated.[32] Pollution from all the industry linked to hydro power has long impaired the Niagara River. Under the terms of the Great Lakes Water Quality Agreement, the Niagara River was declared an Area of Concern (AOC) and was one of the forty-three AOCs requiring a Remedial Action Plan (RAP). The chief concern in the Niagara AOC was toxins. To provide just one prominent example, in the early 1980s the so-called Chippawa Blob was discovered in Canadian waters of the Niagara River, the result of chemicals dumped in the reversed Welland River by the Norton Abrasives Company.[33] Considerable progress has been made since then: two of the three RAP stages have been completed, a number of "beneficial uses" have been restored, and in the fall of 2019 the US Niagara River corridor received a Ramsar designation as a Wetland of International Importance.[34] But much remains to be done, and toxins buried long ago still leach into tributaries and drains heading to the Niagara River. In recent years, radioactive material has been found across Niagara County. For example, in 2018 radioactive soil was discovered in Niagara Falls State Park close to the waterfall.[35]

Flowing Facade

Niagara is still one of the top tourist destinations on Earth. About 22 million people visit every year, according to one estimate, while another pegs attendance at more than 30 million annually. Regardless of which total is correct, far more people visit Niagara Falls today than in any previous era. Yet there is little doubt that Niagara has lost much of its aura, and in the twenty-first century it does not hold sway over our collective imagination as it once did. This is due in part to changing cultural tastes and sensibilities. But that is only part of the story. Another reason for Niagara's decline is more straightforward: onlookers are gazing at only a fraction of the water volume that had captivated so many prior to the industrial age. In other words, ever since we began siphoning off its water and changing its shape, Niagara Falls has been objectively less impressive, though it is still stunning enough for many people to want to see it, or at least say that they have seen it along with the casinos and carnivalesque attractions.

Niagara's history reveals the many ways in which borders impact the development and governance of natural resources. North American governments were willing to sacrifice the natural sublime to benefit industrial development – or turn it into a manufactured or calculated sublime.[36] The history of public hydroelectric developments, water diversions, and engineering manipulations at Niagara Falls highlights the differences as well as similarities between the United States and Canada when it comes to national and regional attitudes toward not only the great cataract but also the environment, progress, technology, and identity. There was a reciprocal and dialectical relationship between waterscapes and culture, for Niagara was imbued with new associations while simultaneously creating and altering these meanings. As a result, changing cultural attitudes about the beauty and sublimity of Niagara Falls cannot be separated from material changes to the cataract.

Patrick McGreevy has referred to Niagara Falls as a "wall of mirrors" that reflects back different cultural meanings to the two countries.[37] Niagara is not only a boundary between two nations and between nature and culture but also a "boundary object" that is interpreted differently by various communities while maintaining a common identity.[38] Scholars have shown that Niagara resonated with American nationalism during the nineteenth century. The Falls became a symbol of the nation itself, something that separated the New World from the Old, as well as of America's burgeoning power, liberty, and manifest destiny. By the late nineteenth century, however, the meaning of Niagara for Americans was in flux. For some it was declining as a symbol of natural splendour, for others it was the epitome of the technological sublime, testifying to the fact that American ingenuity could permit even the wildest natural features to be controlled.

In the twentieth century, Niagara Falls became equally, if not more so, intertwined with Canadian identity. This was not only because of Niagara's declining significance in the American collective consciousness but also because of Canadian hydraulic nationalism springing from that nation's embrace of hydro power, technology, and the larger Horseshoe Falls.[39] Part of this was the unique Canadian formulation of the links between liberal society and energy that came out of (central) Canada's much greater dependence on homegrown hydro power. America's decreasing association with Niagara Falls might even be said to have a physical landscape manifestation: the creation of Terrapin Point in the 1950s almost resulted in the United States losing its entire share of Niagara's largest waterfall.

Niagara Falls should be understood as an energy landscape, even an energy sacrifice zone of sorts. Niagara Falls regulation was integral to the evolution of the two countries' respective domestic energy and environmental policies. Niagara developments and the hydro diplomacy that produced them were central to Canadian-American relations and integration, particularly concerning energy, electricity, and natural resources. Cooperatively developing hydroelectricity along the border, and cooperating on reconfiguring a major cataract, is emblematic of the uniqueness of the US-Canada relationship, though it points to the fact that this bilateral relationship is heavily based on practicality, coinciding national interests, and shared environments. Niagara cooperation also symbolized many aspects of America's growing post-1945 empire.[40] This was a new kind of globalized empire that did not require much direct territorial control. Rather, in an age of newly created plastics, synthetic fuels, artificial rubber, novel chemicals, and standardized products (including the English language), the United States expanded its reach and power not just through economic and military might but also through technologies, infrastructure, and networks. This was a more consensual form of colonization since many nations, such as Canada, willingly joined up in order to get a slice of the progress pie.

I am personally conflicted about the history of Niagara Falls. The "preservation" and "improvement" of Niagara Falls were often misleading euphemisms, suggesting that the best way to protect something is to break it before someone else can take it. But the story is also more complicated than easy villains and victims, or simple declensionist narratives. Humanity has received an enormous amount of power in exchange for what is, arguably, a slight reduction in the grandeur of Niagara Falls, at least from the vantage point of the average tourist. And Niagara Falls is still awe-inspiring. While it is easy to critique the engineering mindset, doing so fails to acknowledge how dependent our modern lifestyles are on the products of this mindset, the myriad ways that we all benefit. The engineers did a very impressive job with Niagara Falls: the scenic provisions of the 1950 treaty attained their goal of an unbroken crestline that people still flock to, while allowing maximum power production. The result was a compromise between scenic beauty and electricity generation. The planners and officials believed they were maximizing the greatest public good, and deserve recognition for not *completely* sacrificing the Falls for power development. Of course, public pressure groups deserve much of the credit for holding the feet of governments to the fire when it came to preserving Niagara Falls.

Yet, selecting power generation and certain types of tourism as the main uses of Niagara Falls privileges the industrial and capitalist uses of a water body while circumscribing or foreclosing other uses and values (recreation, fishing, spiritual, and so on). This ecological imperialism finds echo in other types of colonialism, such as the taking of Tuscarora land. Large-scale structural solutions offered more opportunities for experts to display and cement their expertise; for politicians to dispense patronage and win votes; for industrialists to increase profits; for contractors to get a slice of the action. Industrial interests have been the major beneficiaries of cheap hydro power from Niagara Falls. But cheap power also fosters societal beliefs that we can and should have perpetual economic growth, as well as makes possible the consumption rates that now threaten the planet.

Ultimately, what has been done at Niagara Falls feels dishonest for three primary reasons: scale, deception, and commodification. Some landmarks carry special cultural meaning because of their history, or because they provide an experience that is perceived as closer to nature. Niagara Falls is one of those places. Niagara has a sacredness to it, a *genius loci*. As a collective society we may control Niagara Falls, but as individuals we are still at the mercy of its power and fury. In some ways, what we have done to Niagara Falls is just a scaled-up version of what we do in our gardens, our city parks, or our local streams to make them suit our purposes. But scale matters. A specific activity that is relatively innocuous at a micro level can be much more pernicious or damaging when repeated many times or done at a much larger scale. Niagara represents the traditional "hard path" approach to water: a focus on supply-side options, particularly enormous and capital-intensive infrastructure. In the future, sustainable water policies and infrastructure will need to adopt the "soft path," which entails smaller and localized sources; be more cognizant of the water-energy nexus;

and implement water policies and laws based on public trust, water as a commons, and right-to-water principles.[41]

Scholars point out that the tourism industry uses facades and other ways of shaping the tourists' gaze to create the authentic experiences that visitors seek – which perfectly describes Niagara Falls.[42] But can an ersatz waterfall truly be satisfying? The fabrication of Niagara Falls is a shell game meant to deceive, even if some of the end goals are noble. Manipulating the waterfall for the sake of electricity and industry commodifies the Falls. If it is acceptable to manipulate a famous landmark for such ends, it gives licence to manipulate and dominate any other part of nature that suits us. Indeed, since the mid-twentieth century, an era many label as the Anthropocene, humans have so altered the planet's surface that landscape interventions like that which took place at Niagara Falls are now the norm. We have terraformed most of the planet, though not in a way that increases its long-term habitability. Underpinning the engineering of Niagara Falls is an extractive mentality, and changing the Falls helped normalize and encourage the mass commodification and manipulation of nature – to say nothing of the psychological, moral, and spiritual impacts of increasing reliance on technologies, especially large-scale technologies.[43] To replace nature with technology is to lessen our ability to know and experience truth and the sacred, a connection to the transcendent. Niagara Falls is a place marked by hubris and selfishness; we, and the waterfall, need humility and grace.

Notes

Foreword: Iconic Falls, Contrived Landscapes, and Tantalizing Opportunities

1 Louis Hennepin, *A New Discovery of a Vast Country in America* (London: M. Bentley et al. 1698), quoted by Victoria Dickenson, *Drawn from Life: Science and Art in the Portrayal of the New World* (Toronto: University of Toronto Press, 1998), 106–7.
2 Pierre Francois Xavier de Charlevoix, Pehr Kalm, and J. Hector St.John de Crèvecoeur, excerpts from "Historical Accounts of Niagara Falls," https://www.niagarafallsinfo.com/niagara-falls-history/niagara-falls-municipal-history/historical-accounts-of-niagara-falls/. See also Pehr Kalm, *Peter Kalm's Travels in North America: The English Version of 1770,* Vols. 1 and 2, translated by Adolph B. Benson (New York: Dover, 1987) and J. Hector St. John de Crèvecoeur, *Letters from an American Farmer* (New York: E.P. Dutton, 1957 [1782]).
3 Charles Joseph LaTrobe, *The Rambler in North America: 1832–1833,* Vol. 1 (London: R.B. Seeley and W. Burnside, 1835), 72–73, https://www.niagarafallsinfo.com/niagara-falls-history/niagara-falls-municipal-history/historical-accounts-of-niagara-falls/.
4 Isabella Lucy Bird, *The Englishwoman in America* (London: John Murray, 1856), 353.
5 John K. Howat, *Frederic Church* (New Haven: Yale University Press, 2005), 69.
6 Natalie McKnight, "Dickens, Niagara Falls and the Watery Sublime," *Dickens Quarterly,* June 2009.
7 Patrick V. McGreevy, *Imagining Niagara: the Meaning and Making of Niagara Falls* (Amherst, MA: University of Massachusetts Press, 1994).
8 James K. Liston, *Niagara Falls: A Poem, In Three Cantos* (Toronto: Printed and published for the author, by J.H. Lawrence, 1843), 36.
9 Art historians identify a fourth major painting, now lost. *Under Niagara* (122 cm by 183 cm) was apparently painted in 1862 but is known only through lithographs of the original.

10 See Patrick McGreevy, "Imagining the Future at Niagara Falls," *Annals of the Association of American Geographers* 77, 1 (March 1987): 50, which so characterizes the "historical milieu" of the latter part of the nineteenth century.

11 Jill Jonnes, *Empires of Light: Edison, Tesla, Westinghouse, and the Race to Electrify the World* (New York: Random House, 2003).

12 Benjamin Copeland, *Niagara and Other Poems* (Buffalo and New York: The Matthews-Northrup Works, 1904), 11 ("resistless tide") and 12. The "new, totally human" phrase is from McGreevy, "Imagining the Future at Niagara Falls," 50.

13 Karen Dubinsky, *The Second Greatest Disappointment: Honeymooning and Tourism At Niagara Falls* (New Brunswick, NJ: Rutgers University Press, 1999), 33. Henry David Thoreau came this way on his own version of such a tour in the spring of 1861, and noted of Niagara Falls, NY, on May 16: "This is quite a town, with numerous hotels and stores, and with paved streets; and I imagine the Falls will soon be surrounded by a city." See "The First and Last Journeys of Thoreau: Lately Discovered among His Unpublished Journals and Manuscripts" digitized by the Internet Archive, https://archive.org/stream/firstlastjourney02thor/firstlastjourney02thor_djvu.txt.

14 The Canadian Town of Niagara Falls came into being in 1881 when the community of Clifton (itself an amalgamation of the railway division point settlement of Elgin with neighbouring Clifton) adopted that name. A year later, nearby Drummondville became the Village of Niagara Falls. Town and Village amalgamated to form the City of Niagara Falls in 1904.

15 Dubinsky, *Second Greatest,* 93, 71–72.

16 The simple narrative glorifying Lord Dufferin's role was challenged by Gerald Killan, "Mowat and a Park Policy for Niagara Falls," *Ontario History* 70 (June 1975): 115–35. This paragraph abstracts the much fuller discussion in Dubinsky, *Second Greatest,* 83–115; the "beauties " and "factories" quotes are from this source, p. 98, quoting an article in *Saturday Night;* and "enchantment" from p. 93, quoting the *Toronto Globe.*

17 Matthew D. Evenden, *Fish Versus Power: An Environmental History of the Fraser River* (New York: Cambridge University Press, 2004).

18 Daniel Macfarlane "'A Completely Man-Made and Artificial Cataract': The Trans-national Manipulation of Niagara Falls," *Environmental History* 18 (October 2013): 759–84.

19 William Cronon, "The Trouble with Wilderness, or Getting Back to the Wrong Nature," *Environmental History* 1, 1 (January 1996): 7–28.

20 Sarah Pritchard, "Toward an Environmental History of Technology," in *Oxford Handbook of Environmental History,* ed. A.C. Isenberg (New York: Oxford University Press, 2014), 227–58, especially 243–44.

21 For more on this see Daniel Macfarlane, "'As Nearly as May Be': Estimating Ice and Water on the Niagara and St. Lawrence Rivers," *Journal of Historical Geography* 65 (July 2019): 73–84. Tina Loo has engaged these questions in different contexts at some length. For a summary, see Tina Loo, "Questions of Scale," in *The Nature of Canada,* ed. Colin Coates and Graeme Wynn (Vancouver: On Point Press imprint of the UBC Press, 2019), 262–79.

22 Jean-Yves Tizot, "Ebenezer Howard's Garden City Idea and the Ideology of Industrialism," *Cahiers victoriens et édouardiens* 87 (Printemps, 2018).

23 C.B. Purdom, *The Garden City: A Study in the Development of a Modern Town* (London: Dent and Sons, 1913), 141.

24 Ebenezer Howard, *Garden Cities of To-morrow* (London, 1902). Reprinted, edited with a Preface by F.J. Osborn and an Introductory Essay by Lewis Mumford (London: Faber and Faber, [1946]), 56–57.

25 Peter Kropotkin, *Fields, Factories, and Workshops: Industry Combined with Agriculture and Brain Work with Manual Work* (London, Edinburgh, Dublin, New York: Thomas Nelson and Sons, 1912) includes the 1st edition preface. No pagination, online https://lib.anarhija.net/library/petr-kropotkin-fields-factories-and-workshops-or-industry-combined-with-agriculture-and-brain-w.

26 "Preface to Second Edition," Kropotkin, *Fields, Factories, and Workshops.*

27 Amory Lovins, "Energy Strategy: The Road Not Taken?" *Foreign Affairs* (October 1976).

28 Lovins, "Energy Strategy," 77 and 79 (for transmission costs).

29 Ralph Borsodi, advocate of locally based, self-sufficient living argued strongly (in the 1920s) that small-scale, electric-powered manufacturing was no more expensive than large-scale factory production once one took account of reduced distribution costs and the benefits of timing production to meet individual needs rather than promoting mass consumption to benefit producers. Ralph Borsodi, *The Ugly Civilization* (New York: Simon and Schuster, 1929).

30 Lewis Mumford, *Technics and Civilization* (Chicago: University of Chicago Press, 2010, [originally published New York: Harcourt, 1934]), 265.

31 Mumford, *Technics and Civilization,* 266.

32 Lovins, "Energy Strategy," 70.

33 Chris Rhodes, "Peak Oil Is Not a Myth," *Chemistry World* (February 19, 2014).

34 Jonah M. Kessel and Hiroko Tabuchi, "It's a Vast, Invisible Climate Menace. We Made It Visible," *New York Times,* December 12, 2019.

35 Mumford, *Technics and Civilization,* 267.

36 The insurmountable opportunity phrase is often attributed to Walt Kelly's comic strip character Pogo, widely followed in the 1970s – and is so ascribed by Lovins, in "Energy Strategy," 74 – although the Quote Investigator website stops short of a definite attribution, https://quoteinvestigator.com/2017/04/10/opportunity/.

Introduction

1 Other waterfalls with more volume, such as Buyoma Falls, are not nearly the height of Niagara Falls, while other famous waterfalls, such as Victoria or Iguazu Falls, which are higher and wider than Niagara, have substantially lower flow volumes. For the various types of waterfalls, see Richard H. Biesel Jr., *International Waterfall Classification System* (Parker, CO: Outskirts Press, 2006).

2 A number of waterfalls were dammed for hydro power in Northern Europe in the early twentieth century, particularly in the Alps and Scandinavia, such as Sweden's Trollhättan waterfalls. Marc Landry, "Environmental Consequences of the Peace: The Great War, Dammed Lakes, and Hydraulic History in the Eastern Alps," *Environmental History* 20 (2015): 422–48; Eva Jakobsson, *Industrialisering av älvar: Studier kring svensk vattenkraftutbyggnad 1900–1918* (Göteborg: Historiska Institutionen, 1996).

3 For a voluminous collection of writings and artistic works about Niagara Falls, see Charles Mason Dow, *Anthology and Bibliography of Niagara Falls, Vols. I and II* (Albany: State of New York, J.B. Lyon, 1921).

4 Christopher W. Lane, *Impressions of Niagara: The Charles Rand Penney Collection* (Philadelphia: Philadelphia Print Shop, 1993).

5 On the local histories of Niagara Falls and the history of the Niagara River, see Archer Butler Hulbert, *The Niagara River* (New York: G.P. Putnam's Sons, 1908); Donald Braider, *The Niagara (Rivers of America)* (New York: Holt, Rinehart and Winston, 1972); Ralph Greenhill and Thomas Mahoney, *Niagara* (Toronto: University of Toronto Press, 1969); Lisa Aug, *Beyond the Falls* (Niagara Falls, NY: Niagara Books, 1992); Sherman Zavitz, *Niagara Falls: Historical Notes* (St. Catharines, ON: Looking Back Press, 2008); Daniel Dumych, *Images of America: Niagara Falls* (Chicago: Arcadia, 1996); Paul Gromosiak and Christopher Stoianoff, *Images of America: Niagara Falls, 1850–2000* (Charleston, SC: Arcadia, 2012); Daniel Dumych, *Images of America: Niagara Falls, Volume 2* (Chicago: Arcadia, 1998); Robert Higgins, *The Niagara Frontier: Its Place in US and Canadian History* (Kitchener, ON: Upney Editions, 1996); Michael N. Vogel, *Echoes in the Mist: An Illustrated History of the Niagara Falls Area* (Chicago: Windsor, 1991); Kiwanis Club of Stamford, Ontario, *Niagara Falls, Canada: A History of the City and the World Famous Beauty Spot* (Toronto: Ryerson Press, 1967).

6 Elizabeth R. McKinsey, *Niagara Falls: Icon of the American Sublime* (New York: Cambridge University Press, 1985); Jeremy Elwell Adamson et al., *Niagara: Two Centuries of Changing Attitudes, 1697–1907* (New York: Routledge, 1992); Thomas V. Welch, *How Niagara Was Made Free: The Passage of the Niagara Reservation Act in 1885* (Buffalo: Press Union and Times, 1903); *Arcadia Revisited: Niagara Rivers and Falls from Lake Erie to Lake Ontario* (Albuquerque, NM: Buscaglia-Castellani Art Gallery of Niagara University and University of New Mexico Press, 1988); Lane, *Impressions of Niagara.*

7 Yosemite, created in 1864, was actually the first state park, but it was later changed to a national park. Charles M. Dow, *The State Reservation at Niagara: A History* (Albany: J.B. Lyon, 1914); Anne Whiston Spirn, "Constructing Nature: The Legacy of Frederick Law Olmsted," in *Uncommon Ground: Rethinking the Human Place in Nature,* ed. William Cronon (New York: W.W. Norton, 1996), 91–113.

8 Patrick McGreevy, *Imagining Niagara: The Meaning and Making of Niagara Falls* (Amherst, MA: University of Massachusetts Press, 1994); William Irwin, *The New Niagara: Tourism, Technology, and the Landscape of Niagara Falls* (University Park, PA: Pennsylvania State University Press, 1996); Rob Shields, *Places on the Margin: Alternative Geographies of Modernity* (New York: Routledge, 1992).

9 Ginger Strand, *Inventing Niagara: Beauty, Power, and Lies* (New York: Simon and Schuster, 2008); Pierre Berton, *Niagara: A History of the Falls* (Albany: State University of New York Press, 1992); Karen Dubinsky, *The Second Greatest Disappointment: Honeymooning and Tourism at Niagara Falls* (Toronto: Between the Lines, 1999); Linda L. Revie, *The Niagara Companion: Explorers, Artists, and Writers at the Falls, from Discovery through the Twentieth Century* (Waterloo, ON: Wilfrid Laurier University Press, 2003); Ernest Sternberg, "The Iconography of the Tourism Experience," *Annals of Tourism Research* 24, 4 (1997): 951–69; Barry Grant and Joan Nicks, *Covering Niagara: Studies in Local Popular Culture* (Waterloo, ON: Wilfrid Laurier University Press, 2010); Adam Hallett, "Made of the Mist: Nineteenth-Century British and American Views of Niagara," *Literature Compass* 11, 3 (2014): 159–72; Kevin Hutchings, "Romantic Niagara: Environmental Aesthetics, Indigenous Culture, and Transatlantic Tourism, 1794–1850," in *Transatlantic Literary Exchanges, 1790–1870,* ed. Kevin Hutchings and Julia M. Wright (Farnham, UK: Ashgate, 2011), 153–68; Robinder

Kaur Sehdev, "Unsettling the Settler at Niagara Falls: Reading Colonial Culture through the Maid of the Mist" (PhD diss., York University, 2008).

10 John N. Jackson with John Burtniak and Gregory P. Stein, *The Mighty Niagara: One River – Two Frontiers* (Amherst, NY: Prometheus Books, 2003); Hugh J. Gayler, ed., *Niagara's Changing Landscapes* (Ottawa: Carleton University Press, 1994); Patrick McGreevy, *Wall of Mirrors: Nationalism and Perceptions of the Border at Niagara Falls* (Orono, ME: Borderlands Project, 1991); Munroe Eagles, "Organizing across the Canada-US Border: Binational Institutions in the Niagara Region," *American Review of Canadian Studies* 40, 3 (September 2010): 379–94; William H. Siener, "The United Nations at Niagara: Borderlands Collaboration and Emerging Globalism," *American Review of Canadian Studies* 43, 3 (September 2013): 377–93; Shields, *Places on the Margin*. See also John J. Bukowczyk et al., *Permeable Border: The Great Lakes Basin as Transnational Region, 1650–1990* (Pittsburgh: University of Pittsburgh Press, 2005); Shannon Stunden Bower, *Wet Prairie: People, Land, and Water in Agricultural Manitoba* (Vancouver: UBC Press, 2011); Philip Van Huizen, "Building a Green Dam: Environmental Modernism and the Canadian-American Libby Dam Project," *Pacific Historical Review* 79, 3 (August 2010): 418–53; Joseph Taylor III, *Making Salmon: An Environmental History of the Northwest Fisheries Crisis* (Seattle: University of Washington Press, 2001).

11 Cliff Spieler and Tom Hewitt, *Niagara Power: From Joncaire to Moses* (Lewiston, NY: Niagara Power, 1959); Merrill Denison, *The People's Power: The History of Ontario Hydro* (Toronto: McClelland and Stewart, 1960); C.D. Kepner, "Niagara's Water Power," *Niagara Frontier* 15 (1968): 97–105; 16 (1969): 33–41, 75–80; and 17 (1970): 69–79; Ontario Hydro, *Power from Niagara* (Toronto: HEPCO, 1970); H.V. Nelles, *The Politics of Development: Forest, Mines and Hydro-Electric Power in Ontario, 1849–1941* (Toronto: Macmillan, 1974); James Mavor, *Niagara in Politics: A Critical Account of the Ontario Hydro-Electric Commission* (New York: E.P. Dutton, 1925); J. Carr, "Hydro-Electric Power Development in the Niagara Peninsula," in *Industry in the Niagara Peninsula, Proceedings Eleventh Annual Niagara Peninsula History Conference*, ed. John N. Jackson and John Burtniak (St. Catharines, ON: Brock University, 1992), 6–15; Karl Froschauer, *White Gold: Hydroelectric Power in Canada* (Vancouver: UBC Press, 1999); Robert Belfield, "The Niagara Frontier: The Evolution of Electric Power Systems in New York and Ontario, 1880–1935" (PhD diss., University of Pennsylvania, 1981); Robert Belfield, "The Niagara System: The Evolution of an Electric Power Complex at Niagara Falls, 1883–1896," *Proceedings of the IEEE* 64 (1976): 1344–49; Robert Belfield, "Technology Transfer and Turbulence: The Evolution of an International Energy Complex at Niagara Falls, 1896–1906," *HSTC Bulletin: Journal of the History of Canadian Science, Technology and Medicine/HSTC Bulletin: revue d'histoire des sciences, des techniques et de la médecine au Canada* 5, 2 (1981): 69–98; Norman Ball, *The Canadian Niagara Power Company Story* (Erin, ON: Boston Mills Press, 2005); Ken Glennon, *Hard Hats of Niagara: The Niagara Power Project* (South Bend, IN: Dog Ear, 2011); Brett Gawronski, Jana Kasikova, Lynda Schneekloth, and Thomas Yots, *The Power Trail: History of Hydro-Electricity at Niagara* (Buffalo: Brian Meyer, 2014); Edward Dean Adams, *Niagara Power*, 2 vols. (Niagara Falls, NY: Niagara Falls Power Company, 1927); Steven Lubar, "Transmitting the Power of Niagara: Scientific, Technological and Cultural Contexts of an Engineering Decision," *IEEE Technology and Society Magazine*, March 1989, 11–18; Daniel Dumych, *The Canadian Niagara Power Company – One Hundred Years, 1892–1992* (Niagara Falls, ON: Canadian Niagara Power Company, 1992); Thomas Stieve, "National

Identity in Niagara Falls, Canada and the United States," *Tijdschrift voor Economische en Sociale Geografie* 96, 1 (2005): 3–14.

12 Martha Moore Trescott, *The Rise of the American Electrochemicals Industry, 1880–1910* (New York: Praeger, 1981); Alice Mah, *Industrial Ruination, Community and Place: Landscapes and Legacies of Urban Decline* (Toronto: University of Toronto Press, 2012).

13 Some exceptions are Jackson with Burtniak and Stein, *The Mighty Niagara;* Strand's *Inventing Niagara;* Spirn's "Constructing Nature."

14 Niagara's "beauty versus power" has echoes of the titular tension in Matthew Evenden's *Fish versus Power: An Environmental History of the Fraser River* (New York: Cambridge University Press, 2004).

15 Modern state formation is fundamentally about creating the conditions for capitalist development, which is, in turn, fundamentally about extracting value from nature. Jason W. Moore and Raj Patel, *A History of the World in Seven Cheap Things: A Guide to Capitalism, Nature and the Future of the Planet* (Oakland: University of California Press, 2017); Andreas Malm, *Fossil Capital: The Rise of Steam Power and the Roots of Global Warming* (London: Verso Books, 2016).

16 Christopher Armstrong and H.V. Nelles called this the "internal pluralism of the state" in *Wilderness and Waterpower: How Banff National Park Became a Hydro-Electric Storage Reservoir* (Calgary: University of Calgary Press, 2013). See also Elsbeth Heaman, *A Short History of the State in Canada* (Toronto: University of Toronto Press, 2015).

17 The little academic historical work that has been produced on PASNY/NYPA includes: Robert Lifset, *Power on the Hudson: Storm King Mountain and the Emergence of Modern Environmentalism* (Pittsburgh: University of Pittsburgh Press, 2014); Rock Brynner, *Natural Power: The New York Power Authority's Origins and Path to Clean Energy* (New York: Cosimo, 2016). See also Raymond Edward Petersen, "Public Power and Private Planning: The Power Authority of the State of New York" (PhD diss., City University of New York, 1990).

18 At the same time, in the quest to recognize the influence of a non-human nature, it is possible to go too far in the opposite direction by overascribing intentionality and imaginative powers to non-sentient forces and erasing the types of autonomy that do exist between humanity and the rest of the world. For example, Sheila Jasanoff makes the point that only humans can "imagine" in her "Future Imperfect: Science, Technology, and the Imagination of Modernity," in *Dreamscapes of Modernity: Sociotechnical Imaginaries and the Fabrication of Power,* ed. Sheila Jasanoff and Sang-Hyun Kim (Chicago: University of Chicago Press, 2015), 17.

19 I say it is unnatural while recognizing that many of the distinctions between "artificial" (made by humans) and "natural" (that which is not altered by humans) are problematic and can lead to a flawed dualism.

20 When it comes to theorizing about the melding of water, technology, energy, and infrastructure, Richard White's "organic machine" probably still stands as the most influential concept: Richard White, *The Organic Machine: The Remaking of the Columbia River* (New York: Hill and Wang, 1995). Other examples of works on water that combine environmental history with science and technology studies include: Joy Parr, *Sensing Changes: Technologies, Environments, and the Everyday, 1953–2003* (Vancouver: UBC Press, 2009); Sara B. Pritchard, *Confluence: The Nature of Technology and the Remaking of the Rhône* (Cambridge, MA: Harvard University Press, 2011); Martin Melosi, *The Sanitary City: Urban Infrastructure in*

America from Colonial Times to the Present (Baltimore: Johns Hopkins University Press, 2000); Joel Tarr, *The Search for the Ultimate Sink: Urban Pollution in Historical Perspective* (Akron, OH: University of Akron Press, 1996); Ashley Carse, *Beyond the Big Ditch: Politics, Ecology, and Infrastructure at the Panama Canal* (Cambridge, MA: MIT Press, 2014); Mark Cioc, *The Rhine: An Eco-Biography, 1815–2000* (Seattle: University of Washington Press, 2002); David Blackbourn, *The Conquest of Nature: Water, Landscape, and the Making of Modern Germany* (New York: W.W. Norton, 2007); David Biggs, *Quagmire: Nation-Building and Nature in the Mekong Delta* (Seattle: University of Washington Press, 2010); Theodore Steinberg, *Nature Incorporated: Industrialization and the Waters of New England* (Amherst, MA: University of Massachusetts Press, 1994); Daniel Schneider, *Hybrid Nature: Sewage Treatment and the Contradictions of the Industrial Ecosystem* (Cambridge, MA: MIT Press, 2013); Peter Coates, *A Story of Six Rivers* (London: Reaktion Books, 2013); Eva Jakobsson, "Industrialization of Rivers: A Water System Approach to Hydropower Development," *Knowledge, Technology, and Policy* 14, 4 (Winter 2002): 41–56.

21 Works from the 1990s by the likes of Richard White, William Cronon, Joel Tarr, Martin Melosi, Mark Fiege, and Jeffrey Stine can be considered the vanguard of scholarly work on hybrid natures (e.g., organic machine, second nature), as well as Bruno Latour, *We Have Never Been Modern* (Cambridge, MA: Harvard University Press, 1993). The surge of scholarship combining environmental and technological history, which has been given a number of different titles in addition to "envirotech," is well represented in two edited collections: Martin Reuss and Stephen H. Cutcliffe, eds., *The Illusory Boundary: Environment and Technology in History* (Charlottesville, VA: University of Virginia Press, 2010); and Dolly Jørgensen, Finn Arne Jørgensen, and Sara B. Pritchard, eds., *New Natures: Joining Environmental History with Science and Technology Studies* (Pittsburgh: University of Pittsburgh Press, 2013). The introductions and contributions to these edited collections outline the envirotech field, but see also Edmund Russell et al., "The Nature of Power: Synthesizing the History of Technology and Environmental History," *Technology and Culture* 52 (April 2011): 246–59; and Sara Pritchard, "Toward an Environmental History of Technology," in *Oxford Handbook of Environmental History*, ed. Andrew C. Isenberg (New York: Oxford University Press, 2014), 227–58.

22 On technopolitics, see Gabrielle Hecht and Paul N. Edwards, "The Technopolitics of Cold War: Towards a Transregional Perspective," in *Essays on Twentieth Century History*, ed. Michael Adas (Philadelphia: Temple University Press, 2010), 271–314. A 2013 state-of-the-field essay identified hybridity as a "defining tendency of recent scholarship in American environmental history": Paul S. Sutter, "The World with Us: The State of American Environmental History," *Journal of American History* 100, 1 (2013): 94–119.

23 Thomas Zeller, "Aiming for Control, Haunted by Its Failure: Towards an Envirotechnical Understanding of Infrastructures," *Global Environment* 10, 1 (2017): 215. On energy infrastructure see Christopher Jones, *Routes of Power: Energy and Modern America* (Cambridge, MA: Harvard University Press, 2014); Andrew Needham, *Power Lines: Phoenix and the Making of the Modern Southwest* (Princeton, NJ: Princeton University Press, 2015).

24 On envirotechnical systems/regimes, see Pritchard, *Confluence.* On large technological systems, see Thomas P. Hughes, "The Evolution of Large Technological Systems," in *The Social Construction of Technological Systems: New Directions in the Sociology and History of Technology*, ed. Deborah G. Douglas, Wiebe E. Bijker, and Thomas P. Hughes (Cambridge, MA: MIT Press, 1987), 51–82; Belfield, "The Niagara System." One could also invoke the concept of "virtual river," used by both Richard White and Ellen Wohl: Ellen Wohl, *Rivers*

in the Landscape: Science and Management (Hoboken, NJ: Wiley-Blackwell, 2014); White, *The Organic Machine*, 106–9. To borrow another term from science and technology studies, we might say that the waterfall was transmuted into a technological artifact: Dolly Jørgensen, "Artifacts and Habitats," in *The Routledge Companion to the Environmental Humanities*, ed. Ursula Heise, Jon Christensen, and Michele Niemann (New York: Routledge, 2017), 246–57; Langdon Winner, *The Whale and the Reactor: A Search for Limits in an Age of High Technology* (Chicago: University of Chicago Press, 1988); David Noble, *America by Design: Science, Technology, and the Rise of Corporate Capitalism* (London: Oxford University Press, 1979); Donna Haraway, *Simians, Cyborgs, and Women: The Reinvention of Nature* (New York: Routledge, 1990). The term *assemblage* draws from Actor-Network Theory; see Bruno Latour, *Reassembling the Social* (Oxford: Oxford University Press, 2007); Anna Lowenhaupt Tsing, *The Mushroom at the End of the World: On the Possibility of Life in Capitalist Ruins* (Princeton, NJ: Princeton University Press, 2015).

25 Sara Pritchard and Thomas Zeller assert that "industrial processes were embedded within, and ultimately dependent upon, natural resources, environmental processes, and ecosystems": Sara B. Pritchard and Thomas Zeller, "The Nature of Industrialization," in Reuss and Cutcliffe, *The Illusory Boundary*, 70. In a similar vein, Liza Piper identifies an assimilative and integrative industrial process in the Canadian subarctic, where "nature, economy, and society each adapted to one another": Liza Piper, *The Industrial Transformation of Subarctic Canada* (Vancouver: UBC Press, 2009), 10. In the context of another dammed Canadian river and the resulting reservoir, Dolly Jørgensen discusses competing notions of "natural" in "Competing Ideas of 'Natural' in a Dam Removal Controversy," *Water Alternatives* 10, 3 (2017): 840–52.

26 Donald Worster's *Rivers of Empire*, which details the emergence of a "hydraulic society" in the US West, rightfully remains a touchstone for water historians. Donald Worster, *Rivers of Empire: Water, Aridity, and the Growth of the American West* (New York: Pantheon, 1985), 7.

27 Daniel Macfarlane, "Natural Security: Canada-US Environmental Diplomacy," in *Undiplomatic History: Rethinking Canada in the World*, ed. Asa McKercher and Philip Van Huizen (Montreal and Kingston: McGill-Queen's University Press, 2019), 107–36; Kurkpatrick Dorsey, *The Dawn of Conservation Diplomacy: US-Canadian Wildlife Protection Treaties in the Progressive Era* (Seattle: University of Washington Press, 1998); Kurkpatrick Dorsey and Mark Lytle, "Forum: New Directions in Diplomatic and Environmental History," *Diplomatic History* 32, 4 (2008): 517–646.

28 Daniel Macfarlane, "Watershed Decisions: The St. Lawrence Seaway and Sub-National Water Diplomacy," *Canadian Foreign Policy Journal* 21, 3 (2015): 212–23.

29 On sociotechnical imaginaries, see Jasanoff and Kim, *Dreamscapes of Modernity*. On cultural landscapes, see Simon Schama, *Landscape and Memory* (Toronto: Random House, 1995); UNESCO definitions of type of cultural landscapes, http://whc.unesco.org/en/culturallandscape/#1.

30 Canadian and American historiography is replete with notions of the river narrative and aquatic symbolism. Jean Manore, "Rivers as Text: From Pre-Modern to Post-Modern Understandings of Development, Technology and the Environment in Canada and Abroad," in *A History of Water, Vol. 3: The World of Water*, ed. Terje Tvedt and Terje Oestigaard (London: I.B. Tauris, 2006), 229. The outstanding and comprehensive *A History of Water* series should be consulted by all water scholars. On hydraulic nationalism in the Great Lakes–St. Lawrence basin, see Daniel Macfarlane, *Negotiating a River: Canada, the US,*

and the Creation of the St. Lawrence Seaway (Vancouver: UBC Press, 2014); Andrew Biro, "Half-Empty or Half-Full?" in *Eau Canada: The Future of Canada's Water,* ed. Karen Bakker (Vancouver: UBC Press, 2006), 321–33.

31 The link between nation building and riverine environments has a long lineage in the Canadian context, including the meta-historical Staples and Laurentian interpretations. Harold Innis, *The Fur Trade in Canada: An Introduction to Canadian Economic History* (New Haven, CT: Yale University Press, 1930); Donald Creighton, *The Empire of the St. Lawrence: A Study of Commerce and Politics* (Toronto: Macmillan, 1956); Christopher Armstrong, Matthew Evenden, and H.V. Nelles, *The River Returns: An Environmental History of the Bow* (Montreal and Kingston: McGill-Queen's University Press, 2009).

32 H.R. Rice, "More Canadian Kilowatts at Niagara, Part 1," *Compressed Air Magazine* 58, 8 (August 1953): 210–17; "Hydro as Myth" is the name of a chapter in Nelles, *The Politics of Development.*

33 Andrew Watson, "Coal in Canada," in *Powering Up Canada: The History of Power, Fuel, and Energy from 1600,* ed. Ruth Sandwell (Montreal and Kingston: McGill-Queen's University Press, 2016), 213–50; Daniel French, *When They Hid the Fire: A History of Electricity and Invisible Energy in America* (Pittsburgh: University of Pittsburgh Press, 2017).

34 "Niagara Falls Geology: Facts and Figures," Niagara Parks, https://www.niagaraparks.com/visit-niagara-parks/plan-your-visit/niagara-falls-geology-facts-figures/.

35 Adams, *Niagara Power,* 1:20.

36 On the geology and physical geography of Niagara Falls, see Jackson with Burtniak and Stein, *The Mighty Niagara;* Gayler, *Niagara's Changing Landscapes;* Irving H. Tesmer, ed., *Colossal Cataract: The Geological History of Niagara Falls* (Albany: State University of New York Press, 1981); Glenn C. Forrester, *The Falls of Niagara* (New York: D. Van Nostrand, 1928).

37 Aug, *Beyond the Falls,* 94. For an extended analysis of the changing ecology of the Niagara region, see Patricia Eckel, *Ecological Restoration and Power at Niagara Falls* (St. Louis, MO: Botanical Services, St. Louis, 2016). On the ecological history of the landscapes of the Great Lakes basin (especially Southern Ontario), see John Riley, *The Once and Future Great Lakes Country: An Ecological History* (Montreal and Kingston: McGill-Queen's University Press, 2013).

38 Writing in the early twenty-first century, the authors of *The Mighty Niagara* reported that the area "supported an unusual variety of plant and animal life: 2,200 plants species, nearly 400 species of birds, and 47 species of reptiles and amphibians." Jackson with Burtniak and Stein, *The Mighty Niagara,* 373.

Chapter 1: Harnessing Niagara

1 National Register of Historic Places Inventory – Nomination Form, "Adams Power Plant Transformer House" (1978), http://large.stanford.edu/courses/2013/ph240/iskhakov2/docs/75001212.pdf.

2 Jackson et al., *The Mighty Niagara,* 69–74; S.D. Scott and P.K. Scott, *The Niagara Reservation Archaeological and Historical Resource Survey* (Albany: New York Office of Parks, Recreation, and Historic Preservation, Historic Sites Bureau, March 1983); Patricia Eckel, *Botanical Heritage of Islands at the Brink of Niagara Falls* (St. Louis, MO: Botanical Services, St. Louis, 2013).

3 W.B. Turner, "The Early Settlement of Niagara," in Gayler, *Niagara's Changing Landscapes,* 192.
4 Spieler and Hewitt, *Niagara Power,* 4.
5 For a collection of travel writing and accounts of Niagara Falls over many centuries, see Dow, *Anthology and Bibliography of Niagara Falls.*
6 Daniel Macfarlane and Peter Kitay, "Hydraulic Imperialism: Hydro-electric Development and Treaty 9 in the Abitibi Region," *American Review of Canadian Studies* 47, 3 (Fall 2016): 380–97. In the Niagara context, see also Sehdev, "Unsettling the Settler at Niagara Falls."
7 On the legend of the Maid of the Mist, see L.R. Grol, *Tales from the Niagara Peninsula: A Legend of the Maid of the Mist* (Fonthill, ON: Fonthill Studio, 1975).
8 On the Hudson River School, see Tricia Cusack, *Riverscapes and National Identities* (Syracuse, NY: Syracuse University Press, 2009); McKinsey, *Niagara Falls.* According to Claire Perry, "in a startling display of creative hubris, Church also advocated remodeling the Falls themselves, maintaining that judicious changes would give them a more balanced appearance. 'The natural formation of the rocks seemed to invite some artistic treatment especially by cutting channels for the purpose of forming picturesque cascades.'" Claire Perry, *The Great American Hall of Wonders: Art, Science and Invention in the Nineteenth Century* (New York: Giles, 2011), 73.
9 Dubinsky, *The Second Greatest Disappointment,* 37.
10 Ibid., 60–61. The term "allowably indigenous" comes from Chelsea Vowel, *Indigenous Writes: A Guide to First Nations, Metis and Inuit Issues in Canada* (Winnipeg: Highwater Press, 2016), 68.
11 On hydraulic power at Niagara Falls: Spieler and Hewit, *Niagara Power;* Kepner, "Niagara's Water Power"; Belfield, "Technology Transfer and Turbulence"; Adams, *Niagara Power;* Terry Reynolds, *Stronger Than a Hundred Men: A History of the Vertical Water Wheel* (Baltimore: Johns Hopkins University Press, 2002). For a list of factories and companies using power from the Falls at the turn of the century, see Niagara Falls Power Company, *Niagara Falls Power: Its Application and Uses on the Niagara Frontier* (Buffalo: Courier, 1901).
12 On the bridges of Niagara, see Ralph Greenhill, *Spanning Niagara: The International Bridges, 1848–1962* (Seattle: University of Washington Press, 1985); Irwin, "Bridge to a New Niagara," chap. 2 in *The New Niagara.*
13 Gail E.H. Evans, "Storm over Niagara: A Catalyst in Reshaping Government in the United States and Canada During the Progressive Era," *Natural Resources Journal* 32 (Winter 1992): 36.
14 Hulbert, *The Niagara River,* 76.
15 This quote comes from artist B. Kroupa and is reproduced in Dow, *Anthology and Bibliography of Niagara Falls,* 2:1132.
16 Irwin, *The New Niagara,* 71.
17 Irwin, *The New Niagara;* Welch, *How Niagara Was Made Free.*
18 Patricia Jasen, *Wild Things: Nature, Culture, and Tourism in Ontario, 1790–1914* (Toronto: University of Toronto Press, 1995); Patricia Jasen, "Romanticism, Modernity, and the Evolution of Tourism on the Niagara Frontier, 1790–1850," *Canadian Historical Review* 72 (1991): 283–318.
19 Dubinsky stresses that the famed proprietor of Table Rock House, Saul Davis, was Jewish and employed black staff. Dubinsky, *The Second Greatest Disappointment,* 89–92. On

American tourism in this era, see also John Sears, *Sacred Places: American Tourist Attractions in the Nineteenth Century* (New York: Oxford University Press, 1989).

20 Berton, *Niagara,* 185.

21 Mavor, *Niagara in Politics,* 16.

22 For more detail on this legislative process, see Alfred Runte, "Beyond the Spectacular: The Niagara Falls Preservation Campaign," *New York Historical Society Quarterly* 57 (January 1973): 30–50.

23 Welch, *How Niagara Was Made Free;* Dow, *The State Reservation at Niagara.*

24 Gawronksi et al., *The Power Trail,* 103.

25 Irwin, *The New Niagara,* 84; Sandra Olsen and Francis Kowsky, *The Distinctive Charms of Niagara Scenery: Frederick Law Olmsted and the Niagara Reservation* (Niagara Falls, NY: Buscaglia-Castellani Art Gallery, 1985). For an insightful critique of the planning for the new reservation, see Eckel, *Ecological Restoration and Power at Niagara Falls.*

26 Gerald Killan, "Mowat and a Park Policy for Niagara Falls, 1873–1887," *Ontario History* 70, 2 (June 1978): 115–36. On the history of Ontario's Niagara parkland, see George Seibel, *Ontario's Niagara Parks* (Toronto: Niagara Parks Commission, 1991).

27 Killan, "Mowat and a Park Policy for Niagara Falls."

28 Ibid., 125.

29 Seibel, *Ontario's Niagara Parks.*

30 On the historical ecology of this area, see Riley, *The Once and Future Great Lakes Country,* 326.

31 Way, *Ontario's Niagara Parks,* 50.

32 The parks commission was also permitted to charge small entrance fees for guided tours and other offerings, such as the elevator down to the base of the Falls. Karen Dubinsky, in *The Second Greatest Disappointment,* 106, writes that this mixing of public- and private-sector tourism resulted in a "curious and intriguing relationship" that put them alternatively in alliance and in competition.

33 On future-oriented and technologically utopian schemes at Niagara Falls, see William Irwin, "Electricity's Throne: Niagara Falls and the Utopian Impulse," chap. 5 of *The New Niagara;* Patrick McGreevy, "The Future at Niagara," chap. 5 of *Imagining Niagara.* On anticipatory, or imaginative, geographies in the Great Lakes–St. Lawrence basin see Eric D. Olmanson, *The Future City on the Inland Sea: A History of Imaginative Geographies of Lake Superior* (Athens, OH: Ohio University Press, 2007); Jennifer Bonnell, *Reclaiming the Don: An Environmental History of Toronto's Don River Valley* (Toronto: University of Toronto Press, 2014).

34 Way, *Ontario's Niagara Parks,* 59.

35 Marvin McInnis states that "Canada was arguably the most successful exploiter of the new technology of the Second Industrial Revolution" since that country was lagging behind other industrialized nations at the time. Furthermore, Canada's rapid rate of industrialization in this period, roughly 1897 to the First World War, was spurred by the rapid expansion of a homegrown engineering class. Marvin McInnis, "Engineering Expertise and the Canadian Exploitation of the Technology of the Second Industrial Revolution," in *Technology and Human Capital in Historical Perspective,* ed. Jonaas Ljungberg (New York: Palgrave Macmillan, 2004), 49; Thomas Parke Hughes, *Networks of Power: Electrification in Western Society, 1880–1930* (Baltimore: Johns Hopkins University Press, 1993); David Nye, *Electrifying America: Social Meanings of a New Technology, 1880–1940* (Cambridge, MA: MIT Press, 1992); David E. Nye, *American Technological Sublime* (Cambridge, MA:

MIT Press, 1994); Harold L. Platt, *The Electric City: Energy and the Growth of the Chicago Area, 1880–1930* (Chicago: University of Chicago Press, 1991); Paul W. Hirt, *The Wired Northwest: The History of Electric Power, 1870s–1970s* (Lawrence: University Press of Kansas, 2012); Julie A. Cohn, *The Grid: Biography of an American Technology* (Cambridge, MA: MIT Press, 2017).

36 Thomas Evershed, *Water-Power at Niagara Falls to Be Successfully Utilized. The Niagara River Hydraulic Tunnel, Power and Sewer Co.* (Buffalo: Matthew, Northup, Art-Printing Works, Office of the *Buffalo Morning Express,* 1886).

37 Belfield notes that the Niagara system relied on Swiss design advances in hydroelectric technology: Belfield, "The Niagara System."

38 On the war of currents, see literature such as Adams, *Niagara Power;* Hughes, *Networks of Power;* Bernard Carlson, *Tesla: Inventor of the Electrical Age* (Princeton, NJ: Princeton University Press, 2013); Jill Jonnes, *Empires of Light: Edison, Tesla, Westinghouse and the Race to Electrify the World* (New York: Random House, 2003).

39 Contrary to expectations, industries that wanted cheap hydro power did establish themselves directly in the vicinity of Niagara. The Pittsburgh Reduction Company (which was established in 1886 and in 1907 became the Aluminum Company of America, or ALCOA) had set up operations at Niagara, and eventually moved all its operations there. By the eve of the First World War, ALCOA was the largest single producer of aluminum, and the single largest user of electricity, in the world.

40 Lubar, "Transmitting the Power of Niagara."

41 On the history of the Niagara Falls Power Company and the Adams station see Adams, *Niagara Power.* On the history of hydroelectricity in the 1880s and 1890s on the US side of Niagara, see Louis Cassier Co., *The Harnessing of Niagara* (New York and London: Cassiers Magazine, 1895); Gawronksi et al., *The Power Trail.*

42 It took three years to build and twenty-eight people reportedly died in its construction. Charles H. Werner, "The Niagara Falls Tunnel," *Cassiers Magazine* 2, 8 (June 1892): 75. For an overview of the Adams station, see Paul Gromosiak, "A Brief History of the Edward Dean Adams Power Plant," n.d., http://www.teslaniagara.org/wp-content/uploads/2014/12/261389503-History-of-the-Adams-Power-Plant-PG.pdf.

43 In conjunction, the Niagara Development Company, part of the Niagara Falls Power Company, built a planned worker's community, called "Echota" (Cherokee for "Town of Refuge"), on reclaimed swampland to the east of the Adams station.

44 There were almost thirty electrochemical companies in operation at Niagara Falls by 1910. Trescott, *The Rise of the American Electrochemicals Industry,* 39. Niagara Falls also attracted concerns such as Henry D. Perky's Shredded Wheat factory. Irwin, "'The Wonder of the Age': Shredded Wheat and the Natural Food Company's Model Factory," chap. 7 in *The New Niagara.*

45 This quote is from Garnault Agassiz's September 1912 article in the *National* magazine titled "Mighty Thunderer." Quoted in Dow, *Anthology and Bibliography of Niagara Falls,* 2:1046.

46 On this company's history, see Norman Ball, *The Canadian Niagara Power Company Story;* Dumych, *The Canadian Niagara Power Company.*

47 The first electrical power was generated on the Ontario side of the Niagara peninsula in 1886, when the St. Catharines Electric Light and Power Company built a small DC station at a lock on the second Welland Canal. Other small electricity generation took place at another lock on the canal and in Merriton. Jackson et al., *The Mighty Niagara,* 206–7.

48 The company had secured an exclusive franchise in 1892 from the Province of Ontario, and had intended to create a power monopoly on both sides of Niagara Falls. However, it held off on developing a generating station because the technology was evolving so quickly. Eventually, it gave up its exclusive franchise in 1899, and other companies built generating stations, imitating the company's path-breaking technology. Norman Ball avers that "there was some merit in Canadian Niagara's claim that it had to wait until the technology had been proven in the United States before it could proceed," and "in the long run, the loss of the monopoly rights probably saved the company from being absorbed by Ontario Hydro." Ball, *The Canadian Niagara Power Company Story,* 37–38. See also Carl A.S. Hall, "Electrical Utilities in Ontario Under Private Ownership, 1890–1914" (PhD diss., University of Toronto, 1968).

49 Hirt, *The Wired Northwest,* 54.

50 The Village and Town of Niagara Falls, Ontario, became one city on January 1, 1904, with a population of under 7,000 people. By 1910, this city was home to about 9,000 people, while Niagara Falls, NY, had a population of more than 30,000. Factories also began to spring up on the Ontario side to take advantage of the cheap electricity: American Cyanamid, Canadian Carborundum, Norton Abrasives, Shredded Wheat, Dominion Chain, International Silver, Oneida Community, Ontario Paper Company, McGlashan-Clarke, and Ohio Brass. Kiwanis Club of Stamford, Ontario, *Niagara Falls, Canada,* 46.

51 This company had a transmission line running to Fort Erie and then across to Buffalo.

52 Water is taken from the river about a mile above the crest of the Horseshoe Falls and just above the rim of the first cascade of the upper rapids. From thence three pipelines, each about 6,300 feet long, lead to the power station situated at the base of the cliff below the Horseshoe Falls. The plant operated at an average head of 180 feet and featured a total installed capacity of 208,200 horsepower.

53 The water is diverted from the river at Tempest Point (midway between the headworks of the Ontario Power Company and the Canadian Niagara Power Company) directly to the power station, and from there a tailrace tunnel about 2,000 feet long carries the water to the base of the Horseshoe Falls. The plant operates under an average head of 130 feet and has a total installed capacity of 164,500 horsepower. On the history of electricity in Toronto, see Robert M. Stamp, *Bright Lights, Big City: The History of Electricity in Toronto* (Toronto: Ontario Association of Archivists, 1991).

54 Nelles, *The Politics of Development,* 217.

55 Belfield, "Technology Transfer and Turbulence," 69.

56 Firms like Shawinigan Company and Montreal Light, Heat, and Power Company soon adopted the Niagara system. Belfield, "The Niagara System"; Louis C. Hunter, *A History of Industrial Power in the United States, 1780–1930,* vol. 1, *Waterpower in the Century of the Steam Engine* (Charlottesville, VA: University Press of Virginia, 1979).

57 On historic energy transitions, see Vaclav Smil, *Energy and Civilization: A History* (Cambridge, MA: MIT Press, 2017).

58 Though Niagara transmission was not the first (electricity transmission had already taken place in Germany, California, Colorado, and Oregon), it captured the North American public imagination, in no small part because of its potential and what it represented. Donald C. Jackson, *Building the Ultimate Dam: John S. Eastwood and the Control of Water in the West* (Norman, OK: University of Oklahoma Press, 2005): 47–54; Hirt, *The Wired Northwest,* 27.

59 Lewis Mumford wrote in 1934 that "electrification marked the beginning of a new age," which he referred to as "the 'neotechnic phase' of civilizations." French, *When They Hid the Fire,* 115–16, 145.

Chapter 2: Saving Niagara

1 For more on this concept, see Daniel Macfarlane and Andrew Watson, "Hydro Democracy: Water Power and Political Power in Ontario," *Scientia Canadensis* 40, 1 (2018): 1–18; Timothy Mitchell, *Carbon Democracy: Political Power in the Age of Oil* (New York: Verso, 2011). For a rigorous history of Canada's energy usage, see Richard W. Unger and John Thistle, *Energy Consumption in Canada in the 19th and 20th Centuries: A Statistical Outline* (Rome: Consiglio Nazionale delle Ricerche [CNR], Instituto di Studi sulle Società del Mediterraneo [ISSM], 2013).

2 Archives of Ontario (hereafter AO), RG 38, no. 105, Ontario Power Commission, Hydro Electric Power Commission (January 1, 1904, to December 31, 1918): Geo Y. Wisner, "Lake Erie Regulation," The Board of Engineers of Deep Waterways (n.d.). See also Arthur M. Woodford, *Charting the Inland Seas: A History of the US Lake Survey* (Detroit: US Army Corps of Engineers, Detroit District, 1991); Peter Neary, "Grey, Bryce, and the Settlement of Canadian-American Differences, 1905–1911," *Canadian History Review* 49, 4 (1968): 357–80.

3 This proposal owed much to the esteemed hydraulic engineer George Y. Wisner, who was part of the International Waterways Commission, the IJC's predecessor.

4 The Chicago Diversion was the first large-scale diversion out of the Great Lakes basin and refers to the volume of water allowed to be withdrawn by the Sanitary and Ship Canal. This volume has legally changed several times. The canal reversed the flow of the Chicago River *away* from Lake Michigan, and thus out of the Great Lakes watershed, eventually to the Mississippi, in order to provide sewage disposal for the Windy City. Canada and the American Great Lakes states strenuously objected to the Chicago Diversion, arguing that it would lower levels throughout the Great Lakes, hurting not only shipping interests but also hydroelectric development at places like Niagara Falls. It became a thorn in the side of US-Canada environmental diplomacy, shaping and disrupting interstate politics and foreign relations for more than a century.

5 "New Plans for Niagara Development," *Municipal Journal and Public Works* 41 (July–December 1916): 164.

6 Irwin, *The New Niagara,* 219–20.

7 The term "intermittent waterfall" was used by Thomas H. Norton in a 1916 paper read to the American Electro Chemical Society. See Dow, *Anthology and Bibliography of Niagara Falls,* 2:1051.

8 Evans, "Storm over Niagara: A Study of the Interplay of Cultural Values, Resource Politics, and Environmental Policy, in an International Setting, 1670s–1950" (PhD diss., University of California at Santa Barbara, 1991).

9 For just one of many examples, see Charles M. Dow, "How to Protect Niagara Falls," *The Outlook,* January 27, 1906, 179–90.

10 AO, RG 38, no. 107, Hydro Electric Power Commission (January 1, 1921, to December 31, 1923): General Agreement: Progress Report, International Waterways Commission, December 1, 1905; "The British Ambassador to the Secretary of State," November 13, 1905,

Document 415, Government of the United States, *Foreign Relations of the United States (FRUS) 1905* (Washington, DC: US Government Printing Office, 1905), https://history.state.gov/historicaldocuments/frus1905/d415.

11 Ernest Morrison, *J. Horace McFarland – A Thorn for Beauty* (Harrisburg: Pennsylvania Historical and Museum Commission, 1995). On McFarland, see also Julian Chambliss, "Perfecting Space: J. Horace McFarland and the American Civic Association," *Pennsylvania History: A Journal of Mid-Atlantic Studies* 77, 4 (Autumn 2010): 486–97.

12 Evans, "Storm over Niagara: A Study of the Interplay of Cultural Values, Resource Politics, and Environmental Policy, in an International Setting, 1670s–1950," 42; see also Morrison, *J. Horace McFarland.*

13 On Roosevelt and the environment, see Ian Tyrell, *Crisis of the Wasteful Nation: Empire and Conservation in Theodore Roosevelt's America* (Chicago: University of Chicago Press, 2015); Douglas Brinkley, *The Wilderness Warrior: Theodore Roosevelt and the Crusade for America* (New York: Harper, 2010).

14 Morrison, *J. Horace McFarland,* 111.

15 Evans, "Storm over Niagara: A Study of the Interplay of Cultural Values, Resource Politics, and Environmental Policy, in an International Setting, 1670s–1950," 44.

16 N.F. Dreisziger, "The Campaign to Save Niagara Falls and the Settlement of United States–Canadian Differences, 1906–1911," *New York History* 55, 4 (October 1974): 446.

17 AO, RG 38, no. 108, International Waterways Commission (January 1, 1903, to December 31, 1915): Address of Charles M. Dow (1905/6).

18 Samuel P. Hays suggests that Burton was responsible for the failure of the multi-purpose water development approach "more than any other single man." Samuel P. Hays, *Conservation and the Gospel of Efficiency: The Progressive Conservation Movement, 1890–1920* (Pittsburgh: University of Pittsburgh Press, 1959), 93. A more favourable portrait of Burton is offered in Donald Pisani, "Water Planning in the Progressive Era: The Inland Waterways Commission Reconsidered," *Journal of Policy History* 18, 4 (2006): 389–418.

19 Alvin Gluek, "The Lake Carriers' Association and the Origins of the International Waterways Commission," *The Inland Seas: Quarterly Journal of the Great Lakes Historical Society* 36 (1980): 236–45.

20 The final report of the IWC was released during debate on the Burton bill, reportedly because McFarland was still putting a bug in Roosevelt's ear, which may have influenced matters. Morrison, *J. Horace McFarland,* 112–14; Evans, "Storm over Niagara: A Study of the Interplay of Cultural Values, Resource Politics, and Environmental Policy, in an International Setting, 1670s–1950," 514–17.

21 Joint Resolution of March 3, 1909, extending the Burton Act to June 29, 1911; Joint Resolution of August 22, 1911, further extending the above-mentioned act to March 1, 1912; Joint Resolution of April 5, 1912, further extending the act to March 4, 1913. Subsequent attempts, such as the Cline Bill, unsuccessfully tried to further extend the Burton Act. N.F. Dreisziger contends that the Burton Act drew American pressure for Niagara measures away from the IWC, allowing it to eventually reach a fair and wide-ranging settlement. Dreisziger, "The Campaign to Save Niagara Falls," 446 and 457, fn 43.

22 Ernest Morrison also notes that the ACA's grassroots mail-in campaign on behalf of Niagara preservation resulted in 6,585 letters from the concerned public to the secretary of war. Morrison, *J. Horace McFarland,* 118–19.

23 Much of this section on the IWC is derived from N.F. Dreisziger, "The International Joint Commission of the United States and Canada, 1895–1920: A Study in Canadian-American

Relations" (PhD diss., University of Toronto, 1974). Dreisziger offers a multi-causal story of the IWC's origins, with an emphasis on Niagara Falls developments. Alvin Gluek contends that development at Sault St. Marie was the more important factor, particularly in the call in the 1902 Rivers and Harbors Act for what would become the IWC. However, Dreisziger's argument, including the emphasis on Niagara Falls, is more convincing. Alvin Gluek, "The Lake Carriers' Association"; Alvin Gluek, "The Lake Carriers' Association and the Origins of the International Waterways Commission," *The Inland Seas: Quarterly Journal of the Great Lakes Historical Society* 37 (1981): 27–31.

24 International Joint Commission (hereafter IJC), Docket 64, Niagara Falls Reference, box 114, 64-8-4:1, US Congressional Bills and Proceedings: Reports Upon the Existing Water-Power Situation at Niagara Falls, so far as concerns the Diversion of Water on the American Side, by the American Members of the International Waterways Commission and Captain Charles W. Kutz, Corps of Engineers, US Army, 1906; IJC, Docket 64, Niagara Falls Reference, box 105, 64-1-1-1: References in the Compiled Reports of the International Waterways Commission to the 36,000–20,000 cfs Diversion at Niagara Falls, and to the 10,000 cfs Allowance for the Chicago Drainage Canal.

25 Canadian Niagara Power Company (9,500 cfs); Ontario Power Company (12,000 cfs); Electrical Development Company (11,200 cfs); Niagara Falls Park Railway Company (1,500 cfs); Welland Canal (1,800 cfs). They also agreed that the Chicago Diversion should be limited to 10,000 cfs.

26 Technically it was titled "Treaty between the United States and Great Britain Relating to Boundary Waters, and Questions Arising between the United States and Canada."

27 On the history of the IJC, see Dreiziger, "The International Joint Commission"; William R. Willoughby, *The Joint Organizations of Canada and the United States* (Toronto: University of Toronto Press, 1979); Robert Spencer, John Kirton, and Kim Richard Nossal, eds., *The International Joint Commission Seventy Years On* (Toronto: University of Toronto Centre for International Studies, 1981). On the creation of the IJC, see in particular chapters by David Whorley and Meredith Denning in Daniel Macfarlane and Murray Clamen, eds., *The First Century of the International Joint Commission* (Calgary: University of Calgary Press, 2020).

28 An earlier version of the Boundary Waters Treaty, the so-called Clinton-Gibbons draft, had included Lake Erie when it limited diversions to 36,000:18,500 cfs, whereas the final version of the BWT stipulated only that Lake Erie's level not be appreciably affected. David Whorley, "From IWC to BWT: Canada-US Institution Building, 1902–1909," in Macfarlane and Clamen, *The First Century of the International Joint Commission,* 56.

29 *Treaty between the United States and Canada Concerning Boundary Waters: Hearings and Proceedings Before the Senate Committee on Foreign Relations,* 60th Congress (January–February 1909), 269–74.

30 Ibid. According to Don Courtney Piper, Root "implied that the reason why all references to the diversion were omitted was that Canada obtained a diversion of 36,000 cfs from the Niagara River and the United States only 20,000 cfs." Don Courtney Piper, *The International Law of the Great Lakes: A Study of Canadian–United States Cooperation* (Durham, NC: Duke University Press, 1967), 94.

31 For a detailed discussion of the elaborate political maneuvering that led to the creation of Ontario Hydro, perhaps the best source is the chapter "Power Politics" in H.V. Nelles's *The Politics of Development,* though other studies on Ontario Hydro provide differing perspectives on this commission's origins. It is worth pointing out that the Whitney Conservative

government, on taking office in 1905, cancelled a contract negotiated "during the last hours" of the previous Liberal government that would have granted the Electrical Development Company the rights "to all the remaining waterpowers at Niagara Falls." Nelles, *The Politics of Development,* 258.

32 H.V. Nelles has written that provincial control of natural resources (hydro power, logging, mining) was more likely to favour the interests of the industrial community than those of the general populace – in Nelles's oft-quoted formulation, the Ontario government all too often became a "client" of the business community. Nelles, *The Politics of Development,* 495. On logging, see also Mark Kuhlberg, *In the Power of the Government: The Rise and Fall of Newsprint in Ontario,* 1894–1932 (Toronto: University of Toronto Press, 2015); Michael Stamm, *Dead Tree Media: Manufacturing the Newspaper in Twentieth-Century North America* (Baltimore: Johns Hopkins Press, 2018). On the role of transnational progressivism and antimonopolism in the creation of Ontario Hydro see Mark Sholdice, "The Ontario Experiment: Hydroelectricity, Public Ownership, and Transnational Progressivism, 1906–1939" (PhD. diss., University of Guelph, 2019). Sholdice points out that many of the drivers behind the creation of Ontario Hydro, such as Sir Adam Beck, never adopted electricity in their own business practices. Differing views of Adam Beck and Ontario Hydro are forwarded in Merrill Denison, *The People's Power;* James Mavor, *Niagara in Politics: A Critical Account of the Hydro-Electric Commission* (New York: E.P. Dutton, 1925); W.R. Plewman, *Adam Beck and the Ontario Hydro* (Toronto: The Ryerson Press, 1947).

33 Denison, *The People's Power,* 41. See also Nelles, *The Politics of Development;* Neil B. Freeman, *The Politics of Power: Ontario Hydro and Its Government, 1906–1995* (Toronto: University of Toronto Press, 1996).

34 Evans, "Storm over Niagara: A Study of the Interplay of Cultural Values, Resource Politics, and Environmental Policy, in an International Setting, 1670s–1950," 31, 53.

35 New York enacted a new Water Power Act in 1921 that favoured private developments. For a succinct overview of the various legislation and laws, see Roscoe C. Martin, "Water Programs," chap. 2 in *Water for New York: A Study in State Administration of Water Resources* (Syracuse: Syracuse University Press, 1960).

36 A decision by the Privy Council in 1898 had given ownership of the beds of all inland waterways to Ontario, though this was complicated by the federal government's jurisdiction over navigation. Christopher Armstrong, *The Politics of Federalism: Ontario's Relations with the Federal Government, 1867–1942* (Toronto: University of Toronto Press), 73. See also Janet Martin-Nielsen, "South over the Wires: Hydroelectricity Exports from Canada, 1900–1925," *Water History* 1 (2009): 109–29.

37 The 1907 act would undergo minor modifications in 1925, 1937, and 1955, with the duty abolished in 1963. See Martin-Neilsen and Perlgut for more on the 1907 Act and subsequent modifications. The 1955 amendment to the Fluid and Electricity Export Act brought in natural gas, while the 1963 changes allowed for the export of large blocks of electricity. A.E.D. Grauer, "The Export of Electricity from Canada," in *Canadian Issues: Essays in Honour of Henry F. Angus,* ed. R.M. Clark (Toronto: University of Toronto Press, 1961), 248–85; Mark Perlgut, *Electricity across the Border: The U.S.-Canadian Experience* (New York: C.D. Howe Research Institute, 1978).

38 Armstrong, *The Politics of Federalism,* 72.

39 For one example of the usage of this term, see John Lyell Harper, *The Suicide of the Horseshoe Fall* (Niagara Falls: G.H. Courter, 1916).

40 Niagara Parks Commission (NPC), *Thirty-Third Annual Report of the Commissioners of the Queen Victoria Niagara Falls Park* (Toronto: L.K. Cameron, 1918).
41 Government of the State of New York, State Reservation at Niagara (hereafter SRN), *Fifth Annual Report, State Reservation at Niagara, 1888* (Albany, NY: The Argus Company, 1888); G.K. Gilbert, *Rate of Recession of Niagara Falls,* Department of the Interior, United States Geological Survey, Bulletin No. 306 (Washington, DC: US Government Printing Office, 1907).
42 Way, *Ontario's Niagara Parks,* 47; NPC, *Sixth Annual Report of the Commissioners of the Queen Victoria Niagara Falls Park* (Toronto: L.K. Cameron, 1891). Note that prior to 1927, the Niagara Parks Commission was titled the Queen Victoria Niagara Falls Park Commission, but for simplicity and continuity I will use "Niagara Parks Commission" and "NPC."
43 Patricia Eckel notes that the long-time owners of Goat Island, the Porter family, are often hailed as its protector, seemingly ignoring the family's documented attempts to develop or sell the island, and other adjoining islands, to industrialists. Eckel, *Ecological Restoration and Power at Niagara Falls,* 29.
44 SRN, *Seventh Annual Report, State Reservation at Niagara, 1890* (Albany, NY: J.B. Lyon, 1891).
45 The visitor was Captain Enys. Dow, *Anthology and Bibliography of Niagara Falls,* 1:80–81.
46 Seibel, *Ontario's Niagara Parks,* 20.
47 AO, RG 38, no. 108, International Waterways Commission (January 1, 1903, to December 31, 1915): Address of Charles M. Dow, President of the Commissioners of the State Reservation at Niagara to The International Commission Appointed to Investigate Concerning the Conditions and Uses of the Waters Adjacent to the Boundary Line Between the United States and Canada (1905/6).
48 Likewise, the percentage of the river's flow passing over the American Falls varied widely: some said 5 percent, others said 15 percent, and so on. This was part of the quantification of water by the emerging scientific field of hydrology, which reduced water to its measurable volumes and scientific properties in the service of making it legible and intelligible.
49 IJC, Docket 64, box 106, 64-2-7-1: Dedication Ceremonies, Niagara (September 28, 1957): "Preservation of Niagara Falls," Message from the President of the United States transmitting a letter from the Secretary of Water, submitting additional information concerning the operation of the United States Lake Survey from June 29, 1906, to June 29, 1911, December 7, 1911.
50 Buffalo History Museum Archives (hereafter BHMA), Oversize TC 401, U54 1911: Letter of the Chief of Engineers, January 21, 1909, submitting to the Secretary of War the report of Maj. Charles Keller of November 30, 1908. For the detailed report and the extensive engineering studies on which it was based, see the following US Army Corps of Engineers series in the US National Archives: United States Government, National Archives and Records Administration II (hereafter NARA II), RG 77, Office of the Chief of Engineers, Lake Survey Books – Niagara Falls, 4 boxes, 10/3/1970–6/30/1976.
51 Kiwanis Club of Stamford, Ontario, *Niagara Falls, Canada,* 46, 368.
52 NPC, *Twenty-Ninth Annual Report of the Commissioners of the Queen Victoria Niagara Falls Park* (Toronto: L.K. Cameron, 1914).
53 NPC, *Eighteenth Annual Report of the Commissioners of the Queen Victoria Niagara Falls Park* (Toronto: L.K. Cameron, 1903).

54 For more detail on ice control at Niagara Falls, see Daniel Macfarlane, "'As Nearly as May Be': Estimating Ice and Water on Niagara and St. Lawrence Rivers," *Journal of Historical Geography* 65 (July 2019): 73–84.
55 Adams, *Niagara Power,* 2: 147.
56 This committee was terminated in 1913. SRN, *Twenty-Ninth Annual Report, State Reservation at Niagara, 1912* (Albany, NY: J.B. Lyon Company, 1913).
57 IJC, Progress Report of the IJC on the Reference by the United States and Canada in "The Pollution of Boundary Waters," January 16, 1914, 45. Quoted in Jamie Benidickson, "The IJC and Water Quality in the Bacterial Age," in Macfarlane and Clamen, *The First Century of the International Joint Commission.*
58 The Town of Niagara Falls, with permission from the Queen Victoria Niagara Falls Park Commission, had built new waterworks in the 1890s. This was necessary because diversions had lowered Niagara River water levels at the shoreline where the water intakes were situated. The new waterworks replaced the previous "unsightly structures which had long defaced the shore-line above the cataract" but were demolished in 1937. Way, *Ontario's Niagara Parks,* 47.
59 To further improve the level at the water intakes, in 1908 the Niagara Falls Water Commission deposited limestone blocks into the gap between Randolph's tower and the shoreline. NPC, *Twentieth Annual Report of the Commissioners of the Queen Victoria Niagara Falls Park* (Toronto: L.K. Cameron, 1905).
60 NPC, *Twenty-First Annual Report of the Commissioners of the Queen Victoria Niagara Falls Park* (Toronto: L.K. Cameron, 1906).
61 NPC, *Twenty-Seventh Annual Report of the Commissioners of the Queen Victoria Niagara Falls Park* (Toronto: L.K. Cameron, 1912).
62 In addition to the large generating stations and the International Railway Company (for its electric railway between Queenston and Chippawa). The City of Niagara Falls also generated small amounts of power for waterworks pumps.
63 Federal intervention was required to let the Niagara utilities break their power supply contracts with non-essential consumers. Col. Charles Keller, *The Power Situation during the War* (Washington, DC: US Government Printing Office, by authority of the Secretary of War, 1921).
64 SRN, *Thirty-Fifth Annual Report, State Reservation at Niagara, 1918* (Albany, NY: J.B. Lyon, 1919).
65 Keller, *The Power Situation during the War.*
66 Froschauer, *White Gold.*
67 See Armstrong, *The Politics of Federalism,* 76–82.
68 On coal in North America, see Watson, "Coal in Canada"; Jones, *Routes of Power;* Peter Schulman, *Coal and Empire: The Birth of Energy Security in Industrial America* (Baltimore: Johns Hopkins University Press, 2015).
69 See Nye, *Electrifying America.*
70 SRN, *Thirty-Fourth Annual Report, State Reservation at Niagara, 1917* (Albany, NY: J.B. Lyon, 1918). Between the 1906 Burton Act and the 1920 Federal Water Power Act, Congress passed seven separate resolutions concerning Niagara diversions and the War Department issued nine revocable permits. David L. Nass, *Public Policy and Public Works: Niagara Falls Redevelopment as a Case Study* (Chicago: Public Works Historical Society, 1979), 18.
71 Library and Archives Canada (hereafter LAC), RG 11, vol. 4214, file 534-4, Hunter (Public Works) to Under-Sec of State Department of External Affairs (DEA), May 10, 1921.

72 These were Public Resolution No. 45 of January 19, 1917, in force until July 1, 1917; Public Resolution No. 33 of June 29, 1918, in force until July 1, 1919; and Public Resolution No. 3 of July 12, 1919, in force until June 10, 1920. In 1920, the Federal Power Act and a reconstituted Federal Power Commission came into effect. The FPC became the Federal Energy Regulatory Commission in 1977.

73 Hays, *Conservation and the Gospel of Efficiency,* 239–40. On the history of federal water resources see Beatrice Holmes, *A History of Federal Water Resources Programs, 1800–1960* (Washington, DC: US Department of Agriculture, 1972); Beatrice Holmes, *History of Federal Water Resources Programs and Policies, 1961–70* (Washington, DC: US Department of Agriculture, 1969); Donald C. Swain, *Federal Conservation Policy, 1921–1933;* David P. Billington, Donald C. Jackson, and Martin V. Melosi, *The History of Large Federal Dams: Planning, Design, and Construction in the Era of Big Dams* (Denver: US Department of the Interior, Bureau of Reclamation, 2005).

74 Donald Swain mentions that the Niagara Falls Power Company brought pressure to bear on the FPC. In 1930, the Federal Water Power Act made the FPC an independent agency, and the Federal Power Act of 1935 gave the FPC the responsibility for regulation of interstate whole power rates. Donald C. Swain, *Federal Conservation Policy, 1921–1933* (Berkeley: University of California Press, 1963), 113–15; Donald Pisani, *Water and American Government: The Reclamation Bureau, National Water Policy and the West, 1902–1935* (Oakland: University of California Press, 2002); John L. Neufeld, *Selling Power: Economics, Policy, and Electric Utilities before 1940* (Chicago: University of Chicago Press, 2016).

75 Two other companies, the International Paper Company and the Pettebone-Cataract Company, had both rented small volumes of water for generation (up to 730 cfs and 262.6 cfs, respectively), but the water rights of both were leased for use at the Schoellkopf generating station in 1918 and 1925, with the 1925 lease bought out in 1947. However, these water rentals became a bone of contention with the Federal Power Commission and the courts. See Charles P. Schwartz Jr., "Niagara Mohawk v. FPC: Have Private Water Rights Been Destroyed by the Federal Power Act?" *University of Pennsylvania Law Review* 102 (1953): 31–79.

76 Hirt, *The Wired Northwest,* 168.

77 Ibid., 187.

78 Martin-Nielsen, "South over the Wires," 126–27. See also Grauer, "The Export of Electricity from Canada," 276; Cohn, *The Grid.*

79 Jean Manore notes that the transmission line to Toronto from Niagara was itself an engineering marvel. Jean Manore, *Cross Currents: Hydro-Electricity and the Engineering of Northern Ontario* (Waterloo, ON: Wilfrid Laurier University Press, 1999), 71. Belfield, "Technology Transfer and Turbulence," 214–16.

80 By reducing hydroelectric exports from Ontario, the US Burton Act had seemingly backfired by giving Ontario Hydro a "comparative advantage in subsequent Niagara-based developments." Belfield, "Technology Transfer and Turbulence," 83, 91.

81 Keith Fleming, *Power at Cost: Ontario Hydro and Rural Electrification, 1911–1958* (Montreal and Kingston: McGill-Queen's University Press, 1991), 7–9; Norman Ball, *Building Canada: A History of Public Works* (Toronto: University of Toronto Press, 1988), 179.

82 According to Christopher Armstrong, "Beck procured from the Ontario cabinet an order-in-council in June 1914 formally dividing up the 36,000 cubic feet per second of water which Canada could divert at the Falls for power purposes." Armstrong, *The Politics of Federalism,* 76–82.

83 "Canadian Niagara was granted 8,225 second-feet, the Ontario Power Company 11,180 second-feet, and the Electrical Development Company, 9,985 second-feet. The remaining 6,500 second-feet were thus available for future use." Ibid., 77.
84 In 1913, Ontario Hydro had requested that the Ontario government allow this, but it was not until 1916–17 that the necessary legislation was enacted. See Freeman, *The Politics of Power.*
85 The 1917 amendment of the Power Commission Act allowed Ontario Hydro to acquire private companies through share purchases. For more detail, see Freeman, *The Politics of Power,* 48–50; Seibel, *Ontario's Niagara Parks,* 158–66.
86 NPC, *Thirty-Fourth Report of the Commissioners of the Queen Victoria Niagara Falls Park* (Toronto: L.K. Cameron, 1919).
87 On Ontario Hydro and rural electrification, see Fleming, *Power at Cost.*
88 The Ontario Niagara Development Act of 1917, which supplemented the 1916 act of the same name, allowed Ontario Hydro to issue bonds. Freeman, *The Politics of Power,* 46–47; Denison, *The People's Power,* 127.
89 Engineers had proposed plans that could utilize the entire head from Lake Erie to Lake Ontario. Glenys Biggar, *Ontario Hydro's History and Description of Hydro-Electric Generating Stations* (Toronto: Ontario Hydro, 1991), 251.
90 The Welland River was also known as Chippawa Creek. The canal traversed farmland, roads, and railways, which necessitated property acquisition and other infrastructural changes, such as seventeen new bridges. Seibel, *Ontario's Niagara Parks,* 102–4; Way, *Ontario's Niagara Parks,* 93; Hydro-Electric Power Commission of Ontario, *Hydro-Electric Power in the Niagara District* (Toronto: HEPCO, July 1920).
91 The Hamilton and Lake Erie Power Company had first proposed reversing the flow of the Welland River, though it planned to develop power at Ball's Falls where the Jordan River went over the Niagara Escarpment. Dumych, *The Canadian Niagara Power Company.*
92 In the vicinity of the Whirlpool, for instance, Ontario Hydro paid $83,541 for 322 acres from ten different property owners. Hydro-Electric Power Commission of Ontario Archives (hereafter HEPCO), "H.E.P.C. Property – Whirlpool – Queenston Forebay" (n.d.).
93 James Hull, "Raising Standards: Public Works and Industrial Practice in Interwar Ontario," *Scientia Canadensis* 25 (June 2003): 7–30. On concrete standards, see also Amy Slaton, *Reinforced Concrete and the Modernization of American Building, 1900–1930* (Baltimore: Johns Hopkins University Press, 2003).
94 It required the excavation of over 30 million cubic yards of material, with 340,000 cubic yards of concrete poured. Biggar, *Ontario Hydro's History,* 251; Kiwanis Club of Stamford, Ontario, *Niagara Falls, Canada,* 372–73.
95 HEPCO, Letter from Gaby (Chief Engineer, HEPCO) to Jackson (Park Superintendent), Re Experimental Hydraulic Work, April 24, 1918; HEPCO, R.S. Lea (Consulting Engineer, HEPCO), "Report of Progress on Queenston-Chippawa Power Development, January 31 to April 9, 1921" (Toronto: Clarkson W. James, 1921).
96 The parks commission had granted permission for these model studies on its property, but Ontario Hydro did not do itself any favours by failing to remove the model apparatus after the first summer, which meant there was no water for the bathing pool. Cooper also arranged for informational and personnel exchanges between the Soviets and Ontario Hydro. Belfield, "The Niagara Frontier," 393, 496n12.
97 Cooper criticized many other aspects of the design for the power works and connected infrastructure. Hugh L. Cooper and Co., "Final Report on Chippawa Hydro-Electric Co.,

October 22, 1920," in HEPCO, Lea, *Report of Progress on Queenston-Chippawa Power Development, January 31 to April 9, 1921.*

98 W.R. Plewman, *Adam Beck and the Ontario Hydro* (Toronto: Ryerson Press, 1947), 275–76.

99 IJC, Docket 64, Niagara Falls Reference, box 106, 64-3-1-1: Correspondence, Vol. 1, Re Niagara River, May 26, 1953.

100 The Gregory Commission both praised and condemned the commission and its chairman. See Denison, *The People's Power,* 132–34; Freeman, *The Politics of Power,* 55–58.

101 Twenty-five-cycle power was the standard at the time for the entire Niagara system: "Thus Ontario Hydro was something less than the monolithic superpower system it appeared to be, since it could not supply its Niagara power to markets near Toronto where 60-cycle power was in use." Christopher Armstrong and H.V. Nelles, *Monopoly's Moment: The Organization and Regulation of Canadian Utilities, 1830–1930* (Philadephia: Temple University Press, 1986), 316.

102 By the end of 1922, Ontario Hydro "was producing 1,855,900,415 kilowatt-hours of energy annually at nineteen hydroelectric facilities, while purchasing 437,307,185 kilowatt-hours from private sources." Fleming, *Power at Cost,* 9. For technical specifications on Ontario Hydro generating stations, see Biggar, *Ontario Hydro's History.*

103 NARA II, RG 59, Decimal File, 1945–49, box 3306, file 711.4216 Ni/149: Note from the Secretary of State to the British Ambassador, February 3, 1923; LAC, RG 11, vol. 4214, file 534-4: Note from the Secretary of State to the British Ambassador, February 3, 1923.

104 Irwin, *The New Niagara,* 206.

105 Melanie Hall, "Niagara Falls: Preservation and the Spectacle of Anglo-American Accord," in *Towards World Heritage: International Origins of the Preservation Movement, 1870–1930,* ed. Melanie Hall (Farnham, UK: Ashgate, 2009), 45–66.

106 Granted, given that Niagara was a prime tourist destination that was frequently photographed, there is some visual evidence going back to the nineteenth century.

Chapter 3: Negotiating Niagara

1 From a letter that Helen Keller wrote in 1893, quoted in Robert Higgins, ed., *An Awful Grandeur: Niagara in Quotes* (Kitchener, ON: Upney Editions, 1998). A wide-ranging collection of pre-1921 writings about Niagara Falls, including diary entries, can be found in Dow, *Anthology and Bibliography of Niagara Falls.*

2 This did not always ensure water in front of the scenic tunnels and in 1924 they were extended by 180 feet, for a total length of 980 feet. NPC, *Thirty-Ninth Annual Report of the Commissioners of the Queen Victoria Niagara Falls Park* (Toronto: L.K. Cameron, 1924).

3 NARA II, RG 59, Decimal File, 1945–49, box 3306, file 711.4216 Ni/157: Note from the Secretary of State to the British Ambassador, February 3, 1923.

4 "Establishment of a Joint Board of Control to Supervise the Diversion of Waters from the Niagara River," The Secretary of State to the British Ambassador (Geddes), February 3, 1923, 711.4216 Ni/149, *FRUS 1923,* vol. 1, *Canada* (Washington, DC: US Government Printing Office, 1923), 498–500; LAC, RG 25, G-1, vol. 1347, file 267: Note from US Secretary of State, February 3, 1923.

5 LAC, RG 11, vol. 4214, file 534-4: Kellogg to Chilton, October 20, 1925. The US Army Corps of Engineers records in the national archives feature a microfiche with slides comparing the effect of water diversions on the Niagara River and Falls documented by

investigations: NARA II, RG 77, 519398: Photographs Comparing the Water Diversion Investigations of the Niagara River, 1927.

6 Premier Ferguson also mentioned Ontario's desire to start the Ogoki–Long Lac diversions in this correspondence: LAC, RG 25, G-1, vol. 1437, file 1228, pt. I: Premier of Ontario (Ferguson) to Minister of Interior (Stewart), November 30, 1925.

7 Government of Canada, Department of Mines, Geological Survey, *The Niagara Falls Survey of 1927* (Ottawa: F.A. Acland, 1930).

8 While there was much misinformation in Samuel Wyer's smear campaign, within the Ontario Hydro rate system larger users and utilities did cross-subsidize domestic consumers to an extent. Furthermore, Ontario Hydro pooled its electricity province-wide and did not offer lower rates to industries and concerns that purchased in bulk, as was often done on the New York side. Samuel S. Wyer, *Niagara Falls: Its Power Possibilities and Preservation* (Washington, DC: Smithsonian Institution, 1925).

9 Ibid. See also Richard Lowitt, "Ontario Hydro: A 1925 Tempest in an American Teapot," *Canadian Historical Review* 49, 3 (September 1968): 267–74. It did not help matters that water levels in Lake Erie, and thus in the Niagara River, were very low in 1925 and 1926; by 1929, they would rebound and attain higher levels not seen in half a century, but then would be diminished again between 1930 and 1934 before returning to higher levels during the war years. NPC, *Forty-Seventh Annual Report of the Commissioners of the Queen Victoria Niagara Falls Park* (Toronto: L.K. Cameron, 1932).

10 Wyer, *Niagara Falls.*

11 NPC, *Fortieth Annual Report of the Commissioners of the Queen Victoria Niagara Falls Park* (Toronto: L.K. Cameron, 1925). For more on Niagara Falls illumination history see Seibel, *Ontario's Niagara Parks,* 120–23. During the 1930s, the parkway would be further enhanced under the direction of Ontario politician T.B. McQueston, who with his penchant for roads, parkways, and recreational facilities might be thought of as Ontario's analog to Robert Moses. Joan Coutu, "Vehicles of Nationalism: Defining Canada in the 1930s," *Journal of Canadian Studies* 37 (Spring 2002): 199.

12 Nye, *American Technological Sublime,* 171.

13 LAC, RG 25, G-1, vol. 1438, file 1228, pt. II: Stewart to Ferguson, January 11, 1926; LAC, RG 25-A-2, vol. 741, file 141, microfilm reel T-1757: Memorandum re Conference at Washington on March 18th, 1926 (March 22, 1926); LAC, RG 25, G-1, vol. 1438, file 1228, pt. II: Skelton to US Sec of State, April 27, 1926.

14 LAC, RG 11, vol. 4214, file 534-4: Associated Press, "Great Diversion of Water Sought to Secure Power," December 18, 1928. For more see Daniel Macfarlane, "The Niagara Telecolorimeter," *Environment and Society Portal, Arcadia* (Autumn 2018): 28. Rachel Carson Center for Environment and Society, doi.org/10.5282/rcc/8492.

15 Niagara Power Service, *Niagara the Unconquered* (Buffalo: Buffalo, Niagara, and Eastern Power Corporation, 1929). This model seems to have been built in conjunction with the Niagara Falls Power Company, which in 1925 joined with a number of utilities to form the Buffalo, Niagara, and Eastern Power Corporation. On Harper, see Willi H. Hager, *Hydraulicians in the USA 1800–2000: A Biographical Dictionary of Leaders in Hydraulic Engineering and Fluid Mechanics* (New York: CRC Press/Balkema 2015), 2129.

16 IJC, Docket 64, Niagara Falls Reference, box 114, 64-8-6:1, Special International Niagara Board Interim Report 1927/12/14: "The Preservation of Niagara Falls," Interim Report of the Special International Niagara Board, December 14, 1927.

17 Dwight D. Eisenhower Presidential Library and Archives (hereafter Eisenhower Archives), Staff Files: Records of John S. Bragdon, St. Lawrence Seaway – 1954, box 69, file: St. Lawrence Seaway, Statement of Colonel W.H. Potter, Office Chief of Engineers, US Army, June 27, 1950.

18 Ibid.

19 LAC, RG 25, G-1, vol. 1437, file 1228, pt. I: Memorandum No. 1727 from Canadian Legation (Massey), Washington, to Sec of State DEA, November 14, 1928.

20 Between 1912 and 1924, Canada filed six objections to the Chicago Diversion. A draft treaty on the Chicago Diversion had been discussed between the diplomatic agencies, though no treaty was ultimately signed. LAC, RG 25-A-2, vol. 733, file 113, microfilm reel T-1753: Department of the Interior, Canada, Dominion Water Power and Reclamation Service, "International Problems on Great Lakes – St. Lawrence Deep Waterway, Chicago Diversion, Niagara (with suggested action)" by J.T. Johnson, September 14, 1930.

21 There were rumours that J. Horace McFarland had played a role in swaying the Senate committee toward an adverse opinion. LAC, RG 25-A-2, vol. 741, file 141, microfilm reel T-1757: Massey to Skelton, February 26, 1930.

22 LAC, RG 25, G-1, vol. 1438, file 1228, pt. II: Wrong to Skelton, February 25, 1931.

23 LAC, RG 25, G-1, vol. 1438, file 1228, pt. II: Secretary of State to Embassy in Washington, February 24, 1930. While governor, Roosevelt defended the state's claim to ownership of the Niagara riverbed and control of the water, and negotiated a 1929 agreement with the Niagara Falls Power Company under which this company would obtain a licence from the state and pay a rental fee for the use of the water. This was never implemented, however, since the 1929 Niagara Convention failed. The Niagara Falls Power Company and the State of New York were involved in almost continuous litigation about the company's legal right to water without any payment or rentals going to the state. Eventually, the Feinberg-Daniel bills gave New York the authority to collect rental fees from water-power resources, and the state Water Power and Control Commission fixed an annual rental fee of $1 million for the 20,000 cfs diverted from the Niagara. See Nass, *Public Policy and Public Works,* 19–25.

24 Roosevelt created the Power Authority of the State of New York primarily to publicly develop hydro power on the St. Lawrence River. Since PASNY was a formative influence on Roosevelt's presidential creation of the Tennessee Valley Authority and Bonneville Power Administration, these multi-purpose public projects can trace their lineage, at least in part, back to Ontario Hydro. See Sholdice, "The Ontario Experiment."

25 On Roosevelt's approach to electricity as president, see Philip J. Funigiello, *Toward a National Power Policy: The New Deal and the Electric Utility Industry, 1933–1941* (Pittsburgh: University of Pittsburgh Press, 1973); Richard Hirsh, *Technology and Transformation in the American Electric Utility Industry* (Cambridge, UK: Cambridge University Press, 1989).

26 Hans Huth, *Nature and the American: Three Centuries of Changing Attitudes* (Lincoln, NB: University of Nebraska Press, 1957), 207.

27 The volume of water diverted from the Chicago Sanitary and Ship Canal was effectively limited by a 1930 US Supreme Court decision to 3,200 cfs on an annual basis, though an extension was granted.

28 NARA II, RG 59, box 4047, file 711.42157 SA 29/1288-1/2: Memorandum (by Hickerson), June 23, 1934; President Franklin D. Roosevelt's Office Files, 1933–1945, pt. 4, Subject File, "Memorandum (no author), February 22, 1934."

29 Ontario Hydro undercut its public power stance by underwriting the generation of private power in Quebec, which was to be sent to Ontario. On the contracts with Quebec power producers, see Armstrong, *The Politics of Federalism,* 165–75.

30 In *The Politics of Development,* Nelles has a chapter titled "Hydro as Myth." See also Macfarlane and Watson, "Hydro Democracy."

31 In 1941, oil represented 17 percent, coal 53 percent, and electricity 6 percent of total energy consumed (Unger and Thistle, *Energy Consumption in Canada*). In terms of household rather than total national consumption, data from the Dominion Bureau of Statistics figures show that (unlike the United States) Canadians were consuming very little electricity in their homes before the Second World War. See R.W. Sandwell, *Canada's Rural Majority: Households, Environments, and Economies, 1870–1940* (Toronto: University of Toronto Press, 2016) and in her *Powering Up Canada* see the author's introduction as well as Josh MacFadyen's chapter "Hewers of Wood: A History of Wood Energy in Canada"; Dorotea Gucciardo, "The Powered Generation: Canadians, Electricity, and Everyday Life" (PhD diss., Western University, 2011).

32 LAC, RG 25, vol. 2363, file 1268-B-40, pt. I, St. Lawrence–Niagara River Treaty Proposals (International Correspondence, Feb 17/1938–Jan 31/1941): Survey of Great Lakes–Niagara–St. Lawrence Boundary Waters Negotiations, March 23, 1938.

33 LAC, RG 25, G-1, vol. 1438, file 1228, pt. II: Robbins (US Legation in Ottawa) to Secretary of State, No. 415, March 4, 1935; NARA II, RG 59, file 711.42157 SA 29/1375-1/2: Memorandum to the Secretary of State from FDR, December 23, 1935; Richard Kottman, "Herbert Hoover and the St. Lawrence Treaty of 1932," *New York History* 56 (July 1975): 317–18.

34 For details on these contracts, see Neil B. Freeman, "Injecting the Hydro-Electric Power Commission with Political Sensitivity: The Ferguson Formula, 1924–1943," chap. 4 in *The Politics of Power;* Merrill Denison, "Repudiation and Retrenchment," chap. 23 in *The People's Power.*

35 Under the unratified 1932 treaty, the federal and Ontario governments would have split the cost of the Ogoki–Long Lac diversions.

36 The Montreal Light, Heat and Power Company, which took over the Beauharnois station, was already exporting 75,000 horsepower to ALCOA's Massena plant. NARA II, RG 59, box 4050, file 711.42157 SA 29/1623: Memorandum to the President from Walsh, FPC, Re: Great Lakes–St. Lawrence Treaty, February 15, 1938; Louis-Raphael Pelletier, "The Destruction of the Rural Hinterland: Industrialization of Landscape in Beauharnois County," in *Metropolitan Natures: Environmental Histories of Montreal,* ed. Stéphane Castonguay and Michèle Dagenais (Pittsburgh: University of Pittsburgh Press, 2011), 245–63; Michèle Dagenais, *Montreal, City of Water: An Environmental History* (Vancouver: UBC Press, 2017).

37 The Roosevelt administration figured that private interests were actively undermining a larger agreement covering Great Lakes–St. Lawrence water issues. In 1929, the Supreme Court of Canada had given the provinces the right to develop power, though at the time Prime Minister Mackenzie King claimed his national government had the right to license such projects. The creation of the Royal Commission on Dominion-Provincial Relations, more popularly known as the Rowell-Sirois Commission, would also soon clarify water power on Ontario's major border rivers, allowing Ontario Hydro to proceed with developments on the Niagara, St. Lawrence, and Ottawa Rivers. Historian Christopher Armstrong suspects that Hepburn was the instrument of private power companies. Armstrong, *Politics of Federalism,* 190.

38 Syracuse University Library, Manuscript Collection (hereafter SUL), Robert Moses Papers, box 5, report 1955-6: PASNY, Niagara Power Park and Arterial Development, September 28, 1956.

39 The part that fell was 200 feet long and jutted out 60 feet, but that was but a fragment of the section that fell in the mid-nineteenth century. A large section had also fallen from a point opposite the Niagara Falls Water Works Station in October 1917. NPC, *Forty-Sixth Annual Report of the Commissioners of the Queen Victoria Niagara Falls Park* (Toronto: L.K. Cameron, 1931).

40 NPC, *Forty-Ninth and Fiftieth Combined Annual Report of the Commissioners of the Queen Victoria Niagara Falls Park* (Toronto: L.K. Cameron, 1935).

41 NPC, *Forty-Sixth Annual Report.*

42 Coutu, "Vehicles of Nationalism."

43 This was a submerged concrete spillway diverging outward and downward, 10–12 feet deep and diagonally upstream about 750–900 feet from the end of the powerhouse. For more detail see Ball, *The Canadian Niagara Power Company Story.*

44 An ice boom built in 1918 in New York territory to protect the Niagara Falls Power Company's power station intakes pushed more ice over to the Canadian side of the Niagara River. On scientific expertise and policy-making in North America more generally see Stephen Bocking, *Nature's Experts: Science, Politics, and the Environment* (New Brunswick, NJ: Rutgers University Press, 2004).

45 Canadian officials decided that this weir did not need IJC approval. NPC, *Forty-Ninth and Fiftieth Combined Annual Report;* LAC, RG 25, G-1, vol. 1809, file 968: Johnston to Skelton, October 27, 1936. On O.D. Skelton see Norman Hillmer, *O.D. Skelton: A Portrait of Canadian Ambition* (Toronto: University of Toronto Press, 2016).

46 On the role of technological failure and hostile environments in the Canadian context, see Edward Jones-Imhotep, *The Unreliable Nation: Hostile Nature and Technological Failure in the Cold War* (Cambridge, MA: MIT Press, 2017).

47 Joan Coutu points out that a series of relief sculptures on the side of the Rainbow Bridge "record phases of settlement – a stagecoach, a farm, a rainbow arching over a hayfield – that culminate with depictions of turbines and cog-wheels, which refer to the ultimate domination of nature in the harnessing of the mighty cataracts of the Niagara River in the nearby massive hydroelectric generating stations." Coutu, "Vehicles of Nationalism," 195.

48 On Canadian hydropower during the war, see Matthew Evenden, *Allied Power: Mobilizing Hydro-Electricity during Canada's Second World War* (Toronto: University of Toronto Press, 2015).

49 LAC, MG 26 J4, vol. 210, St. Lawrence (files 1992-3), reels C4280: Memorandum for File Re: Meeting of Cabinet with representatives of the Ontario Legislature, October 3, 1939.

50 LAC, RG 25, vol. 2636, file 1268-D-40C: St. Lawrence River–Niagara River Treaty Proposals (General Correspondence), January 4, 1938, to December 21, 1940: Memorandum by Norman Marr to Mr. Johnston, Re: Possible United States–Canadian Power Interchange Arrangement in connection with St. Lawrence Treaty, November 3, 1939.

51 The Long Lac diversion, completed in 1941, connects the headwaters of the Kenogami River with the Aguasabon River, which naturally discharges into Lake Superior about 250 kilometres east of Thunder Bay, Ontario. The Ogoki diversion, completed in 1943, connects the upper portion of the Ogoki River to Lake Nipigon, and from there flows into Lake Superior. The October 14, 1940, note was followed by supplementary notes on October 31 and November 7, and then additional notes on May 20, 1941, and October 27, 1941. LAC,

RG 25, vol. 3650, file 1268-D-40C: St. Lawrence River–Niagara River Treaty Proposals (General Correspondence), Part 7, June 1/47–Jan 16/48: Letter from Adolf Berle (for the Sec of State) to Christie, October 14, 1940.

52 These two diversions together increase the mean level of each of the Great Lakes as follows: Lake Superior by 6.4 cm (0.21 feet); Lakes Michigan-Huron by 11.3 cm (0.37 feet); Lake Erie by 7.6 cm (0.25 feet); and Lake Ontario by 6.7 cm (0.22 feet).

53 The 1941 agreement also addressed other bilateral water issues in the Great Lakes–St. Lawrence basin. The Ogoki–Long Lac diversions were officially recognized, along the lines of the informal 1940 agreements, and Ontario was covering the entire cost (in the 1932 treaty, the federal government picked up half the tab). For the Chicago Diversion, the limits called for in the 1932 treaty had been enacted in the intervening years, as the diversion had been reduced by over 5,000 cfs, and the 1941 agreement enshrined the new cap and prescribed an arbitral tribunal for cases where the Chicago Diversion was increased without Canadian authorization.

54 LAC, RG 25, vol. 3560, file 1268-K-40C: St. Lawrence–Niagara River Treaty between Canada and United States – Additional Diversion of Water at Niagara Falls, pt. 1 (April 24/41 to Dec 8/41): Letter from Wrong to Robertson, April 24, 1941.

55 HEPCO, Memorandum (by Lindsay) to WL Mackenzie King Re: Remedial Works in Niagara River (n.d., 1943).

56 IJC, Docket 64, Niagara Falls Reference, box 114, file 64-8-5:3, Exchange of Notes Re: Niagara 1941: Exchange of Notes relating to further utilization of water for power purposes at Niagara Falls, November 27, 1941; LAC, RG 25, vol. 3560, file 1268-K-40C: St. Lawrence–Niagara River Treaty between Canada and United States – Additional Diversion of Water at Niagara Falls, pt. 2 (Jan 1/42 to April 30/44): Telegram from Cdn Minister in the US to Sec of State, September 16, 1942; NARA II, RG 77, General Correspondence Relating to Civil Works Projects, 1942–47, box 549, file 825.91 (Niagara R., NY): Hull to McCarthy, October 27, 1941.

57 That included about half the power from the Rankine plant, which had an export contract until 1974. It seems, though I did not find unequivocal information to support this, that the Canadians decided not to protest the continuation of the Rankine plant, as well as an equal distribution of water under the 1950 Niagara Diversion Treaty, since Ontario received much of New York's share under the 1950 treaty until the Robert Moses generating station became operational, which occurred over a decade after the signing of the treaty. IJC, Docket 64, Niagara Falls Reference, box 114, file 64-8-5:4, Exchange of Notes Re: Niagara 1944/05/03: Exchange of Notes – Constituting an Agreement for the Temporary Additional Diversion of Water at Niagara for Power Purposes, May 3, 1944.

58 HEPCO, Niagara River – Diversions of Water for Generation of Power: HEPCO, and Copies of Notes between governments of Canada and US (October 14, 1940, to May 3 1944), December 4, 1947; IJC, Docket 64, Niagara Falls Reference, box 107, file 64-4-2:2, Specifications for Preservation and Enhancement of Niagara Falls, US Flank of Horseshoe Falls, US Corps of Engineers: Preservation and Enhancement of Niagara Falls, International Niagara Falls Engineering Board, March 1, 1953.

59 They were licensed collectively for only 19,725 cfs.

60 The initial 5,000 cfs from the Albany basin was used at Ontario Hydro's Toronto Power Generating Station, and when additional diversions were later authorized, the extra water was in excess of what could be absorbed by the Canadian plants at Niagara. A new DeCew generating station was built, and the additional water from Ogoki–Long Lac was used at

the old and new DeCew plants. The Welland Canal was sourced chiefly from Lake Erie but also received water from the Niagara River via the reversed Welland River. The diversion through the Welland Canal was estimated to lower Lake Erie levels by 0.2 feet, and the Niagara River power diversions lowered the lake by 0.06 feet. The various natural rock ledges, or sills, in the Niagara River minimized the impact of lower Niagara River levels on Lake Erie levels. On the technical aspects of the DeCew plants, see Biggar, *Ontario Hydro's History*, 77–85.

61 HEPCO, USACE Buffalo Office Memorandum (by H. Goldberg), Review of Canadian computations on effect of the Niagara River Weir, July 7, 1943. For more on the American and Army Corps of Engineers involvement in the construction of the weir, see NARA II, RG 77, General Correspondence Relating to Civil Works Projects, 1942–47, box 550, file 825.91 (Niagara, NY).

62 According to Norman Marr, acting controller of the Canadian Department of Mines and Resources, "the effect on Lake Erie of an additional diversion of 2500 cfs at the end of a period of four months has been calculated to result in a total lowering of level of 0.063 feet or three-fourths of one inch. On termination of the additional diversion the lake level should be restored virtually to normal in about eight months' time." LAC, RG 25, vol. 3560, file 1268-K-40C: St. Lawrence–Niagara River Treaty between Canada and United States – Additional Diversion of Water at Niagara Falls, pt. 4 (Jan 1/48 to Nov 30/49): Memorandum, Attention Mr. David Johnson, Re: Further Diversion at Niagara, November 26, 1948; NARA II, RG 77, General Correspondence Relating to Civil Works Projects, 1942–47, box 549, file 800.225 (Niagara R., NY): Olds to Cruise, August 8, 1947.

63 The cost was budgeted at $879,429.38 but in the end was actually $785,446.41. NARA II, RG 59, box 3306, file 711.4216/5-2247: "Niagara Falls River Remedial Works, Submerged Weir: Fifth Progress Report of Construction Sub-Committee as of 31 January 1947," February 28, 1947; NARA II, RG 77, General Correspondence Relating to Civil Works Projects, 1942–47, box 549, file 825.91 (Niagara R., NY): Stone Placing in Niagara River Weir, July 4, 1945.

64 NPC, *Fifty-Seventh Annual Report of the Commissioners of the Queen Victoria Niagara Falls Park* (Toronto: L.K. Cameron, 1943); NPC, *Sixty-Third Annual Report of the Commissioners of the Queen Victoria Niagara Falls Park* (Toronto: L.K. Cameron, 1949); HEPCO, Memorandum to Holden, Re: Niagara River Remedial Weir, February 3, 1944.

65 It was estimated that, before the weir, the Chippawa–Grass Island Pool had been lowered by 0.21 feet by power station diversions. The past dumping of excavated material in the Chippawa–Grass Island Pool likely compensated for some of the diversions from the pool. It is clear that, by this time, at least some experts were aware that natural supply was the bigger factor in determining Great Lakes water levels, and also recognized the slow impact of glacial rebound. HEPCO, "Niagara River Remedial Weir – Affect [sic] on Niagara River Levels," September 23, 1945; HEPCO, "Niagara River, Diversions of Water for Generation of Power: HEPCO," and Copies of Notes between governments of Canada and US: October 14, 1940, to May 3 1944 (December 4, 1947).

66 Part of this stemmed from above-average upper Great Lakes levels and the causeway that was built to aid weir construction; after the war, US authorities removed this causeway. NARA II, RG 77, General Correspondence Relating to Civil Works Projects, 1942–47, box 549, file 825.91 (Niagara R., NY): Removal of Niagara Weir Construction Cableway, May 13, 1946; HEPCO, Riebe (USACE) to Holden, August 3, 1945.

67 HEPCO, Memorandum to Holden, Niagara Weir: Report of Cline and Goldberg, October 9, 1947.
68 HEPCO, USACE Memorandum by Woods to Goethals, Improvement of Goat Island flank of the Horseshoe Falls, October 26, 1942.
69 Dubinsky, *The Second Greatest Disappointment.*
70 In 1945, Southern Ontario had consumed 9.9 billion kW; it would consume 11.8 billion kW in 1949, and estimates put the 1951 requirement at 15.6 billion kW.
71 HEPCO, Law Division: 992.15, "The St. Lawrence Power Project – A Monument to Progress."
72 Annie Michaud, "The Niagara River Remedial Action Plan: 25 Years of Environmental Restoration," Policy Brief 12, Brock University, Niagara Community Observatory, St. Catharines, ON, December 2012.
73 LAC, RG 25, vol. 3560, 1268-D-40C, St. Lawrence River–Niagara River Treaty Proposals – General Correspondence, pt. 9 (Jan 5/49 to Dec 30/49): Memorandum: Niagara Falls Power Diversion, October 13, 1949.
74 Besides the fact that the 1944 notes were never ratified by the US Senate, some notes outlined that the diversions were to end at the cessation of hostilities (and the level of the diversions violated Article V of the Boundary Waters Treaty). LAC, RG 25, vol. 3560, file 1268-K-40C: St. Lawrence–Niagara River Treaty between Canada and United States – Additional Diversion of Water at Niagara Falls, pt. 3 (May 1/44 to Dec 29/47): Memorandum to the Third Political Division, Re: Diversion of Water at Niagara Falls, June 27, 1946.
75 According to a Federal Power Commission report released in September 1949, both countries were taking a total of 87,930 cfs, but it appears that the various treaties and exchanges of notes cumulatively provided for a cap of 89,000 cfs on total diversions. NARA II, RG 59, Decimal File, 1945–49, box 3306, file 711.4216-N51/9-149: Federal Power Commission, "Possibilities for Redevelopment of Niagara Falls for Power," September 1949.
76 LAC, RG 25, vol. 3560, 1268-D-40C, St. Lawrence River–Niagara River Treaty Proposals – General Correspondence, pt. 9 (Jan 5/49 to Dec 30/49): Memorandum for the Secretary to the Cabinet, Re: St. Lawrence Development and Niagara River Diversions, May 26, 1949.
77 NARA II, RG 59, box 3303, files 711.42157 SA 29/10-148 to 711.42157 SA 29/10: State Department Memorandum from Satterthwaite to Hickerson, December 22, 1948; LAC, RG 25, vol. 3560, file 1268-K-40C: St. Lawrence–Niagara River Treaty between Canada and United States – Additional Diversion of Water at Niagara Falls, pt. 3 (May 1/44 to Dec 29/47): Memorandum for Legal Division, Diversion of Water at Niagara Falls, December 28, 1946; IJC, Docket 64, Niagara Falls Reference, box 114, 64-8-5:5, Exchange of Notes Re: DeCew Falls 1948/12/23: Exchange of Notes re DeCew Falls, December 23, 1948.
78 Some consideration was also given to implementing Niagara changes by modifying the Boundary Waters Treaty. It seems worth noting that, as of October 1947, J. Horace McFarland was still writing to government officials about Niagara Falls issues. NARA II, RG 77, General Correspondence Relating to Civil Works Projects, 1942–47, box 549, file 800.225 (Niagara R., NY). Letter to Senator Martin, November 20, 1947.
79 This lobbying likely included advocacy for public power and negotiations about cancelling or buying out export contracts. LAC, RG 25, vol. 3560, file 1268-D-40C: St. Lawrence River–Niagara River Treaty Proposals – General Correspondence, Part 9 (Jan 5/49 to Dec 30, 1949): Memorandum for the Secretary to the Cabinet, RE: St. Lawrence Development and Niagara River Diversions, May 26, 1949.

80 NARA II, RG 84, box 15, Canada, US Embassy, Ottawa: Classified and Unclassified General Records, 1948–1949 (1949: 320.12 to 1949: 322.2), file 322.2: 1949 St. Lawrence Waterway: Letter from St. Laurent to Truman, May 27, 1949.

81 LAC, RG 2, vol. 207, file W-10-3: Water Development, Waterworks, Projects: Niagara Diversion, 1948-59-51: Wrong took Saunders to lunch with Olds of the FPC, June 3, 1949; LAC, RG 19, vol. 4441, file 9810-09, vol. 1: Water – International Joint Commission (IJC) Niagara Falls and River: Canadian Ambassador to US to Sec of State DEA (Wrong to Heeney), WA-1516 of June 1st, Niagara diversions, June 3, 1949.

82 NARA II, RG 59, Decimal File, 1945–49, box 3306, file 711.4216-N51/9-149: Federal Power Commission, "Possibilities for Redevelopment of Niagara Falls for Power," September 1949.

83 The term "nature lovers" was used in NARA II, RG 59, box 3303, files 711.42157 SA 29/10-148 to 711.42157 SA 29/10-3048: Memo for File: Second Meeting on Niagara River Diversion Treaty, October 10, 1949.

84 IJC, Docket 64, Niagara Falls Reference, box 113, 64-8-1:3, Dept. External Affairs Niagara Falls 1953/11/19: E. Robert de Luccia, Chief, Bureau of Power, Federal Power Commission, "Report to Federal Power Commission on Possibilities for Redevelopment of Niagara Falls for Power, Niagara River, New York," Washington, DC, September 1949.

85 The FPC report suggested that 100,000 cfs be guaranteed to go over the Falls during tourist hours, with up to another 100,000 cfs available for power diversions. During nighttime, flows of 50,000 cfs could be diverted. NARA II, RG 59, Decimal File, 1945–49, box 3306, file 711.4216-N51/9-149: Federal Power Commission, "Possibilities for Redevelopment of Niagara Falls for Power," September 1949.

86 LAC, RG 19, file 9810-09, vol. 2: Water – International Joint Commission (IJC) Niagara Falls and River: Diplomatic Note No. 236 (from Harrington, US), October 12, 1949.

87 In 1950, Truman initiated the President's Water Resources Policy Commission in an unsuccessful attempt to bring a more objective and less partisan approach to water development, favouring comprehensive water projects rather than a piecemeal approach. NARA II, RG 59, box 3303, file 711.4216 NI/9-2349: Memorandum for the Secretary of State, September 23, 1949; Elmo Richardson, *Dams, Parks and Politics: Resource Development and Preservation in the Truman-Eisenhower Era* (Lexington, KY: University Press of Kentucky, 1973).

88 That is, Ontario's need for power would likely lead Canada to be "cooperative" and not "drive too hard a bargain": NARA II, RG 59, box 3303, files 711.42157 SA 29/10-148 to 711.42157 SA 29/10: Memo for File: Proposed Treaty with Canada on Niagara River Diversion and Redevelopment, Seventh Meeting, November 10, 1949.

89 NARA II, RG 59, Decimal File, 1945–49, box 3306, file 711.4216/5-2249: Discussion of Tentative Draft of USA-Canada Treaty Relating to Niagara River Diversion, December 9, 1949.

90 LAC, RG 25, vol. 3561, file 1268-K-40C: St. Lawrence–Niagara River Treaty between Canada and United States – Additional Diversion of Water at Niagara Falls, pt. 4 (Jan 1/48 to Nov 30/49): Secretary of State to Canadian Ambassador, October 29, 1949.

91 In February 1950, Canada and the United States quietly exchanged aides-mémoires, the former declaring that it reserved the right to decline the renewal of export contracts once the Americans' facilities allowed them to utilize their full share of the water diversion. LAC, RG 25, vol. 6183, file 1268-K-40, pt. 5.1: St. Lawrence–Niagara River Treaty between Canada and the United States – Additional Diversion of Water at Niagara Falls (29-11-49 to 21-4-50): Letter from Moran (for Heeney) to Deputy Minister of Trade and Commerce,

January 17, 1950; Government of Canada, Department of External Affairs, "Memorandum from Secretary of State for External Affairs to Cabinet, February 18, 1950," Cabinet Document No. 56 50, in *Documents on Canadian External Relations,* vol. 16 (Ottawa: Minister of Public Works and Government Services Canada, 1997), 974; NARA II, RG 59, Decimal File, 1950–54, box 2804, files 611.42322-Ni/1-450 to 611.42322-N/2-22750: *Aide-Mémoire* from Canada to US re: Niagara, February 24, 1950.

92 The DeCew diversion could legitimately have been continued since some of it predated the 1909 BWT and thus could have been grandfathered in like the Chicago Diversion. Additionally, the 800 cfs that New York had diverted for the New York State Barge Canal since before the twentieth century – most of which is for navigation, with a small part for hydroelectric generation – was not included in the US Niagara allotment under the 1950 treaty. However, as a Canadian legal official pointed out, "it could be argued that Article V makes void any permission granted or implied before 1909 for diversion around Niagara Falls. On the other hand, the United States has not, in the last forty years, questioned our right to the DeCew diversion, to my knowledge at least." LAC, RG 25, vol. 6183, file 1268-K-40, pt. 5.1: St. Lawrence–Niagara River Treaty between Canada and the United States – Additional Diversion of Water at Niagara Falls (29-11-49 to 21-4-50): Memorandum for Mr. Blockley, December 1, 1949.

93 NARA II, RG 59, box 3303, file 711.4216/5-2249: Discussion of Tentative Draft of USA-Canada Treaty Relating to Niagara River Diversion, December 9, 1949; Government of Canada, Department of External Affairs, "Extract from Minutes of Meeting of Heads of Divisions, December 12, 1949," in *Documents on Canadian External Relations,* vol. 15 (Ottawa: Minister of Public Works and Government Services Canada, 1995), 980.

94 LAC, RG 25, vol. 3560, file 1268-D-40C: St. Lawrence River–Niagara River Treaty Proposals – General Correspondence, pt. 9 (Jan 5/49 to Dec 30/49): Memorandum for Mr. Menzies, Discussions with Hydro on Niagara Diversions, October 28, 1949.

95 LAC, RG 25, vol. 6183, file 1268-K-40, pt. 5.1: St. Lawrence–Niagara River Treaty between Canada and the United States – Additional Diversion of Water at Niagara Falls (29-11-49 to 21-4-50): Memorandum for Mr. Blockley, December 1, 1949.

96 LAC, RG 25, vol. 3560, file 1268-D-40C: St. Lawrence River–Niagara River Treaty Proposals – General Correspondence, pt. 9 (Jan 5/49 to Dec 30/49): Teletype from Cdn Ambassador to Sec of State DEA, November 14, 1949.

97 LAC, RG 25, vol. 3561, file 1268-K-40C: St. Lawrence–Niagara River Treaty between Canada and United States – Additional Diversion of Water at Niagara Falls, pt. 4 (Jan 1/48 to Nov 30/49): Secretary of State to Canadian Ambassador, November 7, 1949.

98 NARA II, RG 59, Decimal File, 1945–49, box 3306, file 711.4216-NI/5-2247: Transcript of Proceedings: Discussion of Tentative Draft of USA-Canada Treaty Relating to Niagara River Diversions, December 8, 9, 10, 1949.

99 NARA II, RG 59, Decimal File, 1950–54, box 2804, file 611.42322/9-453: Memorandum of Conversation, Dept of State: Draft treaty for the diversion of waters from the Niagara River for power purposes, February 1, 1950.

100 NARA II, RG 59, Box 3303, file 711.4216 NI/10-2749: The Minister in Canada (Armour) to the Secretary of State, No. 441, October 27, 1949.

101 The federal-Ontario agreement actually was not finalized until March 27, 1950, after the Niagara Diversion Treaty was signed. On Ontario's relations with the federal government, see Armstrong, *The Politics of Federalism.*

102 For the purposes of the treaty, the volume of water available was considered the "total outflow from Lake Erie to the Welland Canal and the Niagara River (including the Black Rock Canal) less the amount of water used and necessary for domestic and sanitary purposes and for the service of canals for the purpose of navigation." Via diplomatic letters in 1986, the two federal governments advised the International Niagara Committee that the acceptable metric equivalent of 100,000 cfs would be 2,832 cubic metres per second (m^3/s), and the equivalent of 50,000 cfs would be 1,416 m^3/s.

103 In a 1979 article, B.F. Friesen argued that Canada accepted less than an equal share of diversion from the Niagara River under the 1909 Boundary Waters Treaty and the US advantage stemmed from factors such as the Chicago Diversion, Niagara power exports, the DeCew exemption, New York State Barge Canal diversions, consumptive water use, and the Ogoki–Long Lac diversions. In Friesen's estimation, the 1909 treaty diversion ratios benefited the United States more once the power exports and Chicago Diversion were factored in, and he contends that Canada received the power from only 36 percent of the water but accepted this distribution because it was eager to get the Boundary Waters Treaty into force. B.F. Friesen, "The International Sharing of Niagara River Hydroelectric Power Diversions," *Canadian Water Resources Journal* 4, 4 (1979): 26–38. B.F. Friesen and J.C. Day, "Hydroelectric Power and Scenic Provisions of the 1950 Niagara Treaty," *Water Resources Bulletin – American Water Resources Association* 13, 6 (December 1977): 1175–89. Daniel Macfarlane, "Negotiating Niagara Falls: US-Canada Environmental and Energy Diplomacy," *Diplomatic History* 43, 5 (November 2019): 916–43.

104 The US probably benefited more from using these diversions at Sault Ste. Marie power plants than did Canada since the US plants were larger at Sault St. Marie. There was no agreement in place about Canada receiving a greater share of its Ogoki–Long Lac diversions at the St. Lawrence power works built in the 1950s. The Canadians unsuccessfully fought for the inclusion of a general principle that all water diverted into the Great Lakes system should be the right of whichever territory it had originated from. IJC, Docket 64, Niagara Falls Reference, box 114, 64-2-5:8 to 64-7-1:1, IJC executive session, August 31, 1953.

105 The 1950 treaty flow requirements also overrode the flow limitations in the diplomatic notes from the 1940s.

106 The Department of External Affairs was particularly concerned that Truman would link the Niagara treaty to the St. Lawrence development, or bring up public-versus-private power issues. LAC, RG 25, vol. 6351, file 1268-K-40, pt. 6: St. Lawrence–Niagara River Treaty between Canada and the United States – Additional Diversion of Water at Niagara Falls (May 1/1950–May 31/1950): Letter from Wrong to Pearson, July 11, 1950.

107 Ibid., Moran (for Under-Sec DEA) to Deputy Minister, Department of Transport, May 29, 1950; Memorandum for the Minister, Ratification of the Niagara Diversion Treaty, June 7, 1950.

108 NARA II, RG 84, Canada, US Embassy, Ottawa, Classified General Records, 1950–1961, box 9: Memorandum of Meeting, US Embassy in Ottawa: The Niagara Diversion Treaty and the St. Lawrence Seaway and Power Project, June 21, 1950.

109 NARA II, RG 84, Canada, US Embassy, Ottawa, Classified General Records, box 14, UD 2195C, file 322.2: Niagara Diversion Treaty: Telegram to Secretary of State from Woodward, July 6, 1950; LAC, RG 25, vol. 6351, file 1268-K-40, pt. 6, St. Lawrence–Niagara River Treaty between Canada and the United States – Additional Diversion of Water at

Niagara Falls (May 1/1950–May 31/1950): Memorandum for the Under-Secretary, July 8, 1950.

110 Granted, the State Department official relaying this wrote that "it is *not* believed that any such action is contemplated seriously at this time; Mr. Pearson's and Mr. Hearn's remarks are merely illustrative of a growing Canadian sentiment in the matter" (emphasis in original). NARA II, RG 59, Decimal File, e1950–54, box 2804, file 611.42322 N/8-250: Voluntary Economic Data – Local attitude toward proposed Niagara Diversion Treaty and the direct effect of Treaty provisions upon American-controlled industry on the Niagara Peninsula (Appendix I), August 2, 1950.

111 Private industry generated 94 percent of US electricity as of 1930, but only 75 percent by 1957. Wyatt Well, "Public Power in the Eisenhower Administration," *Journal of Policy History* 20, 2 (2009): 228. On the history of federal water resources, see Holmes, *A History of Federal Water Resources Programs.*

Chapter 4: Empowering Niagara

1 As Bob Johnson points out, once people began burning fossil fuels, the transmission of that power was spatially limited unless it was converted into electricity. See Chapter 4 in Bob Johnson, *Carbon Nation: Fossil Fuels in the Making of American Culture* (Lawrence: University Press of Kansas, 2014).

2 Christopher Jones treats hydroelectricity as a stock and part of the mineral energy regime, whereas R.W. Sandwell emphasizes its renewable and flowing features, and places it within the organic energy regime. Jones, *Routes of Power;* Ruth Sandwell, *Powering Up Canada: The History of Power, Fuel, and Energy from 1600* (Montreal and Kingston: McGill-Queen's University Press, 2016); E.A. Wrigley, *The Path to Sustained Growth: England's Transition from an Organic Economy to an Industrial Revolution* (Cambridge: Cambridge University Press, 2016). On North American energy history see also Matthew Huber, *Lifeblood: Oil, Freedom and the Forces of Capital* (Minneapolis: University of Minnesota Press, 2013); David Nye, *Electrifying America;* Roland Tobey, *Technology as Freedom: The New Deal and the Electrical Modernization of the American Home* (Oakland: University of California Press, 1997).

3 Claudia Goldin and Robert Margo, "The Great Compression: The Wage Structure in the United States at Mid-Century," *Quarterly Journal of Economics* 107 (February 1992): 1–34.

4 According to Robert Gordon, the fruits of the second industrial revolution had "their full effect on productivity years earlier than might have otherwise occurred" because of the Great Depression and the Second World War. Robert Gordon, *The Rise and Fall of American Growth: The US Standard of Living since the Civil War* (Princeton, NJ: Princeton University Press, 2016), 122–28; John R. McNeill and Peter Engelke, *The Great Acceleration: An Environmental History since 1945* (Cambridge, MA: Harvard University Press, 2016).

5 Gordon, *The Rise and Fall of American Growth,* 528.

6 Ibid., 560.

7 Rice, "More Canadian Kilowatts at Niagara, Part 1," 210–17.

8 Manore, *Cross Currents,* 105–12.

9 As of the end of 1950, in Ontario Hydro's Southern Ontario system there were, in operation or under construction, thirty-eight hydroelectric power plants, six steam-electric stations, two small diesel plants, and several plants in Quebec and Ontario from which power was purchased. HEPCO, Letter to Norman Marr, December 12, 1950.

10 HEPCO, Draft of Proposed Paper: "Structural Investigations – Sir Adam Beck Niagara G.S. No. 2, Objectives of Investigation," by W.G. Morrison and A.D. Hogg, September 13, 1955.

11 HEPCO, Memo: "The Stresses in the Tunnel Lining – Sir Adam Beck Niagara Generating Station No. 2," May 23, 1951.

12 HEPCO, 40.1730 F 138: Interim Report: "Rock Stability in the Niagara Region," May 11, 1953.

13 NPC, *Fifty-Eighth Annual Report of the Commissioners of the Queen Victoria Niagara Falls Park* (Toronto: L.K. Cameron, 1944); NPC, *Fifty-Ninth Annual Report of the Commissioners of the Queen Victoria Niagara Falls Park* (Toronto: L.K. Cameron, 1945).

14 In 1951, an observation plaza was added. For a concise history of these scenic tunnels, see Niagara Parks, "The History of Journey Behind the Falls," Niagara Parks, August 14, 2018, https://www.niagaraparks.com/the-history-of-journey-behind-the-falls/.

15 This generating capacity total includes the pumped-storage generating station. "Niagara Power Goes Underground," *Popular Mechanics,* April 1952, 115–17.

16 Ontario Hydro did all the work itself, aside from the tunnels, which were built by the Rayner-Atlas Company and Perini-Walsh and Associates. The tunnels gave way to a canal several miles from the generating stations because the former would have been too hazardous and expensive to build through the glacial silt of the buried St. David's Gorge. Scale models were used for forebay design, as well as to solve turbulence issues where the new canal intersected the old canal right before the forebay. The tunnels were built in segments by innovative methods, including vertical shafts. For an overview of these tunnels and the technological advances involved, see H.R. Rice, "Tunnelers Make Record at Niagara Falls," *Compressed Air Magazine* 59, 6 (June 1954): 169–70; H.R. Rice, "More Canadian Kilowatts at Niagara, Part 1," *Compressed Air Magazine* 58, 8 (August 1953): 210–17; H.R. Rice, "More Canadian Kilowatts at Niagara, Part 2," *Compressed Air Magazine* 58, 9 (September 1953): 244–50; Jesse R. Glaser, "Ontario Hydro Blasts Out Niagara Power Tunnels," *Civil Engineering,* October 1953, 40–44.

17 There was internal debate about where the next powerhouse would be located. Some argued for an alternative location since it was important to disperse their resources in case of a nuclear attack. Ultimately, an additional powerhouse has never been built. HEPCO, "Some Thoughts on Niagara Development," n.d.

18 This canal was enlarged in the mid-1960s. The 12-metre-deep channel of the canal tends to silt up over time. In 1964–65, the canal was drained and debris was removed; it was also widened, the concrete lining was repaired, and gates were installed at each end. This reportedly increased the flow by 184 cubic metres per second (cms). Then, in 1981, some 1,834 cubic metres of concrete, rock, and other debris was dredged out, increasing the water volume about 5 percent, to 617 cms. A March 2018 newspaper article reported that the water flow was back down to about 550 cms, and a rehabilitation project is planned for 2021–22 by Ontario Power Generation that will involve improving the canal (during which the reversed Welland River will temporarily return to its original direction) as well as "refurbishing two of the power station's 10 original generators that have been mothballed for several years." Allan Benner, "OPG to Drain Niagara Power Canal: Welland River Flow Will Temporarily Revert to Normal," *Niagara Falls Review,* March 18, 2018, https://www.niagarafallsreview.ca/news-story/8336021-opg-to-drain-niagara-power-canal/.

19 Compressed air was used to create a wall of air bubbles that almost completely eliminated shock waves. "Air Curtain Fences Blast," *Popular Mechanics,* August 1954, 96–97.

20 The differential between the intakes above the Falls and the powerhouse was 315 feet, and almost the entire drop over that distance could be utilized as head.
21 HEPCO, Holden to Mr. R.W. Osborne, March 12, 1952; HEPCO, "The Economics of the Ontario Hydro Pumped Storage Project," by M. Ward, Director of Planning – Ontario Hydro, 1957.
22 HEPCO, 40-1730 F 138, Pamphlet about Beck No. 2, August 1954.
23 Some of the new residences were constructed in such a way that they could be readily converted into apartments at a later date. HEPCO, Press Release: "'Niagara Falls' Now at Islington," February 1, 1951.
24 HEPCO, Comments on Petition re: Rentals – S.A.B. No. 2, December 21, 1951.
25 Ibid.
26 For a comparison between expropriation and rehabilitation efforts of Ontario Hydro and PASNY as part of the St. Lawrence project, see Chapter 5 of Macfarlane, *Negotiating a River*.
27 HEPCO, SPP Series, C.E. Blee (Chief Engineer, Tennessee Valley Authority) to J.R. Montague (Director of Engineering, HEPCO), February 5, 1952.
28 HEPCO, Advice of Commission Decision – Expropriation of Property in Township of Niagara and Stamford from A.A. Holman, May 26, 1954.
29 HEPCO, Memo for Holden, Meeting of the Stamford Township Council and the H.E.P.C. of Ontario, February 28, 1951.
30 HEPCO, HEPCO to Couillard, Sir Adam Beck–Niagara G.S. No. 2 – Brief by the Township of Niagara, August 12, 1954; HEPCO, Advice of Commission Decision – Negotiation of Agreement with the Township of Niagara, April 20, 1955.
31 HEPCO, Meeting of the Council of the Township of Stamford and the H.E.P.C. of Ontario, February 15, 1951.
32 HEPCO, Press Release: "Hydro and A.F. of L. Sign Master Contract," February 23, 1951. See also Denison, *The People's Power*, 209; Paul C. Weiler, ed., *Mega Projects: The Collective Bargaining Dimension* (Ottawa: Canadian Construction Association, 1986).
33 A statement of understanding signed in March 1958 covered the period from August 1, 1957, to July 31, 1959. HEPCO, Collective Agreement – Ontario Hydro Construction and Construction Allied Council, A.F. of L., November 19, 1953; HEPCO, Statement of Understanding between HEPCO and the Allied Construction Council (A.F. of L.–C.I.O.–C.L.C.) and the Members Unions of the Allied Council, March 4, 1958.
34 HEPCO, Extracts from Article Prepared by Public Relations Divisions – Beck-Niagara G.S. No. 2.
35 These areas were later graded and planted. HEPCO, Rehabilitation – Iroquois, Comments by Saunders, October 13, 1954.
36 Ontario Hydro was given permission by the Niagara Parks Commission. NPC, *Sixty-Sixth Annual Report of the Commissioners of the Queen Victoria Niagara Falls Park* (Toronto: L.K. Cameron, 1952).
37 IJC, Docket 64, Niagara Falls Reference, box 106, 64-3-1-1: Correspondence Vol. 1: Re Niagara River, May 26, 1953.
38 About 550,000 cubic yards of material was spread over approximately 1,900 feet of shoreline between the Queenston-Lewiston Bridge and the Queenston dock, with the remainder deposited south of the bridge at road level along the shore extending twenty feet into the river. HEPCO, Dumping Near Queenston – Execution of Agreement with the Niagara

Parks Commission, September 11, 1952; HEPCO, Sir Adam Beck–Niagara Generating Station No. 2 Niagara River – Proposed Disposal Areas, n.d.

39 The settlements totalled $3,775, ranging from a low award of $75 to a high of $1,000. HEPCO, RE: Fishing in Niagara River near Queenston, February 15, 1954.

40 An Ontario Hydro official internally suggested that the commission's reversal of the Welland River had made the water deeper and faster and it "must have been heavily polluted and the water really unfit for bathing." HEPCO, Memorandum to Holden, Letter dated June 25, 1957, from Clarence Somerville, Deputy Reeve, Village of Chippawa, July 4, 1957.

41 Earlier designs had the parkway running behind the power stations.

42 HEPCO, Memorandum: Sir Adam Beck–Niagara Pumping-Generating Station – Failure of Dyke and Remedial Work, September 26, 1958.

43 As Rock Brynner describes on page 104 of *Natural Power*, PASNY's first operating asset was a seventy-eight-mile transmission line from Taylorville to Massena, which it took over from the federal government in 1951, the same year that the Power Authority Act was amended to add the development of hydro power at Niagara to that of the St. Lawrence. Thomas Dewey reorganized PASNY in 1953, and during his time as governor of New York (1943–54) "PASNY had shifted from the goal of producing and distributing power to a focus on the former, creating an ongoing ideological conflict – between the state and the federal government, and within the federal Department of the Interior – about public versus private development of hydroelectric power." The Truman administration also wanted assurances that if New York was granted the right to generate the power, it would share some with New England. Macfarlane, *Negotiating a River*, 59.

44 Robert Caro reports that Moses also hoped to bring atomic energy development in the state under the aegis of PASNY. Robert Caro, *The Power Broker: Robert Moses and the Fall of New York* (New York: Vintage, 1974), 1062. The "conservation in the truest sense" quote comes from SUL, Robert Moses Papers, box 8: Remarks of Robert Moses at a Luncheon of the New York State Society of Newspaper Editors, Treadway Inn, Niagara Falls, June 20, 1960.

45 Caro summarizes this episode:

> It is impossible to avoid the conclusion that Moses had determined to hound from the state park organization a group of elderly men whose only crime was their refusal to allow him to exercise unbridled power in that organization and to remove them from control of the park they loved, the park that one of them had created, the park to which they had given so much of their lives.

Caro, *The Power Broker*, 253.

46 Hilary Ballon and Kenneth T. Jackson, eds., *Robert Moses and the Modern City* (New York: W.W. Norton, 2007); Phillip Lopate, "A Town Revived, a Villain Redeemed," *New York Times*, February 11, 2007, CY3.

47 Jane Jacobs was likely Moses's most prominent critic during the period when Moses was still active. Jane Jacobs, *The Death and Life of Great American Cities* (New York: Random House, 1961).

48 David Cort, "Robert Moses: King of Babylon," *The Nation*, March 31, 1956.

49 Petersen, "Public Power and Private Planning," 15–17.

50 Power Authority of the State of New York (hereafter PASNY), Meeting Minutes – December 19, 1956: Memorandum to Commissioner Moses, Niagara – Design and Supervision (from Chapin), December 12, 1956.

51 PASNY "completed the sale of bonds aggregating $720,000,000 for the Niagara project by the issuance, on March 23, 1961, of $100,000,000 of bonds at a favorable rate. Because of the advanced stage of construction and the early settlement of most of the construction claims, it was necessary for the Authority to issue 18-month notes in the amount of $15,000,000 at 2 percent interest to cover a temporary deficit in the Niagara Proceeds Account." PASNY, *1961 PASNY Annual Report,* February 19, 1962.

52 PASNY, Meeting Minutes – December 21, 1954: Memorandum to the Members of the Authority, Niagara Report (by Chairman), December 20, 1954.

53 For more detail, see William Leuchtenburg, "The Niagara Compromise," *Current History* 34 (May 1958): 270–74; Nass, *Public Policy and Public Works.*

54 SUL, Robert Moses Papers, box 4: Alexander M. Beebee, "Niagara and Private Enterprise," *New York Herald Tribune,* September 7, 1955.

55 SUL, Robert Moses Papers, box 4: Moses to the Editor of the *New York Herald Tribune, New York Herald Tribune,* September 13, 1955.

56 The controversial defeat of a dam in Hells Canyon in the Pacific Northwest revolved around the public-versus-private development of power. See Karl Boyd Brooks, *Public Power, Private Dams: The Hells Canyon High Dam Controversy* (Seattle: University of Washington Press, 2006).

57 As Sarah Elkind shows, the final arrangement for the Boulder/Hoover Dam decades earlier was also a public-private compromise, and thus could be seen as partially setting the template for public-private sharing of power from Niagara Falls. See Sarah Elkind, "Private Power at Hoover Dam," chap. 4 in *How Local Politics Shape Federal Policy: Business, Power and the Environment in Twentieth Century Los Angeles* (Chapel Hill: University of North Carolina Press, 2011).

58 LAC, RG 25, vol. 6781, file 1268-K-40, pt. 10.1, May 10, 1955, to September 17, 1957, Subject: St. Lawrence–Niagara River Treaty between Canada and the US, Additional Diversion of Water at Niagara: "Niagara versus the Dixiecrats?" *Public Utilities,* September 13, 1956.

59 The Eisenhower cabinet had previously decided that it was not opposed to New York State's development of Niagara power. Government of the United States, *Confidential Files of the Eisenhower White House, Minutes and Documents of the Cabinet Meetings of President Eisenhower (1953–1961),* Reel 1– Cabinet Meetings: "April 24, 1953, 0267" (LexisNexis microfilm, Abilene, KS: Eisenhower Library); Richardson, *Dams, Parks and Politics;* Well, "Public Power in the Eisenhower Administration."

60 For more on the Schoellkopf disaster, see Craig A. Woodworth, "The Schoellkopf Disaster Aftermath in the Niagara River Gorge," *IEEE Energy and Power Magazine* 10, 6 (November-December 2012): 80–96. For a perceptive analysis of another dam disaster, see Norris Hundley and Donald C. Jackson, *Heavy Ground: William Mulholland and the St. Francis Dam Disaster* (Oakland, CA: University of California Press, 2015).

61 Quoted in Don Glynn, "The Collapse of Schoellkopf," *Niagara Falls Gazette,* May 26, 2006, http://www.niagara-gazette.com/news/local_news/the-collapse-of-schoellkopf/article_2e1de7e4-d82e-5d78-b992-ba56d786aac6.html.

62 Ontario Hydro was at the time in the midst of a massive "frequency standardization" program to convert from 25-cycle to 60-cycle power. This program, which started in 1949 and continued for a decade, involved the refitting of hundreds of thousands of appliances and electrical equipment. According to Robert Stamp:

> The decision to build a provincial network on 25-cycle power dated back to 1893 and early developments on the Niagara River. Initially, 25-cycle had possessed certain

advantages: it travelled better over long-distance transmission lines and it proved ideal for such industries as mining and steel-making that used low-speed induction motors. Once transmission problems were solved, however, 60-cycle power soon demonstrated its superiority. It required smaller, cheaper, and more flexible transformers and other equipment. It was better suited for industries using high-speed motors. And for the homeowner, 60-cycle power promised to eliminate the flickering always associated with 25-cycle incandescent and fluorescent lighting. As most of North America embraced 60-cycle power, Ontario seemed out of step.

Stamp, *Bright Lights, Big City,* 49–50.

63 IJC, Docket 64, Niagara Falls Reference, box 113, 64-8-2:3, Application of PASNY to FPC 1956/08/20, License for a Power Project for Development of Niagara: Application of Power Authority of the State of New York to the Federal Power Commission – For a License under Section 4(e) of the Federal Power Act for a Power Project to be located in Niagara County, State of New York for Development of the Niagara River, August 20, 1956; SUL, Robert Moses Papers, box 5: Memorandum in Support of the Application of Power Authority of the State of New York, FPC Project No. 2216, November 14, 1956.

64 LAC, RG 25, vol. 6782, file 12680-K-40, pt. 10.2, Subject: St. Lawrence–Niagara River Treaty between Canada and the USA – Additional Diversion of Water at Niagara: FPC Release No. 8844, Project No. 2116, Opinion No. 98, FPC Rules, November 30, 1956.

65 PASNY, Meeting Minutes – December 19, 1956: Petition for Rehearing of Application of Power Authority of the State of New York to the Federal Power Commission – For a License under Section 4(e) of the Federal Power Act for a Power Project to be located in Niagara County, State of New York for Development of the Niagara River, December 4, 1956.

66 LAC, RG 25, vol. 6782, file 12680-K-40, pt. 10.2, Subject: St. Lawrence–Niagara River Treaty between Canada and the USA – Additional Diversion of Water at Niagara: FPC Release No. 8910, Power Authority of the State of New York Files Petition in D.C. Court of Appeals seeking review of FPC's dismissal of Niagara application, January 9, 1957.

67 LAC, RG 25, vol. 6782, file 12680-K-40, pt. 10.2, Subject: St. Lawrence–Niagara River Treaty between Canada and the USA – Additional Diversion of Water at Niagara: Project No. 2216, Release No. 9397, August 29, 1957.

68 From their intakes, the PASNY conduits could run almost due north, whereas the Ontario Hydro tunnels had to first go west, then north, then back east.

69 PASNY, Meeting Minutes, November 15, 1956.

70 LAC, RG 25, vol. 6782, file 12680-K-40, pt. 11, Subject: St. Lawrence–Niagara River Treaty between Canada and the USA – Additional Diversion of Water at Niagara: Moses to FPC, January 6, 1958.

71 PASNY, Meeting Minutes, March 1, 1957.

72 Chemical Waste Management continues to run the Lake Ontario Ordnance Works, which was the government's designation for the Model City discard complex. See Andrew Jenks, "Model City USA: The Environmental Costs of Victory in World War II and the Cold War," *Environmental History* 12 (July 2007): 552–77; Strand, "The Bomb and Tom Brokaw's Desk," chap. 8 in *Inventing Niagara.*

73 Strand, *Inventing Niagara,* 225.

74 Edmund Wilson, *Apologies to the Iroquois* (New York: Farrar, Straus, and Giroux, 1960), 143.

75 In the end, seventy-six houses were moved from the Alphabet Street area, which received that moniker because the streets were named after letters of the alphabet. PASNY, Meeting

Minutes – August 3, 1960: Niagara Power Project – Niagara Power Project, Disposal of surplus lots and houses in Veteran Heights Development, August 3, 1960.

76 LAC, RG 25, vol. 6782, file 12680-K-40, pt. 11, Subject: St. Lawrence–Niagara River Treaty between Canada and the USA – Additional Diversion of Water at Niagara: IJC Meeting, IJC Docket No. 62: Niagara Falls Reference, April 1957.

77 Aug, *Beyond the Falls,* 196.

78 PASNY, Meeting Minutes – August 9, 1957: House Moving and Relocation at Niagara, August 9, 1957.

79 In September 1961, for example, PASNY purchased land from Niagara Mohawk for $4.75 million (45 acres of land on top of the gorge, 74 acres below the top of the gorge, and 49 acres for the stockpile area). This was in addition to property previously purchased from Niagara Mohawk. PASNY, Meeting Minutes – September 18, 1961: Niagara Power Project, Acquisition of Remaining Niagara Mohawk Property, September 18, 1961.

80 Laurence M. Hauptman, *The Iroquois Struggle for Survival: World War II to Red Power* (Syracuse, NY: Syracuse University Press, 1986), 156; Sehdev, "Unsettling the Settler at Niagara Falls"; Kyle Powys Whyte, "Indigenous Experience, Environmental Justice and Settler Colonialism," in *Nature and Experience: Phenomenology and the Environment,* ed. B. Bannon (Lanham, MD: Rowman and Littlefield, 2016), 157–74.

81 On the Holland Purchase, see William Wyckoff, *The Developer's Frontier: The Making of the Western New York Landscape* (New Haven, CT: Yale University Press, 1988).

82 Hauptman, *The Iroquois Struggle for Survival,* 157. For an insightful analysis of another Mohawk group and the border, see Audra Simpson, *Mohawk Interruptus: Political Life across the Borders of Settler States* (Durham, NC: Duke University Press, 2014).

83 Anthony F.C. Wallace, *Tuscarora: A History* (Albany: SUNY Press, 2012), 133.

84 See Chapter 4 in Nick Estes, *Our History is the Future: Standing Rock versus the Dakota Access Pipeline, and the Long Tradition of Indigenous Resistance* (London: Verso, 2019).

85 Clinton Rickard, *Fighting Tuscarora: The Autobiography of Chief Clinton Rickard,* ed. Barbara Graymont (Syracuse, NY: Syracuse University Press, 1973), 139.

86 Rickard, *Fighting Tuscarora,* 142.

87 Moreover, PASNY would not give them the treaty land rights they desired if they swapped land. Rickard, *Fighting Tuscarora,* 148.

88 SUL, Robert Moses Papers, box 8: Robert Moses, "Tuscarora Fiction and Fact: A Reply to the author of Memoirs of Hecate County and to his Reviewers," 1960.

89 According to Raymond Petersen, "during the hearings on the engineering plans for the Niagara Frontier, the Town of Niagara and City of Niagara Falls were persuaded to withdraw as intervenors in the Federal Power Commission licensing process, leaving the Town of Lewiston and County of Niagara in the position where they appeared to be unreasonably obstructing the project. At the same time the Authority was making the case to the Niagara Frontier region of the potential economic benefit of the project and the costs of the 'obstructive tactics' of the remaining intervenors to the region." Petersen, "Public Power and Private Planning," 20–21.

90 Rickard, *Fighting Tuscarora,* 145.

91 SUL, Robert Moses Papers, box 7: PASNY – "Niagara Desperately Needs More Power," June 9, 1958.

92 Wilson, *Apologies to the Iroquois,* 149–50.

93 PASNY, FPC Opinion No. 217, Opinion and Finding, Project No. 2217, Power Authority of the State of New York, February 2, 1959.
94 PASNY, Meeting Minutes – February 5, 1959: Niagara Power Project – Tuscarora Reservoir, February 5, 1959.
95 SUL, Robert Moses Papers, box 8: Remarks of Robert Moses at a luncheon of the New York State Society of Newspaper Editors, Treadway Inn, Niagara Falls, June 20, 1960.
96 Wallace, *Tuscarora,* 135.
97 The appendices of a PASNY memorandum list the eleven families and gives a breakdown for each, as well as totals: $188,622 for Residence, Including Foundation; $28,573 for Utilities: Water Supply, Sewerage, Electric Service; $29,315 for Access: Roads and Driveways; $25,700 for Cash Advance; and $97,881 for Other: Outbuildings, Grading, Garage, Moving personal items, etc. PASNY, Meeting Minutes, June 20, 1960: Memorandum to Col. William S. Chapin, Relocation of Families – Reservoir Area, June 17, 1960.
98 Wallace, *Tuscarora,* 128.
99 Gawronski et al., *The Power Trail,* 82.
100 Ken Cosentino, "Water Crisis Remains on Tuscarora Reservation," *Niagara Gazette,* October 26, 2019, https://www.niagara-gazette.com/opinion/guest-view-water-crisis-remains-on-tuscarora-reservation/article_45224217-e827-5005-a570-7d83c6c3dada.html.
101 SUL, Robert Moses Papers, box 8: Robert Moses, "Tuscarora Fiction and Fact: A Reply to the author of Memoirs of Hecate County and to his Reviewers," 1960.
102 PASNY released for public consumption a glossy booklet titled "Niagara Power and Local Taxes." SUL, Robert Moses Papers, box 6: PASNY, "Niagara Power and Local Taxes," December 2, 1957.
103 PASNY, Meeting Minutes – February 4, 1957: Niagara Taxes, February 4, 1957.
104 Merritt-Chapman and Scott received a $170.4 million contract for the Moses generating plant and a $66.2 million contract for river intakes and a section of the covered conduits.
105 The total number of workers is taken from Brynner, who also reports that the influx swelled the population of Niagara Falls, New York, to a bit over 100,000 in the 1960 census. Brynner, *Natural Power,* 118.
106 Glennon, *Hard Hats of Niagara.*
107 Ibid., 22.
108 Ibid., 77.
109 PASNY also installed dikes at the downstream tip of Buckhorn Island. Along with deepening the nearshore area, the new shoreline was also intended to constrict the flow of the river.
110 The helicopter pad was removed in the 1990s.
111 Willow Island had been artificially created through construction of an eighteenth-century canal. The Adams plant's impressive portico/door arch was moved to Goat Island, where it now stands.
112 Much of this divided highway was reduced to one lane or removed, and it is no longer named after Moses. In 2018, funding was announced to remove another section of the parkway starting in downtown Niagara Falls and extending almost two miles north.
113 The 282-foot-tall Prospect Point Observation Tower cost $1,268,474 and offers a 14,000-square-foot observation deck protruding about 200 feet above the river.

114 To avoid rock movement problems, the covered conduits that would feed the storage reservoir were built with hinged arches and compressible filler material to provide flexibility. For more detail on the tunnels, see LAC, RG 25, vol. 6782, file 12680-K-2-40, pt. 1, Subject: St. Lawrence–Niagara River Treaty between Canada and the USA – Additional Diversion of Water at Niagara – Clippings: "Niagara Project: Big Job, Big Problems," *Engineering News-Record,* February 26, 1959.

115 PASNY, Meeting Minutes – September 4, 1958: Niagara Project – Relocation of Facilities of Industries, September 4, 1958.

116 Jennifer Read, "Origin of the *Great Lakes Water Quality Agreements:* Concepts and Structures," in *The First Century of the International Joint Commission,* ed. Daniel Macfarlane and Murray Clamen, 347–66; Terence Kehoe, *Cleaning Up the Great Lakes: From Cooperation to Confrontation* (Dekalb: Northern Illinois University Press, 1997); John Hartig, *Burning Rivers: Revival of Four Urban-Industrial Rivers That Caught Fire* (London: Multi-Science, 2010); William McGucken, *Lake Erie Rehabilitated: Controlling Cultural Eutrophication, 1960s–1990s* (Akron, OH: University of Akron Press, 2000).

117 NARA II, RG 59, Decimal File, 1950–54, files 611.42322-N/3-153 to 611.42322-N/12-2353: "Brief History – Water Supply Plant of Niagara Falls, NY," March 15, 1954.

118 IJC, Docket 64, Niagara Falls Reference, box 107, 64-4-5:2, Interim Report of the Engineering Committee of 1952/06/02 Cascade Verification: Memorandum: Summary of "Industrial Wastes Along the Niagara Frontier and their Effect on the International Boundary Waters, by Hayse H. Black" (by C.K. Hurst), October 22, 1954. For a recent and sophisticated study of toxins in the Great Lakes basin, see Nancy Langston, *Sustaining Lake Superior: An Extraordinary Lake in a Changing World* (New Haven, CT: Yale University Press, 2017).

119 For an overview of the state of pollution, and on the contributions of various companies, see NARA II, RG 59, Decimal File, 1945–49, box 3306, files 711.4216 N 1/1-145 to 7.114216 N 1/12-3149: "Industrial Waste Summary Reports: International Joint Commission Boundary Waters Pollution Investigation, Lake Erie–Lake Ontario Section," 1949.

120 This was reportedly the largest bond issue ever up to that point. According to Caro, the success of the bond issue "had been assured by a syndicate of the biggest investment bankers on Wall Street." Caro, *The Power Broker,* 1023.

121 Niagara contracts were signed with Niagara Mohawk Power Corporation; New York State Electric and Gas Corporation; Rochester Gas and Electric Corporation; two electric cooperatives; thirteen villages; the City of Salamanca, NY; and the Public Service Board of Vermont.

122 PASNY, *1962 PASNY Annual Report,* February 1963.

123 PASNY had hired Tuscarora Contractors, a joint construction venture made up of multiple builders, to construct the power plant for just a shade under $40 million. This contractor conglomerate submitted a claim for extra work caused by delays and interferences beyond its control, and PASNY ended up giving a lump sum payment of $2,793,000 in the ensuing settlement. PASNY, Meeting Minutes – February 9, 1961: Niagara Power Project – Increased Costs for Contract N-10 – Construction of Tuscarora Power Plant, February 9, 1961.

124 PASNY, *1961 PASNY Annual Report,* February 1962.

125 PASNY, *1962 PASNY Annual Report,* February 1963.

126 "Niagara Power," *New York Times,* February 10, 1961, 26.

127 The 1,231 parcels of land involved 645 ownerships, not including 222 parcels taken in streets and roads. The number remaining to be settled at the start of 1963 was 181 parcels in 111 ownerships. PASNY, *1962 PASNY Annual Report,* February 1963.

128 Many of the pending cases involved legal title questions to riverfront land because the owners had filled in the river and in effect extended their property into the river.

129 Thirty-four condemnation cases had gone to the Supreme Court; nineteen had been settled, eight had been tried and decided, and seven were still pending. PASNY, *1962 PASNY Annual Report,* February 1963.

130 Since many did not purchase back their houses, PASNY was left with ownership of thirty-one of the houses moved to Veteran Heights, as well as thirty-six vacant lots on this subdivision, and houses it had rented to engineers that were now becoming vacant. Because of the depressed condition of the real estate market, PASNY was forced to sell much of this property at a loss. PASNY, Meeting Minutes – August 3, 1960 – PASNY Minutes: Niagara Power Project, Disposal of surplus lots and houses in Veteran Heights Development: August 3, 1960; PASNY, Meeting Minutes – February 9, 1961: Niagara Power Project, Sale of Houses and Lots Owned by Authority in Veteran Heights Subdivision, February 9, 1961.

131 According to the *1961 PASNY Annual Report* (February 1962):

> At Niagara, power produced at 13.8 kV, is controlled from a switchyard composed of 115 kV, 230 kV and 345 kV sections. Miles of complex circuits and an intricate control system regulate deliveries of power over transmission lines. High-voltage lines connect the Niagara project with the Authority's St. Lawrence project. From Niagara two 345 kV circuits extend to a substation south of Rochester, with one 345 kV circuit continuing east from Rochester to a substation north of Syracuse, thence to the Edic substation north of Utica. There the line connects to a 230 kV facility of Niagara Mohawk which in turn is interconnected with the Authority's Adirondack substation 85 miles south of the St. Lawrence project. From the Adirondack substation, two 230 kV lines owned by the Authority complete the interconnection to the St. Lawrence project at Massena.

132 James Hull, "Watts across the Border: Technology and the Integration of the North American Economy in the Second Industrial Revolution," *Left History* 19, 2 (Fall/Winter 2015/16): 13–31.

133 Martin-Nielsen, "South over the Wires," 126–27.

134 This front door/back door metaphor, and the reasons enumerated in the following sentence, are drawn from McGreevy's *Wall of Mirrors* and *Imagining Niagara.*

135 The term "patriotic topographies" comes from Brian Osborne, "Landscapes, Memory, Monuments and Commemoration: Putting Identity in Its Place," *Canadian Ethnic Studies* 33, 3 (2002): 12.

136 The concept of hydro tourism and this section is primarily derived from Daniel Macfarlane, "Fluid Meanings: Hydro Tourism and the St. Lawrence and Niagara Megaprojects," *Histoire Sociale/Social History* 49, 99 (June 2016): 327–46. See also Donald C. Jackson, *Pastoral and Monumental: Dams, Postcards and the American Landscape* (Pittsburgh: University of Pittsburgh Press, 2013).

137 On the technological sublime, see Nye, *American Technological Sublime.*

138 Caro, *The Power Broker,* 825–26.

139 Kennedy's entire speech can be accessed at "Remarks to the Niagara Power Commission, 4 February 1961," John F. Kennedy Presidential Library and Museum, https://www.jfklibrary.org/Asset-Viewer/Archives/JFKWHA-008.aspx.
140 The Power Vista underwent a $2.3 million renovation in 2001, and more upgrades since then. As of the end of 2010, it had received 6,752,727 visitors, according to a New York Power Authority handout.
141 PASNY, *1962 PASNY Annual Report,* February 1963.
142 Dubinsky, *The Second Greatest Disappointment,* 10.
143 NPC, *Seventy-Third Annual Report of the Commissioners of the Queen Victoria Niagara Falls Park* (Toronto: L.K. Cameron, 1959).
144 To provide just one example, Merrill Denison notes that representatives of the Snowy Mountains Hydro-Electric Authority of Australia came to inspect Ontario Hydro installations and were so impressed that they invited Dr. Otto Holden to examine and report on their plans for a hydroelectric project in southeastern Australia. Denison, *The People's Power,* 252. A number of studies look at the way that hydraulic engineering approaches and technology have been exported or used as part of foreign policy and imperialism, particularly by the United States: Jessica B. Teisch, *Engineering Nature: Water, Development and the Global Spread of American Environmental Expertise* (Chapel Hill: University of North Carolina Press, 2011); Richard P. Tucker, "Containing Communism by Impounding Rivers: American Strategic Interests and the Global Spread of High Dams in the Early Cold War," in *Environmental Histories of the Cold War,* ed. J.R. McNeill and Corinna R. Unger (Cambridge: Cambridge University Press, 2010), 139–64; Nicholas Cullather, "Damming Afghanistan: Modernization in a Buffer State," *Journal of American History* 89 (September 2002): 512–37; David Ekbladh, *The Great American Mission: Modernization and the Construction of an American World Order* (Princeton, NJ: Princeton University Press, 2010); Linda Nash, "Traveling Technology? American Water Engineers in the Columbia Basin and the Helmand Valley," in *Where Minds and Matters Meet: Technology in California and the West,* ed. Volker Janssen (Oakland: University of California Press, 2012), 135–48; Christopher Sneddon, *Concrete Revolution: Large Dams, Cold War Geopolitics, and the US Bureau of Reclamation* (Chicago: University of Chicago Press, 2015).
145 This point about dissemination of engineering expertise is also made in Tina Loo, with Meg Stanley, "An Environmental History of Progress: Damming the Peace and Columbia Rivers," *Canadian Historical Review* 92, 3 (September 2011): 419–21.
146 Tucker, "Containing Communism by Impounding Rivers," 139.
147 The Rockefeller-Moses story is recounted in Caro, *The Power Broker,* 1077–79.
148 Nelles, *The Politics of Development,* 425, 465. See also Paul McKay, *Electric Empire: The Inside Story of Ontario Hydro* (Toronto: Between the Lines, 1983); Jamie Swift and Keith Stewart, *Hydro: The Decline and Fall of Ontario's Electric Empire* (Toronto: Between the Lines, 2004).
149 On fish in the Niagara River, see A.R. Yagi and C. Blott, "Niagara River Watershed Fish Community Assessment (1997 to 2011)," (Ontario Ministry of Natural Resources unpublished report, 2012); David A. Ingleman, Stephen Cox Thomas, and Douglas J. Perrelli, "The Pre-Contact Upper Niagara River Fishery: Shadows of a Changed Environment," *Ontario Archaeology* 92 (2012): 38–73. Lisa Aug suggests that in the late 1970s "government-stocked foreign salmon and trout made the lower Niagara the envy of anglers throughout North America." Aug, *Beyond the Falls,* 15. H.L. Selks, S.J. Walsh, N.M. Burkhead, et al.,

"Conservation Status of Imperiled North American Freshwater and Diadromous Fishes," *Fisheries* 33, 8 (2008): 372–406.

150 Aug, *Beyond the Falls*, 23.

151 Eckel, *Ecological Restoration and Power at Niagara Falls*.

Chapter 5: Disguising Niagara

1 Martin Reuss, "The Art of Scientific Precision: River Research in the United States Army Corps of Engineers to 1945," *Technology and Culture* 40, 2 (April 1999): 292. On the US Army Corps of Engineers, see also Arthur Morgan, *Dams and Other Disasters: A History of the US Army Corps of Engineers in Civil Works* (Boston: Porter Sargent, 1971); Daniel McCool, *River Republic: The Fall and Rise of America's Rivers* (New York: Columbia University Press, 2012).

2 For a more detailed analysis of the Niagara models, see Macfarlane, "Nature Empowered: Hydraulic Models and the Engineering of Niagara Falls." *Technology and Culture* 61, 1 (January 2020): 109–43.

3 IJC, Docket 64, Niagara Falls Reference, box 107, 64-4-2:2, Specifications for Preservation and Enhancement of Niagara Falls, US Flank of Horseshoe Falls, US Corps of Engineers: Preservation and Enhancement of Niagara Falls, International Niagara Falls Engineering Board, March 1, 1953.

4 Lee F. Pendergrass and Bonnie B. Pendergrass, *Mimicking Waterways, Harbors, and Estuaries: A Scholarly History of the Corps of Engineers Hydraulics Laboratory at WES, 1929 to the Present* (Vicksburg, MS: Waterways Experiment Station, US Army Corps of Engineers, 1989). See also Reuss, "The Art of Scientific Precision"; Christine Keiner, "Modeling Neptune's Garden: The Chesapeake Bay Hydraulic Model, 1965–1984," in *The Machine in Neptune's Garden: Historical Studies on Technology and the Marine Environment*, ed. David van Keuren and Helen Rozwadowski (Sagamore Beach, MA: Science History Publications, 2004), 273–314; Ben Fatherree, *The First 75 Years: History of Hydraulic Engineering at the Waterways Experiment Station* (Washington, DC: US Army Corps of Engineers, Engineer Research and Development Center, 2004); Herbert D. Vogel, "Practical River Laboratory Hydraulics," *Transactions of the American Society of Civil Engineers* 100 (1935): 118–84; Hunter Rouse and Simon Ince, *History of Hydraulics* (Iowa City: Iowa Institute of Hydraulic Research, 1957); Hunter Rouse, *Hydraulics in the United States, 1776–1976* (Iowa City: Iowa Institute of Hydraulic Research, 1976).

5 Pendergrass and Pendergrass, *Mimicking Waterways, Harbors, and Estuaries*, 51–52.

6 IJC, Docket 64, Niagara Falls Reference, box 107, 64-6-2:3, Ontario Hydro NEWS and Pamphlets Re: Niagara: Niagara River Remedial Works, November 20, 1952.

7 Ontario Hydro, *Power from Niagara*.

8 Pendergrass and Pendergrass, *Mimicking Waterways, Harbors, and Estuaries*.

9 HEPCO, Press Release: "Niagara Falls" Now at Islington, February 1, 1951.

10 The model required about 550 tons of sand, 9,000 board feet of plywood, and 25 cubic yards of concrete for the surface. One cubic foot of water in the model was equivalent to 167,000 cubic feet in the river, and 1 minute of flow in the model was equivalent to 46 minutes in the river. HEPCO, Press Release: "Niagara Falls" Now at Islington, February 1, 1951; "Niagara Power Goes Underground"; IJC, Docket 64, Niagara Falls Reference, box 107, 64-4-2:2, Specifications for Preservation and Enhancement of Niagara Falls, US Flank

of Horseshoe Falls, US Corps of Engineers: Preservation and Enhancement of Niagara Falls, International Niagara Falls Engineering Board, March 1, 1953.

11 IJC, Docket 64, Niagara Falls Reference, box 107, 64-4-2:2, Specifications for Preservation and Enhancement of Niagara Falls, US Flank of Horseshoe Falls, US Corps of Engineers: Preservation and Enhancement of Niagara Falls, International Niagara Falls Engineering Board, March 1, 1953.

12 Ibid.

13 IJC, Docket 64, Niagara Falls Reference, box 107, 64-4-4:1C, Copies of First Progress Report: International Niagara Falls Engineering Board, First Progress Report, March 31, 1951.

14 IJC, Docket 64, "Report to the Governments of the United States of America and Canada on Remedial Works Necessary to Preserve and Enhance the Scenic Beauty of the Niagara Falls and River," May 1953.

15 HEPCO, Memorandum for Files: Meeting of the IJC and INFEB at Waterways Experiment Station, Vicksburg, Mississippi (8 October 1951), October 19, 1951.

16 HEPCO, Holden to Schull, November 14, 1951.

17 HEPCO, Niagara River: Summary of the Events Leading to the Production of a New Bed Contour Map and the New Water Surface Contour Map of the Cascade Area, February 5, 1952.

18 This consisted of shining a searchlight beam across the water and by transit determining the water levels at various points along the beam. Knowing the position of the light and the direction of the beam and knowing the position and elevation of the instruments, the location of any point could be computed from the horizontal angle, while the level could be calculated from the observed vertical angle. HEPCO, Niagara River: Summary of the Events Leading to the Production of a New Bed Contour Map and the New Water Surface Contour Map of the Cascade Area, February 5, 1952.

19 IJC, Docket 64, Niagara Falls Reference, box 107, 64-4-4:4, Fourth Progress Report: United States and Canadian Model Studies of the Niagara River and Falls – Cascades Verification, Interim Report of the Working Committee of the International Niagara Falls Engineering Board, Prepared by the Waterways Experiment Station, Vicksburg, Miss., USA, and The Hydro-Electric Commission of Ontario, Toronto, Ontario, Canada, June 2, 1952.

20 On standards, see Hull, "Raising Standards"; Maurits Ertsen, *Improvised Planning: Development on the Gezira Plain, Sudan, 1900–1980* (New York: Palgrave Macmillan, 2016).

21 HEPCO, Memorandum for Files: Meeting of the IJC and INFEB at Waterways Experiment Station, Vicksburg, Mississippi (8 October 1951), October 19, 1951.

22 HEPCO, Memo by Holden Re: Niagara River Remedial Works, October 7, 1952.

23 HEPCO, Holden to Marr, May 6, 1952.

24 For another intriguing example of how plant/weed growth was factored into the infrastructural remaking and maintenance of another iconic waterway and "demanding environment," see "Weeds," chap. 12 in Carse, *Beyond the Big Ditch.*

25 See Macfarlane, "'As Nearly as May Be.'"

26 HEPCO, Niagara Remedial Works, Minutes of a Meeting of Working Committee Held at Buffalo on January 31, 1952, February 7, 1952.

27 IJC, Docket 64, Niagara Falls Reference, box 107, 64-4-4:4, Fourth Progress Report: International Niagara Falls Engineering Board, October 1, 1952. It is worth noting that there are several other waterfalls in the world that are "switched on" at specific hours for tourism purposes: Cascata della Marmore near Terni, Italy; Maria Cristina Falls in the

Philippines, and Sweden's Trollhätten Falls. Brian J. Hudson, *Waterfall: Nature and Culture* (London: Reaktion Books, 2012), 200.

28 On hydraulic engineers and cost-benefit analysis see works such as Chapter 7: "US Army Engineers and the Rise of Cost-Benefit Analysis," in Theodore M. Porter, *Trust in Numbers: The Pursuit of Objectivity in Science and Public Life* (Princeton, NJ: Princeton University Press, 1995); Martin Reuss, "Coping With Uncertainty: Social Scientists, Engineers, and Federal Water Resources Planning," *Natural Resources Journal* 32 (1992): 101–35.

29 IJC, Docket 64, Niagara Falls Reference, box 107, 64-4-2:2, Preservation and Enhancement of Niagara Falls: International Niagara Falls Engineering Board, March 1, 1953; IJC, Docket 64, Niagara Falls Reference, box 105, 64-2-5-9: IJC Session – 1952/11/20 Islington, Ont. (with INEB and Committee): Special Meeting of International Joint Commission at Islington, Ontario: Measures to Conserve and Enhance the Scenic Beauties of Niagara, November 20, 1952.

30 IJC, Docket 64, "Report to the Governments of the United States of America and Canada on Remedial Works Necessary to Preserve and Enhance the Scenic Beauty of the Niagara Falls and River," May 1953.

31 Dennis Dack (former speechwriter to Ontario Hydro chairman Robert Saunders), interview by Daniel Macfarlane, Toronto, May 2, 2011.

32 HEPCO, Minutes of the Meeting of the Working Committee of the International Niagara Board of Control at Toronto, Ontario, December 17, 1953; HEPCO, "Tunnels for Kilowatts" (A Proposed Script for the Motion Picture of the Construction of the Twin Hydraulic Tunnels at Sir Adam Beck–Niagara Generating Station No. 2).

33 IJC, Docket 64, Niagara Falls Reference, box 105, 64-2-5-20 (1): Extract from Minutes 1954/04/06 and Executive Session: IJC Semi-Annual Meeting, Niagara River Remedial Works, April 6, 1954.

34 One effective way of saving time and money was to reuse cofferdam material. Approximately 4,500 cubic yards of earth was excavated to form two pockets in the riverbank in which to start the ends of the first stage of the cofferdam. Rock-filled timber cribs, 30 feet wide, were built, 95 feet long for the upstream leg and 65 feet for the downstream leg, with a total volume of approximately 3,000 cubic yards. IJC, Docket 64, Niagara Falls Reference, box 110, 64-7-3:3 (3), INBC – Final Report of Construction of Niagara River Remedial Works – 1960/09/30: Construction of Niagara River Remedial Works, Final Report by the International Niagara Board of Control, September 30, 1960.

35 On the US Army Corps of Engineers and its general history of cofferdam construction, see Patrick O'Bannon, *Working in the Dry: Cofferdams, In-River Construction, and the United States Army Corps of Engineers* (Washington, DC: US Army Corps of Engineers, 2009).

36 HEPCO, Letter from Saunders to Pearson, April 24, 1953; IJC, Docket 64, Niagara Falls Reference, box 114, 64-2-5:8 to 64:7-1:1, Niagara Falls Reference: Memorandum of Meeting between McNaughton and INFEB, April 28, 1953.

37 LAC, RG 19, file 9810-09, vol. 2: Water – International Joint Commission (IJC) Niagara Falls and River: DEA Numbered Letter, Niagara Remedial Works, July 24, 1953.

38 HEPCO, July 27 meeting on Remedial Works to Preserve and Enhance the Scenic Beauty of Niagara Falls, July 29, 1953; IJC, Docket 64, Niagara Falls Reference, box 107, 64-7-2:1, Board of Control – Directive 1953/08/19: Niagara Reference, IJC, Directive Constituting International Niagara Board of Control, August 19, 1953.

39 LAC, RG 25, vol. 6351, 1268-K-40, pt. 8.2, St. Lawrence–Niagara River Treaty between Canada and the United States Additional Diversion of Water at Niagara Falls (July 29,

1953 to January 28, 1954): teletype from Cdn Embassy (LePan) to Under-Sec of State, January 15, 1954.

40 NARA II, RG 59, Decimal File, 1950–54, box 2805, file 611.42322-N/7-2853: Memorandum of Conversation: Diversion of water from Niagara River for power production prior to completion of remedial works, July 28, 1953. McNaughton had been a chief proponent of the all-Canadian St. Lawrence and Columbia approaches. Indeed, McNaughton may be the most underappreciated Canadian foreign policy actor in the post-1945 period, for, in addition to his IJC role, he had positions at the United Nations and on the Permanent Joint Board on Defence.

41 IJC, Docket 64, Niagara Falls Reference, box 110, 64-7-3:3 (3), INBC – Final Report of Construction of Niagara River Remedial Works – 1960/09/30: Construction of Niagara River Remedial Works, Final Report by the International Niagara Board of Control, September 30, 1960.

42 HEPCO, Minutes of Meeting of the International Niagara Board of Control held in Niagara Falls, New York on December 13, 1954.

43 These include piers at Bird Island, piers for the construction of the International Railway and Peace Bridges, channel filling near Squaw Island and Fort Erie, and shoreline fills by private property owners along the upper river. It is estimated that these various channel modifications have led to an 0.4-foot increase in the level of Lake Erie. Frank H. Quinn, "Anthropogenic Changes to Great Lakes Water Levels," *Great Lakes Update* 136 (1999): 1–4.

44 HEPCO, Beck No. 2, Remedial Works: Proposed Observation Area Retaining Wall, Geological Report, May 16, 1955.

45 IJC, Docket 64, Niagara Falls Reference, box 106, 64-3-2: Correspondence 1955: Sutherland to Spence, July 5, 1955.

46 The cost was $12.5 million, $5 million less than initially projected. However, if the extension to the control dam added in the 1960s is factored into the cost, then the project went over budget by a bit.

47 Smaller ceremonies had been held on June 2, 1954, to mark the start of work on the control structure, as well as at the flanks, and another was held when the Terrapin Point fill was turned over to the New York State Division of Parks.

48 The plaque was put on the gate post at the entrance to the parking lot of the control structure and it merely read: "These control works were built and are operated under the direction of the International Joint Commission of the United States and Canada for the preservation and enhancement of the beauty of Niagara Falls pursuant to the Treaty of 1950."

49 IJC, Docket 64, Niagara Falls Reference, box 107, 64-6-3:1, Addresses by General McNaughton: Dedication of Niagara Remedial Works, September 28, 1957.

50 LAC, RG 25, vol. 6780, file 1268-D-40, pt. 52, St. Lawrence Project: General File, 18 September 1957 to 21 April 1958, 28, Address by Alvin Hamilton (Minister of Natural Resources) at Dedication Ceremony, Niagara Falls, 28 September 1957.

51 IJC, Docket 64, Niagara Falls Reference, box 110, 64-7-3:3 (3), INBC – Final Report of Construction of Niagara River Remedial Works – 1960/09/30: Construction of Niagara River Remedial Works, Final Report by the International Niagara Board of Control, September 30, 1960.

52 Ibid.

53 HEPCO, Notes on Meeting with PASNY to Discuss Extension to Niagara River Remedial Works, February 2, 1961.

54 IJC, Docket 74, Niagara River Reference, box 114, 74-1-1: 1 Reference Dated 1961/05/05: Brief to the Governments of Canada and the United States of America on Proposed Extension to Niagara River Remedial Works and on Certain Proposed Operational Procedures by PASNY and HEPCO, March 15, 1961.
55 LAC, RG 25 vol. 6781, file 1268-K-40, 11.2: St. Lawrence–Niagara River Treaty between Canada and the US, Additional Diversion of Water at Niagara, Memorandum: Niagara Reference: Construction of Remedial Works, March 20, 1961.
56 LAC, RG 25, vol. 5027, file 1268-K-40, no. 12, St. Lawrence–Niagara River Treaty Between Canada and the USA – Additional Diversion of Water at Niagara, June 1, 1961 to December 18, 1961: *Niagara Falls Review,* "Remedial Works a Detriment," August 11, 1961.
57 Niagara Falls Public Library (hereafter NFPL) (Ontario), File: Niagara River and Falls – Diversion and Cessation, Adult LHC, vertical file 38080003419098: "Tide Out ... Tide In," August 16, 1962.
58 NFPL (Ontario), File: Niagara River and Falls – Diversion and Cessation, Adult LHC, vertical file 38080003419098: "Whirlpool Goes Backwards from Low Water Level," November 16, 1961; NFPL (Ontario), File: Niagara River and Falls – Diversion and Cessation, Adult LHC, vertical file 38080003419098: River So Low Even Birds and Fish Complaining, November 21, 1961.
59 HEPCO, Letter to Boucher and Patterson, December 9, 1963.
60 If the tour boats are not taken out by the time the winter diversion regime goes into effect, they "will be left high and dry at dockside." Kiwanis Club of Stamford, Ontario, *Niagara Falls, Canada,* 289.
61 PASNY, Meeting Minutes – February 19, 1962: Niagara Power Project, Withdrawal of Request to International Joint Commission for Additional Summertime Water Diversion, February 19, 1962; IJC, Docket 74, Niagara River Reference, box 114, 74-1-1:1, Reference Dated 1961/05/05: Memorandum for File – Niagara Reference, March 1, 1962.
62 PASNY, *1963 PASNY Annual Report.*
63 PASNY also took other steps, such as sinking a plug (a barge) each winter in the diversion dike of Buckhorn Island State Park since the resulting increase in water velocity better facilitated ice movement over the Falls. IJC, Dockets 64, 74, and 75, International Joint Commission Semi-Annual Meeting, Niagara Reference (1950), October 3, 1963.
64 Robert Kohler, *Landscapes and Labscapes: Exploring the Lab-Field Border in Biology* (Chicago: University of Chicago Press, 2002).
65 Caroline Desbiens, *Power from the North: Territory, Identity, and the Culture of Hydroelectricity in Quebec* (Vancouver: UBC Press, 2013); David Massell, *Quebec Hydropolitics: The Peribonka Concessions of the Second World War* (Montreal and Kingston: McGill-Queen's University Press, 2011); Giacomo Parrinello, "Systems of Power: A Spatial Envirotechnical Approach to Water Power and Industrialization in the Po Valley of Italy, 1880–1970," *Technology and Culture* 59, 3 (July 2018): 652–88; Emma Norman, *Governing Transboundary Waters: Canada, the United States, and Indigenous Communities* (New York: Routledge, 2014).
66 IJC, Docket 64, vol. 6781, file 1268-K-40, no. 10, St. Lawrence–Niagara River Treaty between Canada and the US, Additional Diversion of Water at Niagara, May 10, 1955, to September 17, 1957: Niagara Falls Reference: IJC Meeting, October 7, 1959.
67 IJC, Docket 64, Niagara Falls Reference, box 106, 64-2-5-27: International Joint Commission Semi-Annual Meeting, Niagara Reference (1950), October 4–9, 1959. For a brief biographical sketch of Holden, see "Dr. Otter Holden: A Look Back at a Hydro

Pioneer," Ontario Power Generation, accessed April 30, 2018, https://www.opg.com/news-and-media/our- stories/Documents/20170217_Otto_Holden.pdf.

68 HEPCO, Minutes of Meeting of the International Niagara Board of Control held in Niagara Falls, New York, December 13, 1954.

69 HEPCO, Minutes of the Meeting of the Working Committee of the International Niagara Board of Control at Toronto, Ontario, December 17, 1953.

70 NPFL (Ontario), File: Niagara River and Falls – Diversion and Cessation, Adult LHC, vertical file 38080003419098: Niagara Scheme Opposed, October 19, 1961.

71 Ralph Greenhill and Thomas Mahoney note that the noise of the Falls was also a subject of great controversy. Greenhill and Mahoney, *Niagara.* See also Virginia Georgallas, "'The Sublimest Music on Earth': An Aural History of Niagara Falls in the Mid-Nineteenth Century," MusCan (Canadian University Music Society), August 15, 2017, https://www.muscan.org/en/about-us/document-downloads-2/papers/2017-georgallas; Peter Coates, "The Strange Stillness of the Past: Toward an Environmental History of Sound and Noise," *Environmental History* 10 (2005): 636–65; Christopher Caskey, "Listening to a River: How Sound Emerges in River Histories," *Open Rivers: Rethinking Water, Place and Community* 8 (Fall 2017): 146–54; Parr, *Sensing Changes.* There is some speculation that waterfalls such as Niagara produce "negative ions" that produce a sense of well-being or romance. Hudson, *Waterfall,* 75.

72 On uncertainty within hydraulic modelling today, see E. de Rocquigny, Azzedine Soulaimani, and Brian Morse, "Uncertainty in Hydrological and Hydraulic Modeling: Editorial Introduction/Les incertitudes en modelisation hydrologique et hydraulique: Introduction editorial," *Canadian Journal of Civil Engineering* 37, 7 (July 2010): iii. See also Walter G. Vincenti, *What Engineers Know and How They Know It: Analytical Studies from Aeronautical History* (Baltimore: Johns Hopkins University Press, 1990).

73 Pritchard and Zeller, "The Nature of Industrialization."

74 On the materialism of waterways as infrastructure also see works such as Carse, *Beyond the Big Ditch;* Matthew Gandy, *The Fabric of Space: Water, Modernity, and the Urban Imagination* (Cambridge, MA: The MIT Press, 2014); Erik Swyngedouw, *Liquid Power: Contested Hydro-Modernities in Twentieth-Century Spain, 1898–2010* (Cambridge, MA: The MIT Press, 2015).

75 On the blackout, see David Nye, *When the Lights Went Out: A History of Blackouts in America* (Cambridge, MA: MIT Press, 2010).

76 Zeller, "Aiming for Control, Haunted by Its Failure"; Jones-Imhotep, *The Unreliable Nation.*

77 In their article "An Environmental History of Progress" Loo and Stanley point out the engineers' intimate engagement with the Columbia River; White, *The Organic Machine.*

78 James C. Scott, *Seeing Like a State: How Certain Schemes to Improve the Human Condition Have Failed* (New Haven, CT: Yale University Press, 1998), 89–90.

79 Scott lists three elements of what he calls "the most tragic episodes of state development in the late nineteenth and twentieth centuries": in addition to the actual high modernist ideology, he points to the unrestrained use of the immense powers wielded by modern states, and an incapacitated or prostrate civil society. Scott does suggest that high modernism took varied forms, or had staggered developments, across eras and locations. To illustrate, he ascribes a "softer" form of authoritarian high modernism to Tanzanian villagization in the 1970s, while labelling the Brazilian creation of a new capital city from scratch as "ultra-high modernism." Scott identifies three elements: improvement, bureau-

cratic management, and aesthetic dimension. The Tennessee Valley Authority, for Scott, is not fully high modernist. Scott, *Seeing Like a State;* James C. Scott, "High Modernist Social Engineering: The Case of the Tennessee Valley Authority," in *Experiencing the State,* ed. Lloyd Rudolph and John Kurt Jacobsen (New York: Oxford University Press, 2006). See also Edward Tenner, *Why Things Bite Back: Technology and the Revenge of Unintended Consequences* (New York: Alfred A. Knopf, 1996); Paul Josephson, *Industrialized Nature: Brute Force Technology and the Transformation of the Natural World* (Washington, DC: Island Press/Shearwater, 2002); Timothy Mitchell, *Rule of Experts: Egypt, Techno-Politics, Modernity* (Oakland: University of California Press, 2002).

80 For a more elaborate discussion of "negotiated high modernism," see Daniel Macfarlane, "Negotiated High Modernism: Canada and the St. Lawrence Seaway and Power Project," in *Science, Technology, and the Modern in Canada: An Anthology in Honour of Richard Jarrell,* ed. Edward Jones-Imhotep and Tina Adcock (Vancouver: UBC Press, 2018), 326–47.

81 This point is taken from David Pietz's book on the Yellow River. Place-based local knowledge and labour ("walking on two legs," as Pietz labels it) were substituted for technological expertise (though this was partially the result of a lack of expertise, capital, and equipment) in northern Chinese water projects, which led to a number of problems and ecological disturbances, with more than 2,000 dams built during the Mao period eventually failing. David Pietz, *The Yellow River: The Problem of Water in Modern China* (Cambridge, MA: Harvard University Press, 2015), 256–57.

82 Tina Loo, "High Modernism, Conflict, and the Nature of Change in Canada: A Look at *Seeing Like a State,*" *Canadian Historical Review* 97, 1 (Spring 2017): 34–58; Tina Loo, "People in the Way: Modernity, Environment, and Society on the Arrow Lakes," *BC Studies* 142/143 (Summer/Autumn 2004): 161–96; Loo with Stanley, "An Environmental History of Progress."

83 Michael Clarkson estimates that, on average, about twenty-five people per year commit suicide at Niagara Falls. Since the 1960s, four people have gone over the Falls without protection and survived, including Roger Woodward, with the most recent in July 2019. Michael Clarkson, *River of Lost Souls* (N.p.: Amazon Digital Services, 2017); Rob Goodier, "How to Survive an Accidental Plunge over Niagara Falls," *Popular Mechanics,* August 17, 2011, http://www.popularmechanics.com/adventure/outdoors/tips/a6960/how-to-survive-an-accidental-plunge-over-niagara-falls/.

84 NPC, *Seventieth Annual Report of the Commissioners of the Queen Victoria Niagara Falls Park* (Toronto: L.K. Cameron, 1956).

85 Seibel, *Ontario's Niagara Parks,* 47.

86 The confusion stems from the fact that both Terrapin Point and the rock at the base of the waterfall are often shaded the same colour without any indication of relief or height difference. Natural Resources Canada, "The Atlas of Canada – Toporama," https://atlas.gc.ca/toporama/en/index.html; United States Geological Survey (USGS), "The National Map – Data Delivery," https://www.usgs.gov/core-science-systems/ngp/tnm-delivery/. Google Earth Pro also places the border slightly to the west of Terrapin Point. In previous publications, I incorrectly stated that the United States lost its entire share of the Horseshoe Falls, such as in Daniel Macfarlane, "'A Completely Man-Made and Artificial Cataract': The Transnational Manipulation of Niagara Falls," *Environmental History* 18, 4 (October 2013): 776.

Chapter 6: Preserving Niagara

1 Richard Newman, *Love Canal: A Toxic History from Colonial Times to the Present* (New York: Oxford University Press, 2016).

2 According to Michael Vogel, "much of the heart of the old city was razed to make way for the new; the project relocated 295 businesses, 200 families, and about 400 roomers." Vogel, *Echoes in the Mist*, 115.

3 According to David Stradling, "one survey found 215 hazardous waste dumps in Erie and Niagara counties alone. In Niagara Falls, even at some distance from Love Canal, soil samples contained high levels of benzene, groundwater samples contained pesticides such as heptachlor and aldrin, and the air contained tetrachloroethylene. Heavy metals were everywhere in the city. Indeed the region contained so many chemical production sites and so many waste dumps – Niagara County had thirty-eight known industrial landfills – that health officials began to wonder about safety throughout Niagara Falls." David Stradling, *The Nature of New York: An Environmental History of the Empire State* (Ithaca, NY: Cornell University Press, 2010), 216.

4 NFPL (Ontario), File: Niagara River and Falls – Remedial Works, 1970–1999, Adult LHC, vertical file 38080003419197: "American Falls: Death Watch at a Cataract," *Niagara Falls Gazette*, January 31, 1965.

5 On neo-materialism, see Timothy J. LeCain, "Against the Anthropocene: A Neo-Materialist Perspective," *International Journal for History, Culture and Modernity* 3, 1 (2015): 1–28, as well as his *The Matter of History: How Things Create the Past* (Cambridge: Cambridge University Press, 2017).

6 As the affective turn in historical scholarship has demonstrated, emotional responses change over time and are historically contingent. Such is the case at Niagara, evidenced by the changing nature of what constitutes sublimity among the various involved "emotional communities" or "emotional regimes": groups that developed their own unique norms of valuation, expression, style, and shared assumptions, such as tourists, engineers, industrialists, and so on. On emotional communities, see Barbara H. Rosenwein, *Emotional Communities in the Early Middle Ages* (Ithaca, NY: Cornell University Press, 2006); on emotional regimes, see W.M. Reddy, *The Navigation of Feeling: A Framework of the History of Emotions* (Cambridge: Cambridge University Press, 2001). Evidence of such responses to Niagara can be found in diaries, postcards, letters, guestbooks, photographs, art, engineering studies, governmental reports, and so on.

7 LAC, RG 25, 86-4-7:1, American Falls (Niagara), American Falls Engineering Board Final Report – Distribution Vol. 2: Preservation and Enhancement of the American Falls at Niagara, Final Report to the International Joint Commission by the American Falls International Board, June 1974.

8 The escarpment is composed of nearly horizontal layers of sedimentary rock and capped by resistant Lockport Dolomite. Jackson et al., *The Mighty Niagara*, 58–61; Keith J. Tinkler, "Déjà vu: The Downfall of Niagara as a Chronometer, 1845–1941," in Gayler, *Niagara's Changing Landscapes*, 81–110.

9 LAC, RG 25, G-1, vol. 1438, file 1228, pt. II: Supplementary Report of the Special International Niagara Board Re Fall of Rock in the American Falls on January 17, 1931, November 10, 1931.

10 This cavern allowed people to go behind the American Falls. Three people died from a 1920 rockfall at the Cave of the Winds. Elevators to take visitors to the Cave of the Winds

were added in 1925. Because of the danger to the public, the remaining overhang of the Cave of the Winds was blasted off in 1955.

11 According to a 1962 newspaper report, the Falls had gone from 162 feet to 200 feet. NFPL (Ontario), File: Niagara River and Falls – Diversion and Cessation, Adult LHC, vertical file 380800034199098: "Our Falls Are Getting Higher," January 9, 1962; LAC, RG 25, 86-4-4:1, American Falls (Niagara), Plan of Study, Vol. 2: Feasibility Study of a Structure to Control Levels of the Maid of the Mist Pool, Report to the American Falls International Board by H.G. Acres, Consulting Engineers, September 1970.

12 NFPL (Ontario), File: Niagara River and Falls – Remedial Works, 1970–1999, Adult LHC, vertical file 38080003419197: "American Falls: Death Watch at a Cataract," *Niagara Falls Gazette,* January 31, 1965; NFPL (Ontario), File: Niagara River and Falls – Remedial Works, 1970–1999, Adult LHC, vertical file 38080003419197: "Should Niagara Falls Look Like This?" *Niagara Falls Gazette,* September 9, 1973.

13 For example, at a public hearing on the Niagara River in Niagara Falls, New York, there were speakers from the Corps of Engineers, various levels of government, private citizens, and so on, and most were in favour of remedial works and of removing the talus rock. LAC, RG 25, 86-2-1:1, American Falls (Niagara), Transcript of Hearing: Record of Public Hearing on Niagara River, Niagara Falls, New York, for Consideration of Measures Necessary to Preserve and Enhance the Scenic Beauty of the American Falls, January 18, 1966.

14 For a range of newspaper responses, see ibid. The Corps of Engineers was authorized by section 304 of the Rivers and Harbors Act, October 27, 1965.

15 On the Beautiful America initiative, see Martin Melosi, "Lyndon Johnson and Environmental Policy" and Lewis L. Gould, "Lady Bird Johnson and Beautification," both in *The Johnson Years,* vol. 2, *Vietnam, the Environment, and Science,* ed. Robert A. Divine (Lawrence: University Press of Kansas, 1987); Lewis L. Gould, *Lady Bird Johnson: Our Environmental First Lady* (Lawrence: University Press of Kansas, 1999).

16 NFPL (Ontario), File: Niagara River and Falls – Remedial Works, 1969, Adult LHC, vertical file 38080003418637: "Beauty at Niagara," *McKeesport Daily News* (McKeesport, PA), February 18, 1965; LAC, RG 25, 86-2-1:1, American Falls (Niagara), Transcript of Hearing: *Chicago Tribune,* "The Great Society vs. Niagara Falls," February 13, 1965.

17 "Trouble at Niagara," *Milwaukee Journal,* March 5, 1965.

18 On the rust belt, see Steven High, *Industrial Sunset: The Making of North America's Rust Belt* (Toronto: University of Toronto Press, 2003); Tracy Neumann, *Remaking the Rust Belt: The Postindustrial Transformation of North America* (Philadelphia: University of Pennsylvania Press, 2016).

19 NFPL (Ontario), File: Niagara River and Falls – Remedial Works, 1969, Adult LHC, vertical file 38080003418637: "Fall from Beauty," *Democrat and Chronicle* (Rochester, NY), February 4, 1965.

20 See, for example, the "Summary of Public Hearings" on page 11 in the final report of the American Falls International Board to the IJC: LAC, RG 25, 86-4-7:1, American Falls (Niagara), American Falls Engineering Board Final Report – Distribution Vol. 2: Preservation and Enhancement of the American Falls at Niagara, Final Report to the International Joint Commission by the American Falls International Board, June 1974.

21 The Maid of the Mist Boat Company, for example, complained that lower levels would impair its dockage and navigation. The owner of the Spanish Aero Car attraction at the Niagara Whirlpool downstream complained that diverting extra water would tamper with the flow of the whirlpool, which was known to reverse itself when water levels decreased

enough. In response to complaints that the Niagara Rapids experience would suffer, the International Niagara Board of Control countered that the slight decrease in level that could be expected would actually be an improvement because there would be greater "violent surface agitation." LAC, RG 25, 86-3-1:1, American Falls (Niagara), General Correspondence Vol. 1: Letter from Maid of the Mist Steamboat Company to IJC, November 1, 1967; LAC, RG 25, 86-3-1:1, American Falls (Niagara), General Correspondence Vol. 1: International Niagara Board of Control Report, March 12, 1968; LAC, RG 25, 86-2-4:1, American Falls (Niagara), Briefs and Statements 1967/10: PASNY Statement for IJC Public Hearing on October 20, 1967.

22 The two power entities eventually contributed $276,500 each. LAC, RG 25, 86-3-1:1, American Falls (Niagara), General Correspondence Vol. 1: Memorandum to Chairman. Re: Temporary Diversion of Flows from the American Falls, by M.W. Thompson, May 31, 1968.

23 For more detail on the local history of the dewatering see "The Summer of '69: The Dewatering of the American Falls," Niagara Falls Thunder Alley, http://www.niagarafrontier.com/dewater.html.

24 Quoted in William Willingham, *Vital Currents: A History of the North Central Division, US Army Corps of Engineers, 1867–1997* (Washington, DC: US Army Corps of Engineers, 2006), 54.

25 Nuala Drescher, *Engineers for the Public Good: A History of the Buffalo District, US Army Corps of Engineers* (Washington, DC: US Army Corps of Engineers, 1999), 259.

26 T.A. Wilkinson and Captain J.D Guerin, "Dewatering of American (Niagara) Falls Planned for Study," *Engineering Geologist* 4, 3 (August 1969): 1, 4, 6.

27 NFPL (Ontario), File: Niagara River and Falls – Diversion and Cessation, Adult LHC, vertical file 3808000341990 98: "Dry Falls Blamed for Decline," by John Fedoe.

28 This two-volume publication described Niagara Falls as "a superb diamond set in lead." Quoted in Berton, *Niagara,* 181.

29 See page 70 of the AFIB's final report to the IJC: LAC, RG 25, 86-4-7:1, American Falls (Niagara), American Falls Engineering Board Final Report – Distribution Vol. 2: Preservation and Enhancement of the American Falls at Niagara, Final Report to the International Joint Commission by the American Falls International Board, June 1974; LAC, RG 25-A-3-c, vol. 15731, file pt. 5, Boundaries – Water – International Rivers, Bays and Lakes – Niagara River – IJC Niagara Falls, Reference (1970/04/10–1971/06/15): Studies of the Environment of Niagara Falls, Comments by Humphrey Carver, January 2, 1969.

30 Seagram Tower opened in 1962, Oneida Tower in 1964, and Skylon Tower and Niagara International Center tower in 1965.

31 LAC, RG 25-A-3-c, vol. 20851, file pt. 6, Boundaries – Water – International Rivers, Bays and Lakes – Niagara River – IJC Niagara Falls, Reference (1971/06/16–1973/02/15): Memorandum from Robichaud to Sharp, November 1, 1971.

32 Margaret Beattie Bogue, *Fishing the Great Lakes: An Environmental History, 1783–1933* (Madison: University of Wisconsin Press, 2000); Lee Botts and Paul Muldoon, *Evolution of the Great Lakes Water Quality Agreements* (East Lansing: Michigan State University Press, 2005); John Hartig, *Burning Rivers;* Kehoe, *Cleaning Up the Great Lakes;* Jennifer Read, "Addressing 'A Quiet Horror': The Evolution of Ontario Pollution Control Policy in the International Great Lakes, 1909–1972" (PhD diss., University of Western Ontario, 1999); William McGucken, *Lake Erie Rehabilitated;* Claire Campbell, *Shaped by the West Wind: Nature and History in Georgian Bay* (Vancouver: UBC Press, 2005); Nancy Bouchier and

Ken Cruikshank, *The People and the Bay: A Social and Environmental History of Hamilton Harbour* (Vancouver: UBC Press, 2016); Langston, *Sustaining Lake Superior;* Riley, *The Once and Future Great Lakes Country.*

33 Walter Liong-Ting Hang and Joseph P. Salvo, *Ravaged River: Toxic Chemicals in the Niagara* (New York: New York Public Interest Research Group, 1981), 1.

34 LAC, RG 25-A-3-c, vol. 15731, file pt. 5, Boundaries – Water – International Rivers, Bays and Lakes – Niagara River – IJC Niagara Falls, Reference (1970/04/10–1971/06/15) Summary of Inventory Findings and Their Implications (Second Draft), Prepared for the Erie and Niagara Counties Regional Planning Board by Candeub, Fleissig and Associates, May 1971.

35 LAC, RG 25-A-3-c, vol. 20851, file pt. 6, Boundaries – Water – International Rivers, Bays and Lakes – Niagara River – IJC Niagara Falls, Reference (1971/06/16–1973/02/15): Department of External Affairs Background Paper – Niagara Falls Environmental Study, November 10, 1970.

36 LAC, RG 25-A-3-c, vol. 15731, file pt. 5, Boundaries – Water – International Rivers, Bays and Lakes – Niagara River – IJC Niagara Falls, Reference (1970/04/10–1971/06/15): Erie and Niagara Counties Regional Planning Board, Topical Outline of the International Environmental Design Study, January 7, 1971.

37 LAC, RG 25-A-3-c, vol. 15731, file pt. 5, Boundaries – Water – International Rivers, Bays and Lakes – Niagara River – IJC Niagara Falls, Reference (1970/04/10–1971/06/15): Aide Memoire from US Dept of State to Canada DEA, April 7, 1970.

38 LAC, RG 25-A-3-c, vol. 20851, file pt. 6, Boundaries – Water – International Rivers, Bays and Lakes – Niagara River – IJC Niagara Falls, Reference (1971/06/16–1973/02/15): Memorandum by Humphrey Carver, for exploratory discussion meeting in Toronto, April 6, concerning the international interests in the conservation of the beauty of the Falls at Niagara, April 5, 1972.

39 LAC, RG 25-A-3-c, vol. 20851, file pt. 6, Boundaries – Water – International Rivers, Bays and Lakes – Niagara River – IJC Niagara Falls, Reference (1971/06/16–1973/02/15): Department of External Affairs Background Paper – Niagara Falls Environmental Study, November 10, 1970.

40 LAC, RG 25-A-3-c, vol. 15731, file pt. 5, Boundaries – Water – International Rivers, Bays and Lakes – Niagara River – IJC Niagara Falls, Reference (1970/04/10–1971/06/15): Ad Hoc Committee on Niagara Falls Environmental Study, Position Paper, June 14, 1971.

41 LAC, RG 25-A-3-c, vol. 20851, file pt. 6, Boundaries – Water – International Rivers, Bays and Lakes – Niagara River – IJC Niagara Falls, Reference (1971/06/16–1973/02/15): DEA Memorandum from MacLellan to Kingstone, Subject: Niagara Falls Environmental Study Meeting with Mr. Ryan, February 8, 1972.

42 NFPL (Ontario), File: Niagara River and Falls – Remedial Works, 1969, Adult LHC, vertical file 38080003418637: Fissures not shale layers blamed for erosion at falls, August 23, 1969.

43 LAC, RG 25, 86-4-5:1, American Falls (Niagara), Board's Interim Report and Distribution, Vol. 1: Preservation and Enhancement of the American Falls at Niagara, Interim Report to the International Joint Commission by the American Falls International Board, December 1971, Appendix B, 15–16.

44 In terms of combining environmental and emotional history, see Dolly Jørgensen, *Recovering Lost Species in the Modern Age: Histories of Longing and Belonging* (Cambridge, MA: MIT Press, 2019); Finis Dunaway, *Seeing Green: The Use and Abuse of American Environmental Images* (Chicago: University of Chicago Press, 2015); Andrea Gaynor, "Environmental

History and the History of Emotions," Histories of Emotion from Medieval Europe to Contemporary Australia, June 16, 2017, https://historiesofemotion.com/2017/06/16/environmental-history-and-the-history-of-emotions/.

45 The Ontario Hydro chairman denied that this was the case and insisted that the hydraulic model being built in conjunction with PASNY was actually for the purpose of studying ice buildup in the lower river. NFPL (Ontario), File: Niagara River and Falls – Diversion and Cessation, Adult LHC, vertical file 38080003419098: "Deny River Dam Proposed," March 17, 1964; NFPL (Ontario), File: Niagara River and Falls – Remedial Works, 1969, Adult LHC, vertical file 38080003418637: "Engineer Proposes Concrete Dam as Aid to Preserving Falls," *Niagara Falls Evening Review,* November 25, 1911.

46 Acres International, as the firm is now known, was responsible for designing hydro generating plants that, as of Canada's Centennial in 1967, accounted for almost one-third of Canada's installed capacity: Bronwen Ledger, "Acres: Portrait of a Firm," *Canadian Consulting Engineer,* June 1, 2000, https://www.canadianconsultingengineer.com/features/acres-portrait-of-a-firm/.

47 In fact, an Ice Advisory Group was formed and began meeting in 1971. LAC, RG 25, 86-4-4:1, American Falls (Niagara), Plan of Study, Vol. 2: Feasibility Study of a Structure to Control Levels of the Maid of the Mist Pool, Report to the American Falls International Board, H.G. Acres Limited, Consulting Engineers, September 1970.

48 The model cost $51,630 and this was credited toward the amount Ontario Hydro owed for extra water diversions while the American Falls were dewatered.

49 LAC, RG 25, 86-4-5:1, American Falls (Niagara), Board's Interim Report and Distribution, Vol. 1: Preservation and Enhancement of the American Falls at Niagara, Interim Report to the International Joint Commission by the American Falls International Board, December 1971, Appendix B, 19.

50 Ibid., 32.

51 NFPL (Ontario), File: Niagara River and Falls – Remedial Works, 1970–1999, Adult LHC, vertical file 38080003419197: "Some remedies for ailing falls," February 13, 1972. For a tabulation of the results of the survey, see the AFIB's final report to the IJC. LAC, RG 25, 86-2-1:4, American Falls (Niagara), Arrangements for Hearing 1975: Statement of the American Falls International Board, by General Walter Bachus, US Army Corps of Engineers, Chairman of the United States Section of the American Falls International Board, 1975; LAC, RG 25, 86-4-7:1, American Falls (Niagara), American Falls Engineering Board Final Report – Distribution Vol. 2: Preservation and Enhancement of the American Falls at Niagara, Final Report to the International Joint Commission by the American Falls International Board, June 1974; LAC, RG 25, 86-1-7:1, American Falls (Niagara): Preservation and Enhancement of the American Falls, Final Report of the International Joint Commission to United States and Canada, July 1975.

52 NFPL (Ontario), File: Niagara River and Falls – Remedial Works, 1969, Adult LHC, vertical file 38080003418637: "NPC's Gray opposed removal of rocks from base of falls," by Bill Wilkerson, October 25, 1967.

53 LAC, RG 25, 86-2-1:1, American Falls (Niagara), Transcript of Hearing: IJC Public Hearing on Preservation and Enhancement of the American Falls at Niagara, Council Chambers, City Hall, Niagara Falls, NY, March 24, 1972.

54 The statement was made by Jim Henry, an employee of the Buffalo District of the US Army Corps of Engineers and a member of the working committee of the American Falls International Board. NFPL (Ontario), File: Niagara River and Falls – Remedial Works,

1970–1999, Adult LHC, vertical file 38080003419197: "Should Niagara Falls Look Like This?" *Niagara Falls Gazette,* September 9, 1973.

55 LAC, RG 25, 86-1-7:1, American Falls (Niagara): Preservation and Enhancement of the American Falls, Final Report of the International Joint Commission to United States and Canada, July 1975, 34.

56 Matthew Wisnioski, *Engineers for Change: Competing Visions of Technology in 1960s America* (Cambridge, MA: MIT Press, 2012), 3.

57 Ian L. McHarg, *Design with Nature* (New York: Doubleday/Natural History Press, 1969).

58 I make this statement while also keeping in mind the importance of local concerns and quotidian experiences for influencing American environmentalism in this era. Christopher Sellers, *Crabgrass Crucible: Suburban Nature and the Rise of Environmentalism in Twentieth-Century America* (Chapel Hill: University of North Carolina Press, 2012).

59 Though the received understanding of the origins and dating of the American environmental movement has been undergoing revision, standard works on the movement include: Roderick Nash, *Wilderness and the American Mind* (New Haven, CT: Yale University Press, 1965); Samuel Hays, *Beauty, Health and Permanence: Environmental Politics in the United States, 1955–1985* (Cambridge: Cambridge University Press, 1987); Robert Gottlieb, *Forcing the Spring: The Transformation of the American Environmental Movement* (Washington, DC: Island Press, 1993); Benjamin Kline, *First Along the River: A Brief History of the US Environmental Movement* (Lanham, MD: Rowman and Littlefield, 2011); Philip Shabecoff, *A Fierce Green Fire: The American Environmental Movement* (Washington, DC: Island Press, 2003).

60 The criteria included: Improve Safety; Improve Viewing; Enhance Natural Appearance; Compare Cost; Maintain Reversibility/Flexibility; Maintain Natural Process; Maintain Positive Economic Impact on Tourism.

61 LAC, RG 25, 86-1-7:1, American Falls (Niagara): Preservation and Enhancement of the American Falls, Final Report of the International Joint Commission to United States and Canada, July 1975.

62 Ibid.

63 LAC, RG 25, 86-2-1:4, American Falls (Niagara), Arrangements for Hearing 1975: Statement of the American Falls International Board, by General Walter Bachus, US Army Corps of Engineers, Chairman of the United States Section of the American Falls International Board, 1975; LAC, RG 25, 86-4-4:1, American Falls (Niagara), Progress Reports, Vol. 5: Thirteenth Semi-Annual Progress Report to the IJC by the AFIB for the period March 15, 1973, to September 14, 1973.

64 LAC, RG 25, 86-6-2:1, American Falls (Niagara), Press Clippings: "Niagara's Own Way," *Niagara Falls Gazette,* July 28, 1975.

65 For example, a 1977 rock slide at the Great Gorge boardwalk on the Ontario side of the rapids, and a 1979 slide near the Schoellkopf site. Scaling of loose rock takes place annually up and down the cliffs on either side of the lower river: "Rock Scaling along Niagara Gorge," *Niagara Falls Review,* April 18, 2016, http://www.niagarafallsreview.ca/2016/04/18/rock-scaling-along-niagara-gorge.

66 Cresonia Hsieh, "Pardon Us: Niagara Falls Viewing Areas Now Under Construction," *Buffalo News,* June 16, 2016, https://buffalonews.com/2016/06/16/pardon-us-niagara-falls-viewing-areas-now-construction/. See also "Niagara Falls State Park Landscape Improvement Plan," https://parks.ny.gov/parks/attachments/NiagaraFallsHowNYWorksProjectsAreAffectingNiagaraFallsStatePark.pdf (accessed November 13, 2017).

67 In the spring of 2019, I inquired about plans to turn off the American Falls, and a Niagara Falls State Park representative replied that they were "still exploring funding options." Angela Berti, email message to author, May 7, 2019.

68 In the 1950s and 1960s, Table Rock House was renovated and the steel surge tank was removed. The Ontario Parks Commission introduced its People Mover shuttles in the 1970s, and the Spanish Aero Car was modernized in 1985. Starting in late 2018, Table Rock House was given a "refresh" that included refurbished facilities.

69 Jackson with Burtniak and Stein, *The Mighty Niagara,* 232–33.

Conclusion: Fabricating Niagara

1 "Falls Illumination," Niagara Parks, https://www.niagaraparks.com/events/event/falls-illumination/.

2 Strand, *Inventing Niagara,* 197.

3 Leo Marx, *The Machine in the Garden: Technology and the Pastoral Idea in America* (New York: Oxford University Press, 1964).

4 Reuss and Cutcliffe, *The Illusory Boundary.*

5 Sarah Williams Goldhagen, *Welcome to Your World: How the Built Environment Shapes Our Lives* (New York: Harper, 2017).

6 Karl Appuhn, *A Forest on the Sea: Environmental Expertise in Renaissance Venice* (Baltimore: Johns Hopkins University Press, 2009).

7 White, *The Organic Machine,* 64.

8 Yuichi S. Hayakawa and Yukinori Matsukara, "Factors Influencing the Recession Rate of Niagara Falls since the 19th Century," *Geomorphology* 110 (2009): 212–16; Yuichi S. Hayakawa and Yukinori Matsukara, "Stability Analysis of Waterfall Cliff Face at Niagara Falls: An Implication to Erosional Mechanism of Waterfall," *Engineering Geology* 116, 1–2 (2010): 178–83; S.S. Philbrick, "Horizontal Configuration and the Rate of Erosion of Niagara Falls," *Geological Society of America Bulletin* 81 (1970): 3723–32.

9 The studies that the International Joint Commission point to as the most authoritative on the subject contend that decreased erosion at the Horseshoe Falls stems from not just a reduced flow rate over the waterfall but also better stress dispersion across the horizontal profile of the lip. Philbrick, "Horizontal Configuration"; Hayakawa and Matsukara, "Factors Influencing the Recession Rate of Niagara Falls."

10 Hayakawa and Matsukara, the basis for the IJC's figures, indicate that the lip length went from 2,050 feet in 1927 to roughly 2,600 feet in 1950, 2,370 feet in 1964, and back to about 2,500 by the end of the millennium. According to my findings, however, the crest length was 2,600 when the 1950s remedial works, which shortened the crest by 355 feet, began, meaning it would have been 2,245 feet after the completion of the remedial works. But if the crest had a length of 2,370 feet by 1964, it means that the crest length had increased by 125 feet due to erosion in less than a decade. Since this seems exceedingly unlikely, it is more probable that 2,370 feet is an erroneous number. Furthermore, even if 2,370 is accurate, it means that the crest must have lengthened by 130 feet over thirty-six years if its length was 2,600 as of 2000 (which averages out to 3.6 feet per year). This leads to two possible conclusions: erosion since the 1950s has been much more significant than formal studies suggest, and/or past measurements used as baselines were erroneous. "Table 3: Historical changes in lip length of Horseshoe Falls" of Hayakawa and Matsukara, "Factors

Influencing the Recession Rate of Niagara Falls," gives the lengths in metres (1927: 626 m; 1964: 723 m; 2000: 762 m); the conversions into feet are mine.

11 IJC, Docket 64, Niagara Falls Reference, box 107, 64-4-2:2, Specifications for Preservation and Enhancement of Niagara Falls, US Flank of Horseshoe Falls, US Corps of Engineers: Preservation and Enhancement of Niagara Falls, International Niagara Falls Engineering Board, March 1, 1953.

12 The IJC refers questions about erosion rates to the 2009 article by Hayakawa and Matsukara, "Factors Influencing the Recession Rate of Niagara Falls"; Derrick Beach, "Niagara Falls Is Moving," *IJC Newsletter: Water Matters,* November 15, 2018, https://www.ijc.org/en/niagara-falls-moving.

13 Beach, ibid.

14 Even with the current water regime, experts speculate that the waterfall will return to one cataract in two millennia, when the Niagara River recedes upstream past Goat Island. Moreover, after about 15,000 years, the waterfall will have moved approximately four miles, at which point it will hit softer rock that erodes more quickly. These long-term erosion predictions come from Niagara Parks, "Niagara Falls Geology: Facts and Figures," https://www.niagaraparks.com/visit-niagara-parks/plan-your-visit/niagara-falls-geology-facts-figures/.

15 Corey Binns, "Two Studies of Increasing Mist at Niagara Falls Find Two Different Culprits," *New York Times,* July 18, 2006, section F, 3; Conestoga-Rovers and Associates, "Report: Determine if the Ice Boom Has Climatic, Aquatic, Land Management, or Aesthetic Effects, Niagara Power Project, FERC No. 2216" (prepared by Conestoga-Rovers and Associates for New York Power Authority, 2005), http://niagara.nypa.gov/ALP%20working%20documents/finalreports/html/IS39.htm.

16 Ice proved hazardous in other ways on the Niagara frontier: the blizzard of 1977, for example, resulted in the deaths of twenty-nine people. This blizzard also triggered the Love Canal crisis: when the greater-than-normal snow cover melted, it increased the spring leaching of toxins from the old ditch. Steve Henschel, "Looking Back at Niagara's Perfect Storm," *NiagaraThisWeek.com,* January 30, 2017, http://www.niagarathisweek.com/news-story/7092997-looking-back-at-niagara-s-perfect-storm/. For an oral history of the blizzard, see Erno Rossi, *White Death: Blizzard of 77* (Buffalo: 77 Publishing, 1978). On lake effect snow in upstate New York, see Mark Monmonier, *Lake Effect: Tales of Large Lakes, Arctic Winds, and Recurrent Snows* (Syracuse: Syracuse University Press, 2012).

17 The boom, which was initially made of wooden timbers but now consists of several hundred thirty-foot steel pontoons, is usually installed around mid-December, depending on temperatures, and removed in early spring. The boom is installed when the Lake Erie water temperature reaches 4°C or on December 16, whichever comes first. It can be removed when less than 650 square kilometres of ice is present in the eastern section of Lake Erie. Panel on Niagara River Ice Boom Investigations, Water Science and Technology Board Commission on Physical Sciences, Mathematics, and Resources, National Research Council, *The Lake Erie–Niagara River Ice Boom: Operation and Impacts* (Washington, DC: National Academy Press, 1983).

18 Maintenance is the aspect of technological history that is often forgotten in the rush to highlight innovation and novelty, which Andrew L. Russell and Lee Vinsel highlight in their article "After Innovation, Turn to Maintenance," *Technology and Culture* 59, 1 (January 2018): 1–25.

19 The International Niagara Board of Control had issued a new, more flexible directive in 1973 that was implemented by an exchange of notes dated April 17, 1973. More diplomatic notes were exchanged in 1986 advising that for the terms of the 1950 treaty, the acceptable metric equivalent of 100,000 cfs would be 2,832 cubic metres per second (m^3/s) and 50,000 cfs would be 1,416 m^3/s. Deborah H. Lee, Frank H. Quinn, and Anne H. Clites, "Effect of the Niagara River Chippawa Grass Island Pool on Water Levels of Lakes Erie, St. Clair, and Michigan-Huron," *Journal of Great Lakes Research* 24, 4 (1998): 936–48.
20 In November 1983, the NYPA and Ontario Hydro completed a new 345-kilovolt transmission line between the Moses and Beck hydro power complexes.
21 Under the terms of the new fifty-year licence, the NYPA must make at least half the power available to public bodies and non-profit cooperatives under a "preference" scheme, which also requires that 20 percent of the electricity be made available to neighbouring states. The NYPA was also required to undertake eight habitat improvement projects and establish an ecological committee. For more details, see "A Powerful 50 Years at Niagara," *Water Power Magazine,* April 15, 2011.
22 Frank Parlato, "Some Call It Insane Planning a Century without the Use of Local Niagara Hydropower," *Niagara Falls Reporter,* September 29–October 6, 2015, http://niagarafalls reporter.com/Stories/2015/SEP29/NYPA.html.
23 The two pre-existing tunnels are capable of handling approximately 63,566 cfs, with the new tunnel adding approximately 17,657 cfs.
24 For example, Andrei Sedoff, Stephan Schott, and Bryan Karney, "Sustainable Power and Scenic Beauty: The Niagara River Water Diversion Treaty and Its Relevance Today," *Energy Policy* 66 (2014): 526–36.
25 The Welland Canal diverts 9,500 cfs of water from Lake Erie and the New York Barge Canal diverts 1,100 cfs from the Niagara River.
26 The water intakes are higher than the riverbed, so there could be some water that might continue to the Falls, though cofferdams might eliminate this. But even if, say, 5 percent of the Niagara River's flow still dripped over the Falls, it would still be appropriate to contend that Niagara Falls had been turned "off." After all, there were some rivulets going over the American Falls in 1969, but it was still widely characterized as having been turned off. In 2006, *Inventing Niagara* author Ginger Strand interviewed Norm Stressing, the supervisor of operations for the Moses plant, who maintains that the power utilities could turn off Niagara Falls. Glenn C. Forrester, *Niagara Falls and the Glacier* (Hicksville, NY: Exposition Press, 1976), 133–35; Strand, *Inventing Niagara,* 266.
27 One study estimates that hydroelectric dams emit a billion tonnes of greenhouse gases a year: Bridget R. Deemer et al., "Greenhouse Gas Emissions from Reservoir Water Surfaces: A New Global Synthesis," *BioScience* 66, 11 (November 1, 2016): 949–64.
28 Mike Muller, "Hydropower Dams Can Help Mitigate the Global Warming Impact of Wetlands," *Nature,* February 19, 2019, https://www.nature.com/articles/d41586-019-00616-w.
29 This definition is taken from the website Free Flowing Rivers. "Longitudinal" refers to connectivity between upstream and downstream; "lateral" refers to the ability of a river to rise and fall naturally and connect to its floodplains; "temporal" refers to the natural ability of river flows to change intermittently; "vertical" refers to the ability of a river to draw water from or contribute water to underground aquifers and the atmosphere. Free Flowing Rivers, accessed May 12, 2019, http://freeflowingriver.org.

30 For its "connectivity" score, the Niagara River receives "Below threshold," which is the lowest possible category, with "Good connectivity" and "Free-flowing Rivers" the two higher categories. "Maptool," Free Flowing Rivers, http://freeflowingriver.org/maptool/. See also G. Grill et al., "Mapping the World's Free-Flowing Rivers," *Nature: International Journal of Science* 569 (2019): 215–21.

31 I call the term "ecosystem services" potentially flawed because it risks commodifying and reducing non-human nature to only its anthropocentric economic value. A recent study suggests that warming water temperatures because of climate change threaten the Emerald Shiner, a minnow, whose extirpation could potentially have catastrophic reverberations across riverine food webs. Sara Jerving, "A Looming Threat to the Niagara River," *Investigative Post,* September 12, 2018, http://www.investigativepost.org/2018/09/12/a-looming-threat-to-the-niagara-river/.

32 However, most of this loss has been replaced by introduced and invasive species (primarily opportunistic and successional). There are still about 400 different species of vascular plants today, which was the case a century ago. Eckel, *Botanical Heritage;* also Eckel, *Ecological Restoration and Power at Niagara Falls.*

33 Mah, *Industrial Ruination;* see also Liong-Ting Hang and Salvo, *Ravaged River.*

34 For the Canadian section of the AOC, as of May 2019, five beneficial uses have been restored with five impaired beneficial uses remaining. For more information on the Canadian section, see Niagara River Remedial Action Plan at http://ourniagarariver.ca; Michaud, "The Niagara River Remedial Action Plan." The Ramsar Convention on Wetlands of International Importance promotes the conservation and wise use of water-based ecosystems, and the Canadian side is also working toward a Ramsar designation: "Pursuing the Ramsar Designation," Niagara River Remedial Action Plan, http://ourniagarariver.ca/ramsar/.

35 This radioactive slag was likely put down in the early 1960s as a foundation for the Robert Moses Parkway. Thomas J. Prohaska, "Piles of radioactive dirt greet visitors to Niagara Falls State Park," *The Buffalo News,* August 22, 2018, https://buffalonews.com/2018/08/22/radioactive-waste-unearthed-at-niagara-falls-state-park/.

36 The phrase "manufactured" sublime comes from Anne Whiston Spirn, while "calculated" sublime comes from Elizabeth McKinsey in a 2006 documentary. Spirn "Constructing Nature"; Lawrence Hott and Diane Garey, *Niagara Falls* (Arlington, VA: PBS Home Video, 2006), DVD.

37 McGreevy, *Wall of Mirrors.*

38 Susan Leigh Star, "This Is Not a Boundary Object: Reflections on the Origin of a Concept," *Science, Technology, and Human Values* 35, 5 (2010): 601–17. Because of its plastic meanings that vary across communities, Niagara Falls could be considered a large-scale "boundary object." On boundary objects, see Geoffrey C. Bowker and Susan Leigh Star, *Sorting Things Out: Classification and Its Consequences* (Cambridge, MA: MIT Press, 1999). On political boundary terminology, see Joseph Taylor III, "Boundary Terminology," *Environmental History* 13 (July 2008): 454–81.

39 I have previously suggested that Canada selectively engages in "functional" big science, with hydroelectric dams as a prime example, in Macfarlane, "Negotiated High Modernism."

40 Daniel Immerwahr, *How to Hide an Empire: A History of the Greater United States* (New York: Farrar, Straus and Giroux, 2019).

41 Juliet Christian-Smith et al., *A Twenty-First Century US Water Policy* (New York: Oxford University Press, 2012), 297.

42 Dean MacCannell, *The Tourist: A New Theory of the Leisure Class* (Oakland: University of California Press, 1976); John Urry, *The Tourist Gaze: Leisure and Travel in Contemporary Societies,* 2nd ed. (Newbury Park, CA: Sage, 1992).

43 Wendell Berry, *Our Only World: Ten Essays* (Berkeley, CA: Counterpoint, 2015).

Bibliography

Archival Sources

Canada
Archives of Ontario
Hydro-Electric Power Commission of Ontario (now Ontario Power Generation)
Library and Archives Canada
Niagara Falls Public Library (Ontario)
Queen Victoria Niagara Falls Park Commission

United States
Buffalo History Museum Archives
Dwight D. Eisenhower Presidential Library
Harry S. Truman Presidential Library
National Archives and Records Administration II
Niagara Falls Public Library (New York)
Power Authority of the State of New York (now New York Power Authority)
Syracuse University Library

International Joint Commission
International Joint Commission – Canadian Section

Other Sources

Adams, Edward Dean. *Niagara Power.* 2 vols. Niagara Falls, NY: Niagara Falls Power Company, 1927.

Adamson, Jeremy Elwell, Elizabeth McKinsey, Alfred Runte, and John F. Sears. *Niagara: Two Centuries of Changing Attitudes, 1697–1907.* New York: Routledge, 1992.

Appuhn, Karl. *A Forest on the Sea: Environmental Expertise in Renaissance Venice*. Baltimore: Johns Hopkins University Press, 2009.

Arcadia Revisited: Niagara Rivers and Falls from Lake Erie to Lake Ontario. Albuquerque, NM: Buscaglia-Castellani Art Gallery of Niagara University and University of New Mexico Press, 1988.

Armstrong, Christopher. *The Politics of Federalism: Ontario's Relations with the Federal Government, 1867–1942*. Toronto: University of Toronto Press, 1981.

Armstrong, Christopher, and H.V. Nelles. *Monopoly's Moment: The Organization and Regulation of Canadian Utilities, 1830–1930*. Philadephia: Temple University Press, 1986.

–. *Wilderness and Waterpower: How Banff National Park Became a Hydro-Electric Storage Reservoir*. Calgary: University of Calgary Press, 2013.

Armstrong, Christopher, Matthew Evenden, and H.V. Nelles. *The River Returns: An Environmental History of the Bow*. Montreal and Kingston: McGill-Queen's University Press, 2009.

Aug, Lisa. *Beyond the Falls*. Niagara Falls, NY: Niagara Books, 1992.

Ball, Norman. *Building Canada: A History of Public Works*. Toronto: University of Toronto Press, 1988.

–. *The Canadian Niagara Power Company Story*. Erin, ON: Boston Mills Press, 2005.

Ballon, Hilary, and Kenneth T. Jackson, eds. *Robert Moses and the Modern City*. New York: W.W. Norton, 2007.

Belfield, Robert. "The Niagara Frontier: The Evolution of Electric Power Systems in New York and Ontario, 1880–1935." PhD diss., University of Pennsylvania, 1981.

–. "The Niagara System: The Evolution of an Electric Power Complex at Niagara Falls, 1883–1896." *Proceedings of the IEEE* 64 (1976): 1344–49.

–. "Technology Transfer and Turbulence: The Evolution of an International Energy Complex at Niagara Falls, 1896–1906." *HSTC Bulletin: Journal of the History of Canadian Science, Technology and Medicine/HSTC Bulletin: revue d'histoire des sciences, des techniques et de la médecine au Canada* 5, 2 (1981): 69–98.

Benidickson, Jamie. *Levelling the Lake: Transboundary Resource Management in the Lake of the Woods Watershed*. Vancouver: UBC Press, 2019.

Berry, Wendell. *Our Only World: Ten Essays*. Berkeley, CA: Counterpoint, 2015.

Berton, Pierre. *Niagara: A History of the Falls*. Albany: State University of New York Press, 1992.

Biesel, Richard H. Jr. *International Waterfall Classification System*. Parker, CO: Outskirts Press, 2006.

Biggar, Glenys. *Ontario Hydro's History and Description of Hydro-Electric Generating Stations*. Toronto: Ontario Hydro, 1991.

Biggs, David. *Quagmire: Nation-Building and Nature in the Mekong Delta*. Seattle: University of Washington Press, 2010.

Billington, David P., and Donald C. Jackson. *Big Dams of the New Deal Era: A Confluence of Engineering and Politics*. Norman, OK: University of Oklahoma Press, 2006.

Billington, David P., Donald C. Jackson, and Martin V. Melosi. *The History of Large Federal Dams: Planning, Design, and Construction in the Era of Big Dams*. Denver: US Department of the Interior, Bureau of Reclamation, 2005.

Biro, Andrew. "Half-Empty or Half-Full?" In *Eau Canada: The Future of Canada's Water*, edited by Karen Bakker, 321–33. Vancouver: UBC Press, 2006.

Blackbourn, David. *The Conquest of Nature: Water, Landscape, and the Making of Modern Germany.* New York: W.W. Norton, 2007.

Bocking, Stephen. *Nature's Experts: Science, Politics, and the Environment.* New Brunswick, NJ: Rutgers University Press, 2004.

Bogue, Margaret Beattie. *Fishing the Great Lakes: An Environmental History, 1783–1933.* Madison: University of Wisconsin Press, 2000.

Bonnell, Jennifer. *Reclaiming the Don: An Environmental History of Toronto's Don River Valley.* Toronto: University of Toronto Press, 2014.

Botts, Lee, and Paul Muldoon. *Evolution of the Great Lakes Water Quality Agreements.* East Lansing: Michigan State University Press, 2005.

Bouchier, Nancy, and Ken Cruikshank. *The People and the Bay: A Social and Environmental History of Hamilton Harbour.* Vancouver: UBC Press, 2016.

Bowker, Geoffrey C., and Susan Leigh Star. *Sorting Things Out: Classification and Its Consequences.* Cambridge, MA: MIT Press, 1999.

Braider, Donald. *The Niagara (Rivers of America).* New York: Holt, Rinehart and Winston, 1972.

Brinkley, Douglas. *The Wilderness Warrior: Theodore Roosevelt and the Crusade for America.* New York: Harper, 2010.

Brooks, Karl Boyd. *Public Power, Private Dams: The Hells Canyon High Dam Controversy.* Seattle: University of Washington Press, 2006.

Brynner, Rock. *Natural Power: The New York Power Authority's Origins and Path to Clean Energy.* New York: Cosimo, 2016.

Bukowczyk, John J., Nora Faires, David R. Smith, and Randy Widdis. *Permeable Border: The Great Lakes Basin as Transnational Region, 1650–1990.* Pittsburgh: University of Pittsburgh Press, 2005.

Campbell, Claire. *Shaped by the West Wind: Nature and History in Georgian Bay.* Vancouver: UBC Press, 2005.

Carlson, Bernard. *Tesla: Inventor of the Electrical Age.* Princeton, NJ: Princeton University Press, 2013.

Caro, Robert. *The Power Broker: Robert Moses and the Fall of New York.* New York: Vintage, 1974.

Carse, Ashley. *Beyond the Big Ditch: Politics, Ecology, and Infrastructure at the Panama Canal.* Cambridge, MA: MIT Press, 2014.

Caskey, Christopher. "Listening to a River: How Sound Emerges in River Histories." *Open Rivers: Rethinking Water, Place and Community* 8 (Fall 2017): 146–54.

Chambliss, Julian. "Perfecting Space: J. Horace McFarland and the American Civic Association." *Pennsylvania History: A Journal of Mid-Atlantic Studies* 77, 4 (Autumn 2010): 486–97.

Christian-Smith, Juliet, Peter Gleick, Heather Cooley, Lucy Allen, Amy Fanderwarker, and Kate Berry. *A Twenty-First Century US Water Policy.* New York: Oxford University Press, 2012.

Cioc, Mark. *The Rhine: An Eco-Biography, 1815–2000.* Seattle: University of Washington Press, 2002.

Clarkson, Michael. *River of Lost Souls: What We Might Learn from Niagara Falls Suicides.* N.p.: Amazon Digital Services, 2017.

Coates, Peter. *A Story of Six Rivers.* London: Reaktion Books, 2013.

–. "The Strange Stillness of the Past: Toward an Environmental History of Sound and Noise." *Environmental History* 10 (2005): 636–65.

Cohn, Julie A. *The Grid: Biography of an American Technology.* Cambridge, MA: MIT Press, 2017.

Cosentino, Ken. "Water Crisis Remains on Tuscarora Reservation." *Niagara Gazette,* October 26, 2019.

Coutu, Joan. "Vehicles of Nationalism: Defining Canada in the 1930s." *Journal of Canadian Studies* 37 (Spring 2002): 180–203.

Creighton, Donald. *The Empire of the St. Lawrence: A Study of Commerce and Politics.* Toronto: Macmillan, 1956.

Cronon, William. *Nature's Metropolis: Chicago and the Great West.* New York: W.W. Norton, 1992.

Cullather, Nicholas. "Damming Afghanistan: Modernization in a Buffer State." *Journal of American History* 89 (September 2002): 512–37.

Cusack, Tricia. *Riverscapes and National Identities.* Syracuse, NY: Syracuse University Press, 2009.

Dagenais, Michèle. *Montreal, City of Water: An Environmental History.* Vancouver: UBC Press, 2017.

De Rocquigny, E., Assedine Soulaimani, and Brian Morse. "Uncertainty in Hydrological and Hydraulic Modeling: Editorial Introduction/Les incertitudes en modelisation hydrologique et hydraulique: Introduction editorial." *Canadian Journal of Civil Engineering* 37, 7 (July 2010): iii–ix.

Deemer, Bridget R., John A. Harrison, Siyue Li, Jake J. Beaulieu, Tonya DelSontro, Nathan Barros, José F. Bezerra-Neto, Stephen M. Powers, Marco A. dos Santos, and J. Arie Vonk. "Greenhouse Gas Emissions from Reservoir Water Surfaces: A New Global Synthesis." *BioScience* 66, 11 (November 1, 2016): 949–64.

Denison, Merrill. *The People's Power: The History of Ontario Hydro.* Toronto: McClelland and Stewart, 1960.

Desbiens, Caroline. *Power from the North: Territory, Identity, and the Culture of Hydroelectricity in Quebec.* Vancouver: UBC Press, 2013.

Dorsey, Kurkpatrick. *The Dawn of Conservation Diplomacy: U.S.-Canadian Wildlife Protection Treaties in the Progressive Era.* Seattle: University of Washington Press, 1998.

Dorsey, Kurkpatrick, and Mark Lytle. "Forum: New Directions in Diplomatic and Environmental History." *Diplomatic History* 32, 4 (2008): 517–646.

Dow, Charles M. *Anthology and Bibliography of Niagara Falls, Vols. I and II.* Albany: J.B. Lyon, 1921.

–. *The State Reservation at Niagara: A History.* Albany: J.B. Lyon, 1914.

Dreisziger, N.F. "The Campaign to Save Niagara Falls and the Settlement of United States–Canadian Differences, 1906–1911." *New York History* 55, 4 (October 1974): 437–58.

–. "The International Joint Commission of the United States and Canada, 1895–1920: A Study in Canadian-American Relations." PhD diss., University of Toronto, 1974.

Drescher, Nuala. *Engineers for the Public Good: A History of the Buffalo District, US Army Corps of Engineers.* Washington, DC: US Army Corps of Engineers, 1999.

Dubinsky, Karen. *The Second Greatest Disappointment: Honeymooning and Tourism at Niagara Falls.* Toronto: Between the Lines, 1999.

Dumych, Daniel. *The Canadian Niagara Power Company – One Hundred Years, 1892–1992.* Niagara Falls, ON: Canadian Niagara Power Company, 1992.

–. *Images of America: Niagara Falls.* Chicago: Arcadia, 1996.

–. *Images of America: Niagara Falls, Volume 2.* Chicago: Arcadia, 1998.

Dunaway, Finis. *Seeing Green: The Use and Abuse of American Environmental Images.* Chicago: University of Chicago Press, 2015.

Eagles, Munroe. "Organizing across the Canada-US Border: Binational Institutions in the Niagara Region." *American Review of Canadian Studies* 40, 3 (September 2010): 379–94.

Eckel, Patricia. *Botanical Heritage of Islands at the Brink of Niagara Falls.* St. Louis, MO: Botanical Services, St. Louis, 2013.

–. *Ecological Restoration and Power at Niagara Falls.* St. Louis, MO: Botanical Services, St. Louis, 2016.

Elkind, Sarah. *How Local Politics Shape Federal Policy: Business, Power and the Environment in Twentieth Century Los Angeles.* Chapel Hill: University of North Carolina Press, 2011.

Estes, Nick. *Our History is the Future: Standing Rock versus the Dakota Access Pipeline, and the Long Tradition of Indigenous Resistance.* London: Verso, 2019.

Evans, Gail E.H. "Storm over Niagara: A Catalyst in Reshaping Government in the United States and Canada during the Progressive Era." *Natural Resources Journal* 32 (Winter 1992): 27–54.

–. "Storm over Niagara: A Study of the Interplay of Cultural Values, Resource Politics, and Environmental Policy, in an International Setting, 1670s–1950." PhD diss., University of California at Santa Barbara, 1991.

Evenden, Matthew. *Allied Power: Mobilizing Hydro-Electricity during Canada's Second World War.* Toronto: University of Toronto Press, 2015.

–. *Fish versus Power: An Environmental History of the Fraser River.* New York: Cambridge University Press, 2004.

Evershed, Thomas. *Water-Power at Niagara Falls to Be Successfully Utilized. The Niagara River Hydraulic Tunnel, Power and Sewer Co.* Buffalo: Matthew, Northup, Art-Printing Works, Office of the Buffalo Morning Express, 1886.

Fatherree, Ben. *The First 75 Years: History of Hydraulic Engineering at the Waterways Experiment Station.* Washington, DC: US Army Corps of Engineers, Engineer Research and Development Center, 2004.

Fleming, Keith. *Power at Cost: Ontario Hydro and Rural Electrification, 1911–1958.* Montreal and Kingston: McGill-Queen's University Press, 1991.

Forrester, Glenn C. *The Falls of Niagara.* New York: D. Van Nostrand, 1928.

–. *Niagara Falls and the Glacier.* Hicksville, NY: Exposition Press, 1976.

Freeman, Neil B. *The Politics of Power: Ontario Hydro and Its Government, 1906–1995.* Toronto: University of Toronto Press, 1996.

French, Daniel. *When They Hid the Fire: A History of Electricity and Invisible Energy in America.* Pittsburgh: University of Pittsburgh Press, 2017.

Friesen, B.F. "The International Sharing of Niagara River Hydroelectric Power Diversions." *Canadian Water Resources Journal* 4, 4 (1979): 26–38.

Friesen, B.F., and J.C. Day. "Hydroelectric Power and Scenic Provisions of the 1950 Niagara Treaty." *Water Resources Bulletin – American Water Resources Association* 13, 6 (December 1977): 1175–89.

Froschauer, Karl. *White Gold: Hydroelectric Power in Canada.* Vancouver: UBC Press, 1999.

Funigiello, Philip J. *Toward a National Power Policy: The New Deal and the Electric Utility Industry, 1933–1941.* Pittsburgh: University of Pittsburgh Press, 1973.

Gandy, Matthew. *The Fabric of Space: Water, Modernity, and the Urban Imagination.* Cambridge, MA: The MIT Press, 2014.

Gawronski, Brett, Jana Kasikova, Lynda Schneekloth, and Thomas Yots. *The Power Trail: History of Hydro-Electricity at Niagara.* Buffalo: Brian Meyer, 2014.

Gayler, Hugh J., ed. *Niagara's Changing Landscapes.* Ottawa: Carleton University Press, 1994.

Gaynor, Andrea. "Environmental History and the History of Emotions." Histories of Emotion from Medieval Europe to Contemporary Australia, June 16, 2017, https://historiesofemotion.com/2017/06/16/environmental-history-and-the-history-of-emotions/.

Georgallas, Virginia. "'The Sublimest Music on Earth': An Aural History of Niagara Falls in the Mid-Nineteenth Century." MusCan (Canadian University Music Society), August 15, 2017.

Gilbert, G.K. *Rate of Recession of Niagara Falls.* Department of the Interior, United States Geological Survey, Bulletin No. 306. Washington, DC: US Government Printing Office, 1907.

Glaser, Jesse R. "Ontario Hydro Blasts Out Niagara Power Tunnels." *Civil Engineering,* October 1953, 40–44.

Glennon, Ken. *Hard Hats of Niagara: The Niagara Power Project.* South Bend, IN: Dog Ear, 2011.

Gluek, Alvin. "The Lake Carriers' Association and the Origins of the International Waterways Commission." *The Inland Seas: Quarterly Journal of the Great Lakes Historical Society* 36 (1980): 236–45.

Goldhagen, Sarah Williams. *Welcome to Your World: How the Built Environment Shapes Our Lives.* New York: Harper, 2017.

Goldin, Claudia, and Robert Margo. "The Great Compression: The Wage Structure in the United States at Mid-Century." *Quarterly Journal of Economics* 107 (February 1992): 1–34.

Goodier, Rob. "How to Survive an Accidental Plunge over Niagara Falls." *Popular Mechanics,* August 17, 2011, http://www.popularmechanics.com/adventure/outdoors/tips/a6960/how-to-survive-an-accidental-plunge-over-niagara-falls/.

Gordon, Robert. *The Rise and Fall of American Growth: The US Standard of Living since the Civil War.* Princeton, NJ: Princeton University Press, 2016.

Gottlieb, Robert. *Forcing the Spring: The Transformation of the American Environmental Movement.* Washington, DC: Island Press, 1993.

Gould, Lewis L. "Lady Bird Johnson and Beautification." In *The Johnson Years,* Vol. 2, *Vietnam, the Environment, and Science,* edited by Robert A. Divine, 150–81. Lawrence: University Press of Kansas, 1987.

–. *Lady Bird Johnson: Our Environmental First Lady.* Lawrence: University Press of Kansas, 1999.

Grant, Barry, and Joan Nicks. *Covering Niagara: Studies in Local Popular Culture.* Waterloo, ON: Wilfrid Laurier University Press, 2010.

Grauer, A.E.D. "The Export of Electricity from Canada." In *Canadian Issues: Essays in Honour of Henry F. Angus,* edited by R.M. Clark, 248–85. Toronto: University of Toronto Press, 1961.

Greenhill, Ralph. *Spanning Niagara: The International Bridges, 1848–1962.* Seattle: University of Washington Press, 1985.

Greenhill, Ralph, and Thomas Mahoney. *Niagara.* Toronto: University of Toronto Press, 1969.

Grill, G., B. Lehner, M. Thieme, B. Geenen, D. Tickner, F. Antonelli, S. Babu, P. Borrelli, L. Cheng, H. Crochetiere, H. Ehalt Macedo, R. Filgueiras, M. Goichot, J. Higgins, Z. Hogan, B. Lip, M.E. McClain, J. Meng, M. Mulligan, C. Nilsson, J.D. Olden, J.J. Opperman, P. Petry, C. Reidy Liermann, L. Saenz, S. Salinas-Rodriguez, P. Schelle, R.J.P. Schmitt, J. Snider, F. Tan, K. Tockner, P.H. Valdujo, A. van Soesbergen, and C. Zarfl. "Mapping the World's Free-Flowing Rivers." *Nature: International Journal of Science* 569 (2019): 215–21.

Grol, L.R. *Tales from the Niagara Peninsula: A Legend of the Maid of the Mist.* Fonthill, ON: Fonthill Studio, 1975.

Gromosiak, Paul. "A Brief History of the Edward Dean Adams Power Plant." N.d. http://www.teslaniagara.org/wp-content/uploads/2014/12/261389503-History-of-the-Adams-Power-Plant-PG.pdf.

Gromosiak, Paul, and Christopher Stoianoff. *Images of America: Niagara Falls, 1850–2000.* Charleston, SC: Arcadia, 2012.

Gucciardo, Dorotea. "The Powered Generation: Canadians, Electricity, and Everyday Life." PhD diss., Western University, 2011.

Hager, Willi H. *Hydraulicians in the USA 1800–2000: A Biographical Dictionary of Leaders in Hydraulic Engineering and Fluid Mechanics.* New York: CRC Press/Balkema, 2015.

Hall, Carl A.S. "Electrical Utilities in Ontario Under Private Ownership, 1890–1914." PhD diss., University of Toronto, 1968.

Hall, Melanie. "Niagara Falls: Preservation and the Spectacle of Anglo-American Accord." In *Towards World Heritage: International Origins of the Preservation Movement, 1870–1930,* edited by Melanie Hall, 23–44. Farnham, UK: Ashgate, 2009.

Hallett, Adam. "Made of the Mist: Nineteenth-Century British and American Views of Niagara." *Literature Compass* 11, 3 (2014): 159–72.

Haraway, Donna. *Simians, Cyborgs, and Women: The Reinvention of Nature.* New York: Routledge, 1990.

Hartig, John. *Burning Rivers: Revival of Four Urban-Industrial Rivers That Caught Fire.* London: Multi-Science, 2010.

Hauptman, Laurence M. *The Iroquois Struggle for Survival: World War II to Red Power.* Syracuse, NY: Syracuse University Press, 1986.

Hayakawa, Yuichi S., and Yukinori Matsukara. "Factors Influencing the Recession Rate of Niagara Falls since the 19th Century." *Geomorphology* 110 (2009): 212–16.

–. "Stability Analysis of Waterfall Cliff Face at Niagara Falls: An Implication to Erosional Mechanism of Waterfall." *Engineering Geology* 116, 1–2 (2010): 178–83.

Hays, Samuel. *Beauty, Health and Permanence: Environmental Politics in the United States, 1955–1985.* Cambridge: Cambridge University Press, 1987.

–. *Conservation and the Gospel of Efficiency: The Progressive Conservation Movement, 1890–1920.* Pittsburgh: University of Pittsburgh Press, 1959.

Heaman, Elsbeth. *A Short History of the State in Canada.* Toronto: University of Toronto Press, 2015.

Hecht, Gabrielle. *The Radiance of France: Nuclear Power and National Identity after World War II.* Cambridge, MA: MIT Press, 2001.

Hecht, Gabrielle, and Paul N. Edwards. "The Technopolitics of Cold War: Towards a Transregional Perspective." In *Essays on Twentieth Century History*, edited by Michael Adas, 271–314. Philadelphia: Temple University Press, 2010.
Higgins, Robert, ed. *An Awful Grandeur: Niagara in Quotes*. Kitchener, ON: Upney Editions, 1998.
–. *The Niagara Frontier: Its Place in US and Canadian History*. Kitchener, ON: Upney Editions, 1996.
High, Steven. *Industrial Sunset: The Making of North America's Rust Belt*. Toronto: University of Toronto Press, 2003.
Hillmer, Norman. *O.D. Skelton: A Portrait of Canadian Ambition*. Toronto: University of Toronto Press, 2016.
Hirsh, Richard. *Technology and Transformation in the American Electric Utility Industry*. Cambridge, UK: Cambridge University Press, 1989.
Hirt, Paul W. *The Wired Northwest: The History of Electric Power, 1870s–1970s*. Lawrence: University Press of Kansas, 2012.
Holmes, Beatrice. *A History of Federal Water Resources Programs, 1800–1960*. Washington, DC: US Department of Agriculture, 1972.
–. *History of Federal Water Resources Programs and Policies, 1961–70*. Washington, DC: US Department of Agriculture, 1969.
Hott, Lawrence, and Diane Garey. *Niagara Falls*. Arlington, VA: PBS Home Video, 2006. DVD.
Huber, Matthew. *Lifeblood: Oil, Freedom and the Forces of Capital*. Minneapolis: University of Minnesota Press, 2013.
Hudson, Brian J. *Waterfall: Nature and Culture*. London: Reaktion Books, 2012.
Hughes, Thomas P. "The Evolution of Large Technological Systems." In *The Social Construction of Technological Systems: New Directions in the Sociology and History of Technology*, edited by Deborah G. Douglas, Wiebe E. Bijker, and Thomas P. Hughes, 51–82. Cambridge, MA: MIT Press, 1987.
–. *Networks of Power: Electrification in Western Society, 1880–1930*. Baltimore: Johns Hopkins University Press, 1993.
Hulbert, Archer Butler. *The Niagara River*. New York: G.P. Putnam's Sons, 1908.
Hull, James. "Raising Standards: Public Works and Industrial Practice in Interwar Ontario." *Scientia Canadensis* 25 (June 2003): 7–30.
–. "Watts across the Border: Technology and the Integration of the North American Economy in the Second Industrial Revolution." *Left History* 19, 2 (Fall/Winter 2015/16): 13–31.
Hundley, Norris, and Donald C. Jackson. *Heavy Ground: William Mulholland and the St. Francis Dam Disaster*. Oakland: University of California Press, 2016.
Hunter, Louis C. *A History of Industrial Power in the United States, 1780–1930*, Vol. 1, *Waterpower in the Century of the Steam Engine*. Charlottesville, VA: University Press of Virginia, 1979.
Hutchings, Kevin. "Romantic Niagara: Environmental Aesthetics, Indigenous Culture, and Transatlantic Tourism, 1794–1850." In *Transatlantic Literary Exchanges, 1790–1870*, edited by Kevin Hutchings and Julia M. Wright, 153–68. Farnham, UK: Ashgate, 2011.
Huth, Hans. *Nature and the American: Three Centuries of Changing Attitudes*. Lincoln, NB: University of Nebraska Press, 1957.

Hydro-Electric Power Commission of Ontario. *Hydro-Electric Power in the Niagara District.* Toronto: HEPCO, July 1920.
–. *Power from Niagara.* Toronto: HEPCO, 1970.
Immerwahr, Daniel. *How to Hide an Empire: A History of the Greater United States.* New York: Farrar, Straus and Giroux, 2019.
Innis, Harold. *The Fur Trade in Canada: An Introduction to Canadian Economic History.* New Haven, CT: Yale University Press, 1930.
Irwin, William. *The New Niagara: Tourism, Technology, and the Landscape of Niagara Falls.* University Park, PA: Pennsylvania State University Press, 1996.
Jackson, Donald C. *Building the Ultimate Dam: John S. Eastwood and the Control of Water in the West.* Norman, OK: University of Oklahoma Press, 2005.
–. *Pastoral and Monumental: Dams, Postcards and the American Landscape.* Pittsburgh: University of Pittsburgh Press, 2013.
Jackson, John N., and John Burtniak, eds. *Industry in the Niagara Peninsula, Proceedings Eleventh Annual Niagara Peninsula History Conference.* St. Catharines, ON: Brock University, 1992.
Jackson, John N., with John Burtniak and Gregory P. Stein. *The Mighty Niagara: One River – Two Frontiers.* Amherst, NY: Prometheus Books, 2003.
Jacobs, Jane. *The Death and Life of Great American Cities.* New York: Random House, 1961.
Jakobsson, Eva. *Industrialisering av älvar: Studier kring svensk vattenkraftutbyggnad 1900–1918.* Göteborg: Historiska Institutionen, 1996.
–. "Industrialization of Rivers: A Water System Approach to Hydropower Development." *Knowledge, Technology, and Policy* 14, 4 (Winter 2002): 41–56.
Jasanoff, Sheila, and Sang-Hyun Kim. *Dreamscapes of Modernity: Sociotechnical Imaginaries and the Fabrication of Power.* Chicago: University of Chicago Press, 2015.
Jasen, Patricia. "Romanticism, Modernity, and the Evolution of Tourism on the Niagara Frontier, 1790–1850." *Canadian Historical Review* 72 (1991): 283–318.
–. *Wild Things: Nature, Culture, and Tourism in Ontario, 1790–1914.* Toronto: University of Toronto Press, 1995.
Jenks, Andrew. "Model City USA: The Environmental Costs of Victory in World War II and the Cold War." *Environmental History* 12 (July 2007): 552–77.
Jerving, Sara. "A Looming Threat to the Niagara River." *Investigative Post,* September 12, 2018. http://www.investigativepost.org/2018/09/12/a-looming-threat-to-the-niagara-river/.
Johnson, Bob. *Carbon Nation: Fossil Fuels in the Making of American Culture.* Lawrence: University Press of Kansas, 2014.
Jones, Christopher. *Routes of Power: Energy and Modern America.* Cambridge, MA: Harvard University Press, 2014.
Jones-Imhotep, Edward. *The Unreliable Nation: Hostile Nature and Technological Failure in the Cold War.* Cambridge, MA: MIT Press, 2017.
Jonnes, Jill. *Empires of Light: Edison, Tesla, Westinghouse and the Race to Electrify the World.* New York: Random House, 2003.
Jørgensen, Dolly. "Artifacts and Habitats." In *The Routledge Companion to the Environmental Humanities,* edited by Ursula Heise, Jon Christensen, and Michele Niemann, 246–57. New York: Routledge, 2017.
–. "Competing Ideas of 'Natural' in a Dam Removal Controversy." *Water Alternatives* 10, 3 (2017): 840–52.

–. *Recovering Lost Species in the Modern Age: Histories of Longing and Belonging.* Cambridge, MA: MIT Press, 2019.

Jørgensen, Dolly, Finn Arne Jørgensen, and Sara B. Pritchard, eds. *New Natures: Joining Environmental History with Science and Technology Studies.* Pittsburgh: University of Pittsburgh Press, 2013.

Josephson, Paul. *Industrialized Nature: Brute Force Technology and the Transformation of the Natural World.* Washington, DC: Island Press/Shearwater, 2002.

Kehoe, Terence. *Cleaning Up the Great Lakes: From Cooperation to Confrontation.* Dekalb: Northern Illinois University Press, 1997.

Keiner, Christine. "Modeling Neptune's Garden: The Chesapeake Bay Hydraulic Model, 1965–1984." In *The Machine in Neptune's Garden: Historical Studies on Technology and the Marine Environment,* edited by David van Keuren and Helen Rozwadowski, 273–314. Sagamore Beach, MA: Science History, 2004.

Keller, Col. Charles. *The Power Situation during the War.* Washington, DC: US Government Printing Office, by authority of the Secretary of War, 1921.

Kepner, C.D. "Niagara's Water Power." *Niagara Frontier* 15 (1968): 97–105; 16 (1969): 33–41, 75–80; and 17 (1970): 69–79.

Killan, Gerald. "Mowat and a Park Policy for Niagara Falls, 1873–1887." *Ontario History* 70, 2 (June 1978): 115–36.

Kiwanis Club of Stamford, Ontario. *Niagara Falls, Canada: A History of the City and the World Famous Beauty Spot.* Toronto: Ryerson Press, 1967.

Kline, Benjamin. *First Along the River: A Brief History of the US Environmental Movement.* Lanham, MD: Rowman and Littlefield, 2011.

Kohler, Robert E. *Landscapes and Labscapes: Exploring the Lab-Field Border in Biology.* Chicago: University of Chicago Press, 2002.

Kottman, Richard. "Herbert Hoover and the St. Lawrence Treaty of 1932." *New York History* 56 (July 1975): 314–46.

Kuhlberg, Mark. *In the Power of the Government: The Rise and Fall of Newsprint in Ontario, 1894–1932.* Toronto: University of Toronto Press, 2015.

Landry, Marc. "Environmental Consequences of the Peace: The Great War, Dammed Lakes, and Hydraulic History in the Eastern Alps." *Environmental History* 20 (2015): 422–48.

Lane, Christopher W. *Impressions of Niagara: The Charles Rand Penney Collection.* Philadelphia: Philadelphia Print Shop, 1993.

Langston, Nancy. *Sustaining Lake Superior: An Extraordinary Lake in a Changing World.* New Haven, CT: Yale University Press, 2017.

Larkin, Janet Dorothy. *Overcoming Niagara: Canals, Commerce, and Tourism in the Niagara–Great Lakes Borderland Region, 1792–1837.* Albany: SUNY Press, 2018.

Latour, Bruno. *Reassembling the Social.* Oxford: Oxford University Press, 2007.

–. *We Have Never Been Modern.* Cambridge, MA: Harvard University Press, 1993.

LeCain, Timothy. "Against the Anthropocene: A Neo-Materialist Perspective." *International Journal for History, Culture and Modernity* 3, 1 (2015): 1–28.

–. *The Matter of History: How Things Create the Past.* Cambridge: Cambridge University Press, 2017.

Ledger, Bronwen. "Acres: Portrait of a Firm." *Canadian Consulting Engineer,* June 1, 2000. https://www.canadianconsultingengineer.com/features/acres-portrait-of-a-firm/.

Lee, Deborah H., Frank H. Quinn, and Anne H. Clites. "Effect of the Niagara River Chippawa Grass Island Pool on Water Levels of Lakes Erie, St. Clair, and Michigan-Huron." *Journal of Great Lakes Research* 24, 4 (1998): 936–48.

Leuchtenburg, William. "The Niagara Compromise." *Current History* 34 (May 1958): 270–74.

Lifset, Robert. *Power on the Hudson: Storm King Mountain and the Emergence of Modern Environmentalism.* Pittsburgh: University of Pittsburgh Press, 2014.

Liong-Ting Hang, Walter, and Joseph P. Salvo. *Ravaged River: Toxic Chemicals in the Niagara.* New York: New York Public Interest Research Group, 1981.

Loo, Tina."High Modernism, Conflict, and the Nature of Change in Canada: A Look at *Seeing Like a State.*" *Canadian Historical Review* 97, 1 (Spring 2017): 34–58.

–. "People in the Way: Modernity, Environment, and Society on the Arrow Lakes." *BC Studies* 142/143 (Summer/Autumn 2004): 161–96.

Loo, Tina, with Meg Stanley. "An Environmental History of Progress: Damming the Peace and Columbia Rivers." *Canadian Historical Review* 92, 3 (September 2011): 399–427.

Lopate, Phillip. "A Town Revived, a Villain Redeemed," *New York Times,* February 11, 2007, CY3.

Louis Cassier Co. *The Harnessing of Niagara.* New York and London: Cassiers Magazine, 1895.

Lowitt, Richard. "Ontario Hydro: A 1925 Tempest in an American Teapot." *Canadian Historical Review* 49, 3 (September 1968): 267–74.

Lubar, Steven. "Transmitting the Power of Niagara: Scientific, Technological and Cultural Contexts of an Engineering Decision." *IEEE Technology and Society Magazine,* March 1989, 11–18.

MacCannell, Dean. *The Tourist: A New Theory of the Leisure Class.* Oakland: University of California Press, 1976.

Macfarlane, Daniel. "'As Nearly as May Be': Estimating Ice and Water on Niagara and St. Lawrence Rivers." *Journal of Historical Geography* 65 (July 2019): 73–84.

–. "'A Completely Man-Made and Artificial Cataract': The Transnational Manipulation of Niagara Falls." *Environmental History* 18, 4 (October 2013): 759–84.

–. "Fluid Meanings: Hydro Tourism and the St. Lawrence and Niagara Megaprojects." *Histoire Sociale/Social History* 49, 99 (June 2016): 327–46.

–. "Natural Security: Canada-US Environmental Diplomacy." In *Undiplomatic History: Rethinking Canada in the World,* edited by Asa McKercher and Philip Van Huizen, 107–36. Montreal and Kingston: McGill-Queen's University Press, 2019.

–. "Nature Empowered: Hydraulic Models and the Engineering of Niagara Falls," *Technology and Culture* 61, 1 (January 2020): 109–43.

–. "Negotiated High Modernism: Canada and the St. Lawrence Seaway and Power Project." In *Science, Technology, and the Modern in Canada: An Anthology in Honour of Richard Jarrell,* edited by Edward Jones-Imhotep and Tina Adcock, 326–47. Vancouver: UBC Press, 2018.

–. "Negotiating Niagara Falls: US-Canada Environmental and Energy Diplomacy." *Diplomatic History* 43, 5 (November 2019): 916–43.

–. *Negotiating a River: Canada, the US, and the Creation of the St. Lawrence Seaway.* Vancouver: UBC Press, 2014.

–. "The Niagara Telecolorimeter." *Environment and Society Portal, Arcadia* (Autumn 2018): 28. Rachel Carson Center for Environment and Society. doi.org/10.5282/rcc/8492.

–. "Watershed Decisions: The St. Lawrence Seaway and Sub-National Water Diplomacy." *Canadian Foreign Policy Journal* 21, 3 (2015): 212–23.

Macfarlane, Daniel, and Andrew Watson. "Hydro Democracy: Water Power and Political Power in Ontario." *Scientia Canadensis* 40, 1 (2018): 1–18.

Macfarlane, Daniel, and Murray Clamen, eds. *The First Century of the International Joint Commission*. Calgary: University of Calgary Press, 2020.

Macfarlane, Daniel, and Peter Kitay. "Hydraulic Imperialism: Hydro-electric Development and Treaty 9 in the Abitibi Region." *American Review of Canadian Studies* 47, 3 (Fall 2016): 380–97.

Mah, Alice. *Industrial Ruination, Community and Place: Landscapes and Legacies of Urban Decline*. Toronto: University of Toronto Press, 2012.

Malm, Andreas. *Fossil Capital: The Rise of Steam Power and the Roots of Global Warming*. London: Verso Books, 2016.

Manore, Jean. *Cross Currents: Hydro-Electricity and the Engineering of Northern Ontario*. Waterloo, ON: Wilfrid Laurier University Press, 1999.

–. "Rivers as Text: From Pre-Modern to Post-Modern Understandings of Development, Technology and the Environment in Canada and Abroad." In *A History of Water*, Vol. 3, *The World of Water*, edited by Terje Tvedt and Terje Oestigaard, 229–53. London: I.B. Tauris, 2006.

Martin-Nielsen, Janet. "South over the Wires: Hydroelectricity Exports from Canada, 1900–1925." *Water History* 1 (2009): 109–29.

Marx, Leo. *The Machine in the Garden: Technology and the Pastoral Idea in America*. New York: Oxford University Press, 1964.

Massell, David. *Quebec Hydropolitics: The Peribonka Concessions of the Second World War*. Montreal and Kingston: McGill-Queen's University Press, 2011.

Mavor, James. *Niagara in Politics: A Critical Account of the Ontario Hydro-Electric Commission*. New York: E.P. Dutton, 1925.

McCool, Daniel. *River Republic: The Fall and Rise of America's Rivers*. New York: Columbia University Press, 2012.

McGreevy, Patrick. *Imagining Niagara: The Meaning and Making of Niagara Falls*. Amherst, MA: University of Massachusetts Press, 1994.

–. *Wall of Mirrors: Nationalism and Perceptions of the Border at Niagara Falls*. Orono, ME: Borderlands Project, 1991.

McGucken, William. *Lake Erie Rehabilitated: Controlling Cultural Eutrophication, 1960s–1990s*. Akron, OH: University of Akron Press, 2000.

McHarg, Ian L. *Design with Nature*. New York: Doubleday/Natural History Press, 1969.

McInnis, Marvin. "Engineering Expertise and the Canadian Exploitation of the Technology of the Second Industrial Revolution." In *Technology and Human Capital in Historical Perspective*, edited by Jonaas Ljungberg, 49–78. New York: Palgrave Macmillan, 2004.

McKay, Paul. *Electric Empire: The Inside Story of Ontario Hydro*. Toronto: Between the Lines, 1983.

McKinsey, Elizabeth R. *Niagara Falls: Icon of the American Sublime*. New York: Cambridge University Press, 1985.

McNeill, John R., and Peter Engelke. *The Great Acceleration: An Environmental History since 1945*. Cambridge, MA: Harvard University Press, 2016.

Melosi, Martin. *The Sanitary City: Urban Infrastructure in America from Colonial Times to the Present.* Baltimore: Johns Hopkins University Press, 2000.

Michaud, Annie. "The Niagara River Remedial Action Plan: 25 Years of Environmental Restoration." Policy Brief 12, Brock University, Niagara Community Observatory, St. Catharines, ON, December 2012.

Mitchell, Timothy. *Carbon Democracy: Political Power in the Age of Oil.* New York: Verso, 2011.

–. *Rule of Experts: Egypt, Techno-Politics, Modernity.* Oakland: University of California Press, 2002.

Molburg, J.C., J.A. Kavicky, and K.C. Picel. "The Design, Construction, and Operation of Long-Distance High Voltage Electricity Transmission Technologies." Oak Ridge, TN: Argonne National Library, US Department of Energy, November 2007.

Monmonier, Mark. *Lake Effect: Tales of Large Lakes, Arctic Winds, and Recurrent Snows.* Syracuse: Syracuse University Press, 2012.

Moore, Jason W., and Raj Patel. *A History of the World in Seven Cheap Things: A Guide to Capitalism, Nature and the Future of the Planet.* Oakland: University of California Press, 2017.

Morgan, Arthur. *Dams and Other Disasters: A History of the U.S. Army Corps of Engineers in Civil Works.* Boston: Porter Sargent, 1971.

Morrison, Ernest. *J. Horace McFarland – A Thorn for Beauty.* Harrisburg, PA: Pennsylvania Historical and Museum Commission, 1995.

Muller, Mike. "Hydropower Dams Can Help Mitigate the Global Warming Impact of Wetlands." *Nature,* February 19, 2019. https://www.nature.com/articles/d41586-019-00616-w.

Murphy, Michelle. "Chemical Infrastructures of the St. Clair River." In *Toxicants, Health and Regulation since 1945,* edited by Soraya Boudia and Nathalie Jas, 103–15. London: Pickering and Chatto, 2013.

Nash, Linda. "Traveling Technology? American Water Engineers in the Columbia Basin and the Helmand Valley." In *Where Minds and Matters Meet: Technology in California and the West,* edited by Volker Janssen, 135–48. Oakland: University of California Press, 2012.

Nash, Roderick. *Wilderness and the American Mind.* New Haven, CT: Yale University Press, 1965.

Nass, David L. *Public Policy and Public Works: Niagara Falls Redevelopment as a Case Study.* Chicago: Public Works Historical Society, 1979.

Neary, Peter. "Grey, Bryce, and the Settlement of Canadian-American Differences, 1905–1911." *Canadian History Review* 49, 4 (1968): 357–80.

Needham, Andrew. *Power Lines: Phoenix and the Making of the Modern Southwest.* Princeton, NJ: Princeton University Press, 2015.

Nelles, H.V. *The Politics of Development: Forest, Mines and Hydro-Electric Power in Ontario, 1849–1941.* Toronto: Macmillan, 1974.

Neumann, Tracy. *Remaking the Rust Belt: The Postindustrial Transformation of North America.* Philadelphia: University of Pennsylvania Press, 2016.

Newman, Richard. *Love Canal: A Toxic History from Colonial Times to the Present.* New York: Oxford University Press, 2016.

Niagara Parks. "The History of Journey behind the Falls." Niagara Parks, August 14, 2018. https://www.niagaraparks.com/the-history-of-journey-behind-the-falls/.

Niagara Parks Commission (NPC). *Annual Reports of the Commissioners of the Queen Victoria Niagara Falls Park.* Various reports. Toronto: L.K. Cameron, 1891–1959.

Niagara Power Service. *Niagara the Unconquered.* Buffalo: Buffalo, Niagara, and Eastern Power Corporation, 1929.

Norman, Emma. *Governing Transboundary Waters: Canada, the United States, and Indigenous Communities.* New York: Routledge, 2014.

Nye, David. *American Technological Sublime.* Cambridge, MA: MIT Press, 1994.

–. *Electrifying America: Social Meanings of a New Technology, 1880–1940.* Cambridge, MA: MIT Press, 1992.

–. *When the Lights Went Out: A History of Blackouts in America.* Cambridge, MA: MIT Press, 2010.

O'Bannon, Patrick. *Working in the Dry: Cofferdams, In-River Construction, and the United States Army Corps of Engineers.* Washington, DC: US Army Corps of Engineers, 2009.

Olmanson, Eric D. *The Future City on the Inland Sea: A History of Imaginative Geographies of Lake Superior.* Athens, OH: Ohio University Press, 2007.

Olsen, Sandra, and Francis Kowsky. *The Distinctive Charms of Niagara Scenery: Frederick Law Olmsted and the Niagara Reservation.* Niagara Falls, NY: Buscaglia-Castellani Art Gallery, 1985.

Osborne, Brian. "Landscapes, Memory, Monuments and Commemoration: Putting Identity in Its Place." *Canadian Ethnic Studies* 33, 3 (2002): 618–42.

Panel on Niagara River Ice Boom Investigations, Water Science and Technology Board Commission on Physical Sciences, Mathematics, and Resources, National Research Council. *The Lake Erie–Niagara River Ice Boom: Operation and Impacts.* Washington, DC: National Academy Press, 1983.

Parlato, Frank. "Some Call It Insane Planning a Century without the Use of Local Niagara Hydropower." *Niagara Falls Reporter,* September 29–October 6, 2015. http://niagarafallsreporter.com/Stories/2015/SEP29/NYPA.html.

Parr, Joy. *Sensing Changes: Technologies, Environments, and the Everyday, 1953–2003.* Vancouver: UBC Press, 2009.

Parrinello, Giacomo. "Systems of Power: A Spatial Envirotechnical Approach to Water Power and Industrialization in the Po Valley of Italy, 1880–1970," *Technology and Culture* 59, 3 (July 2018): 652–88.

Pelletier, Louis-Raphael. "The Destruction of the Rural Hinterland: Industrialization of Landscape in Beauharnois County." In *Metropolitan Natures: Environmental Histories of Montreal,* edited by Stéphane Castonguay and Michèle Dagenais, 187–210. Pittsburgh: University of Pittsburgh Press, 2011.

Pendergrass, Lee F., and Bonnie B. Pendergrass. *Mimicking Waterways, Harbors, and Estuaries: A Scholarly History of the Corps of Engineers Hydraulics Laboratory at WES, 1929 to the Present.* Vicksburg, MS: Waterways Experiment Station, US Army Corps of Engineers, 1989.

Perlgut, Mark. *Electricity across the Border: The U.S.-Canadian Experience.* New York: C.D. Howe Research Institute, 1978.

Perry, Claire. *The Great American Hall of Wonders: Art, Science and Invention in the Nineteenth Century.* New York: Giles, 2011.

Petersen, Raymond Edward. "Public Power and Private Planning: The Power Authority of the State of New York." PhD diss., City University of New York, 1990.

Philbrick, S.S. "Horizontal Configuration and the Rate of Erosion of Niagara Falls." *Geological Society of America Bulletin* 81 (1970): 3723–32.

Pietz, David. *The Yellow River: The Problem of Water in Modern China.* Cambridge, MA: Harvard University Press, 2015.

Piper, Don Courtney. *The International Law of the Great Lakes: A Study of Canadian–United States Cooperation.* Durham, NC: Duke University Press, 1967.

Piper, Liza. *The Industrial Transformation of Subarctic Canada.* Vancouver: UBC Press, 2009.

Pisani, Donald. *Water and American Government: The Reclamation Bureau, National Water Policy and the West, 1902–1935.* Oakland: University of California Press, 2002.

–. "Water Planning in the Progressive Era: The Inland Waterways Commission Reconsidered." *Journal of Policy History* 18, 4 (2006): 389–418.

Platt, Harold L. *The Electric City: Energy and the Growth of the Chicago Area, 1880–1930.* Chicago: University of Chicago Press, 1991.

Plewman, W.R. *Adam Beck and the Ontario Hydro.* Toronto: Ryerson Press, 1947.

Porter, Theodore M. *Trust in Numbers: The Pursuit of Objectivity in Science and Public Life.* Princeton, NJ: Princeton University Press, 1995.

"A Powerful 50 Years at Niagara." *Water Power Magazine,* April 15, 2011. https://www.waterpowermagazine.com/features/featurea-powerful-50-years-at-niagara/.

Pritchard, Sara B. *Confluence: The Nature of Technology and the Remaking of the Rhône.* Cambridge, MA: Harvard University Press, 2011.

–. "Toward an Environmental History of Technology." In *Oxford Handbook of Environmental History,* edited by Andrew C. Isenberg, 227–58. New York: Oxford University Press, 2014.

Quinn, Frank H. "Anthropogenic Changes to Great Lakes Water Levels." *Great Lakes Update* 136 (1999): 1–4.

RBC Blue Water Project. *2017 RBC Canadian Water Attitudes Study.* Toronto: RBC, 2017. http://www.rbc.com/community-sustainability/_assets-custom/pdf/CWAS-2017-report.pdf.

Read, Jennifer. "Addressing 'A Quiet Horror': The Evolution of Ontario Pollution Control Policy in the International Great Lakes, 1909–1972." PhD diss., University of Western Ontario, 1999.

Reddy, W.M. *The Navigation of Feeling: A Framework of the History of Emotions.* Cambridge: Cambridge University Press, 2001.

Reuss, Martin. "The Art of Scientific Precision: River Research in the United States Army Corps of Engineers to 1945." *Technology and Culture* 40, 2 (April 1999): 292–323.

–. "Coping With Uncertainty: Social Scientists, Engineers, and Federal Water Resources Planning." *Natural Resources Journal* 32 (1992): 101–35.

Reuss, Martin, and Stephen H. Cutcliffe, eds. *The Illusory Boundary: Environment and Technology in History.* Charlottesville, VA: University of Virginia Press, 2010.

Revie, Linda L. *The Niagara Companion: Explorers, Artists, and Writers at the Falls, from Discovery through the Twentieth Century.* Waterloo, ON: Wilfrid Laurier University Press, 2003.

Reynolds, Terry. *Stronger Than a Hundred Men: A History of the Vertical Water Wheel.* Baltimore: Johns Hopkins University Press, 2002.

Rice, H.R. "More Canadian Kilowatts at Niagara, Part 1." *Compressed Air Magazine* 58, 8 (August 1953): 210–17.

–. "More Canadian Kilowatts at Niagara, Part 2." *Compressed Air Magazine* 58, 9 (September 1953): 244–50.

–. "Tunnelers Make Record at Niagara Falls." *Compressed Air Magazine* 59, 6 (June 1954): 169–70.

Richardson, Elmo. *Dams, Parks and Politics: Resource Development and Preservation in the Truman-Eisenhower Era.* Lexington, KY: University Press of Kentucky, 1973.

Rickard, Clinton. *Fighting Tuscarora: The Autobiography of Chief Clinton Rickard.* Edited by Barbara Graymont. Syracuse, NY: Syracuse University Press, 1973.

Riley, John. *The Once and Future Great Lakes Country: An Ecological History.* Montreal and Kingston: McGill-Queen's University Press, 2013.

"Rock Scaling along Niagara Gorge." *Niagara Falls Review,* April 18, 2016. http://www.niagarafallsreview.ca/2016/04/18/rock-scaling-along-niagara-gorge.

Rosenwein, Barbara H. *Emotional Communities in the Early Middle Ages.* Ithaca, NY: Cornell University Press, 2006.

Rossi, Erno. *White Death: Blizzard of 77.* Buffalo: 77 Publishing, 1978.

Rouse, Hunter, and Simon Ince. *History of Hydraulics.* Iowa City: Iowa Institute of Hydraulic Research, 1957.

–. *Hydraulics in the United States, 1776–1976.* Iowa City: Iowa Institute of Hydraulic Research, 1976.

Runte, Alfred. "Beyond the Spectacular: The Niagara Falls Preservation Campaign." *New York Historical Society Quarterly* 57 (January 1973): 30–50.

Russell, Andrew L., and Lee Vinsel. "After Innovation, Turn to Maintenance." *Technology and Culture* 59, 1 (January 2018): 1–25.

Russell, Edmund, James Allison, Thomas Finger, John K. Brown, Brian Balogh, and W. Bernard Carlson. "The Nature of Power: Synthesizing the History of Technology and Environmental History." *Technology and Culture* 52 (April 2011): 246–59.

Sandwell, Ruth. *Canada's Rural Majority: Households, Environments, and Economies, 1870–1940.* Toronto: University of Toronto Press, 2016.

–, ed. *Powering Up Canada: The History of Power, Fuel, and Energy from 1600.* Montreal and Kingston: McGill-Queen's University Press, 2016.

Schama, Simon. *Landscape and Memory.* Toronto: Random House, 1995.

Schneider, Daniel. *Hybrid Nature: Sewage Treatment and the Contradictions of the Industrial Ecosystem.* Cambridge, MA: MIT Press, 2013.

Schulman, Peter. *Coal and Empire: The Birth of Energy Security in Industrial America.* Baltimore: Johns Hopkins University Press, 2015.

Schwartz, Charles P. Jr. "Niagara Mohawk v. FPC: Have Private Water Rights Been Destroyed by the Federal Power Act?" *University of Pennsylvania Law Review* 102 (1953): 31–79.

Scott, James C. "High Modernist Social Engineering: The Case of the Tennessee Valley Authority." In *Experiencing the State,* edited by Lloyd I. Rudolph and John Kurt Jacobsen, 3–52. New York: Oxford University Press, 2006.

–. *Seeing Like a State: How Certain Schemes to Improve the Human Condition Have Failed.* New Haven, CT: Yale University Press, 1998.

Scott, S.D., and P.K. Scott. *The Niagara Reservation Archaeological and Historical Resource Survey.* Albany: New York Office of Parks, Recreation, and Historic Preservation, Historic Sites Bureau, March 1983.

Sears, John. *Sacred Places: American Tourist Attractions in the Nineteenth Century.* New York: Oxford University Press, 1989.

Sedoff, Andrei, Stephan Schott, and Bryan Karney. "Sustainable Power and Scenic Beauty: The Niagara River Water Diversion Treaty and Its Relevance Today." *Energy Policy* 66 (2014): 526–36.

Sehdev, Robinder Kaur. "Unsettling the Settler at Niagara Falls: Reading Colonial Culture through the Maid of the Mist." PhD diss., York University, 2008.

Seibel, George. *Ontario's Niagara Parks.* Toronto: Niagara Parks Commission, 1991.

Selks, H.L, S.J. Walsh, N.M. Burkhead, et al. "Conservation Status of Imperiled North American Freshwater and Diadromous Fishes." *Fisheries* 33, 8 (2008): 372–406.

Sellers, Christopher. *Crabgrass Crucible: Suburban Nature and the Rise of Environmentalism in Twentieth-Century America.* Chapel Hill: University of North Carolina Press, 2012.

Shabecoff, Philip. *A Fierce Green Fire: The American Environmental Movement.* Washington, DC: Island Press, 2003.

Sheriff, Carol. *The Artificial River: The Erie Canal and the Paradox of Progress, 1817–1862.* New York: Hill and Wang, 1997.

Shields, Rob. *Places on the Margin: Alternative Geographies of Modernity.* New York: Routledge, 1992.

Sholdice, Mark. "The Ontario Experiment: Hydroelectricity, Public Ownership, and Transnational Progressivism, 1906–1939." PhD diss., University of Guelph, 2019.

Siener, William H. "The United Nations at Niagara: Borderlands Collaboration and Emerging Globalism." *American Review of Canadian Studies* 43, 3 (September 2013): 377–93.

Simpson, Audra. *Mohawk Interruptus: Political Life across the Borders of Settler States.* Durham, NC: Duke University Press, 2014.

Slaton, Amy. *Reinforced Concrete and the Modernization of American Building, 1900–1930.* Baltimore: Johns Hopkins University Press, 2003.

Smil, Vaclav. *Energy and Civilization: A History.* Cambridge, MA: MIT Press, 2017.

Sneddon, Christopher. *Concrete Revolution: Large Dams, Cold War Geopolitics, and the US Bureau of Reclamation.* Chicago: University of Chicago Press, 2015.

Spencer, Joseph William Winthrop. *The Falls of Niagara. Their Evolution and Varying Relations to the Great Lakes: Characteristics of the Power, and the Effects of Its Diversion, 1905–1906, Geological Survey of Canada.* Ottawa: King's Printer, 1907.

Spencer, Robert, John Kirton, and Kim Richard Nossal, eds. *The International Joint Commission Seventy Years On.* Toronto: University of Toronto Centre for International Studies, 1981.

Spieler, Cliff, and Tom Hewitt. *Niagara Power: From Joncaire to Moses.* Lewiston, NY: Niagara Power, 1959.

Spirn, Anne Whiston. "Constructing Nature: The Legacy of Frederick Law Olmsted." In *Uncommon Ground: Rethinking the Human Place in Nature,* edited by William Cronon, 91–113. New York: W.W. Norton, 1996.

Stamm, Michael. *Dead Tree Media: Manufacturing the Newspaper in Twentieth-Century North America.* Baltimore: Johns Hopkins University Press, 2018.

Stamp, Robert M. *Bright Lights, Big City: The History of Electricity in Toronto.* Toronto: Ontario Association of Archivists, 1991.

Star, Susan Leigh. *Sorting Things Out: Classification and Its Consequences.* Cambridge, MA: MIT Press, 1999.

–. "This Is Not a Boundary Object: Reflections on the Origin of a Concept." *Science, Technology, and Human Values* 35, 5 (2010): 601–17.

Steinberg, Theodore. *Nature Incorporated: Industrialization and the Waters of New England.* Amherst, MA: University of Massachusetts Press, 1994.

Sternberg, Ernest. "The Iconography of the Tourism Experience." *Annals of Tourism Research* 24, 4 (1997): 951–69.

Stieve, Thomas. "National Identity in Niagara Falls, Canada and the United States." *Tijdschrift voor Economische en Sociale Geografie* 96, 1 (2005): 3–14.

Stradling, David. *The Nature of New York: An Environmental History of the Empire State.* Ithaca, NY: Cornell University Press, 2010.

Strand, Ginger. *Inventing Niagara: Beauty, Power, and Lies.* New York: Simon and Schuster, 2008.

Stunden Bower, Shannon. *Wet Prairie: People, Land, and Water in Agricultural Manitoba.* Vancouver: UBC Press, 2011.

Sutter, Paul S. "The World with Us: The State of American Environmental History." *Journal of American History* 100, 1 (2013): 94–119.

Swain, Donald C. *Federal Conservation Policy, 1921–1933.* Berkeley: University of California Press, 1963.

Swift, Jamie, and Keith Stewart. *Hydro: The Decline and Fall of Ontario's Electric Empire.* Toronto: Between the Lines, 2004.

Swyngedouw, Erik. *Liquid Power: Contested Hydro-Modernities in Twentieth-Century Spain, 1898–2010.* Cambridge, MA: The MIT Press, 2015.

Tarr, Joel. *The Search for the Ultimate Sink: Urban Pollution in Historical Perspective.* Akron, OH: University of Akron Press, 1996.

Taylor, Joseph III. *Making Salmon: An Environmental History of the Northwest Fisheries Crisis.* Seattle: University of Washington Press, 2001.

Teisch, Jessica B. *Engineering Nature: Water, Development and the Global Spread of American Environmental Expertise.* Chapel Hill: University of North Carolina Press, 2011.

Tenner, Edward. *Why Things Bite Back: Technology and the Revenge of Unintended Consequences.* New York: Alfred A. Knopf, 1996.

Tesmer, Irving H., ed. *Colossal Cataract: The Geological History of Niagara Falls.* Albany: State University of New York Press, 1981.

Tobey, Ronald. *Technology as Freedom: The New Deal and the Electrical Modernization of the American Home.* Oakland: University of California Press, 1997.

Trescott, Martha Moore. *The Rise of the American Electrochemicals Industry, 1880–1910.* New York: Praeger, 1981.

Tsing, Anna Lowenhaupt. *The Mushroom at the End of the World: On the Possibility of Life in Capitalist Ruins.* Princeton, NJ: Princeton University Press, 2015.

Tucker, Richard P. "Containing Communism by Impounding Rivers: American Strategic Interests and the Global Spread of High Dams in the Early Cold War." In *Environmental Histories of the Cold War,* edited by John R. McNeill and Corinna R. Unger, 139–64. Cambridge: Cambridge University Press, 2010.

Tyrell, Ian. *Crisis of the Wasteful Nation: Empire and Conservation in Theodore Roosevelt's America.* Chicago: University of Chicago Press, 2015.

Unger, Richard W., and John Thistle. *Energy Consumption in Canada in the 19th and 20th Centuries: A Statistical Outline.* Rome: Consiglio Nazionale delle Ricerche (CNR), Instituto di Studi sulle Società del Mediterraneo (ISSM), 2013.

Urry, John. *The Tourist Gaze: Leisure and Travel in Contemporary Societies.* 2nd ed. Newbury Park, CA: Sage, 1992.

Van Huizen, Philip. "Building a Green Dam: Environmental Modernism and the Canadian-American Libby Dam Project." *Pacific Historical Review* 79, 3 (August 2010): 418–53.

Vincenti, Walter G. *What Engineers Know and How They Know It: Analytical Studies from Aeronautical History.* Baltimore: Johns Hopkins University Press, 1990.

Vogel, Michael N. *Echoes in the Mist: An Illustrated History of the Niagara Falls Area.* Chicago: Windsor, 1991.

Vowel, Chelsea. *Indigenous Writes: A Guide to First Nations, Metis and Inuit Issues in Canada.* Winnipeg: Highwater Press, 2016.

Wallace, Anthony F.C. *Tuscarora: A History.* Albany: SUNY Press, 2012.

Way, Ronald. *Ontario's Niagara Parks.* Toronto: Niagara Parks Commission, 1960.

Weiler, Paul C., ed. *Mega Projects: The Collective Bargaining Dimension.* Ottawa: Canadian Construction Association, 1986.

Welch, Thomas V. *How Niagara Was Made Free: The Passage of the Niagara Reservation Act in 1885.* Buffalo: Press Union and Times, 1903.

Well, Wyatt. "Public Power in the Eisenhower Administration." *Journal of Policy History* 20, 2 (2009): 227–62.

White, Richard. *The Organic Machine: The Remaking of the Columbia River.* New York: Hill and Wang, 1995.

Whyte, Kyle Powys. "Indigenous Experience, Environmental Justice and Settler Colonialism." In *Nature and Experience: Phenomenology and the Environment,* edited by B. Bannon, 157–74. Lanham, MD: Rowman and Littlefield, 2016.

Wilkinson, T.A., and Captain J.D. Guerin. "Dewatering of American (Niagara) Falls Planned for Study." *Engineering Geologist* 4, 3 (August 1969): 1, 4, 6.

Willingham, William. *Vital Currents: A History of the North Central Division, US Army Corps of Engineers, 1867–1997.* Washington, DC: US Army Corps of Engineers, 2006.

Willoughby, William R. *The Joint Organizations of Canada and the United States.* Toronto: University of Toronto Press, 1979.

Wilson, Edmund. *Apologies to the Iroquois.* New York: Farrar, Straus, and Giroux, 1960.

Winner, Langdon. *The Whale and the Reactor: A Search for Limits in an Age of High Technology.* Chicago: University of Chicago Press, 1988.

Wisnioski, Matthew. *Engineers for Change: Competing Visions of Technology in 1960s America.* Cambridge, MA: MIT Press, 2012.

Wohl, Ellen. *Rivers in the Landscape: Science and Management.* Hoboken, NJ: Wiley-Blackwell, 2014.

Woodford, Arthur M. *Charting the Inland Seas: A History of the US Lake Survey.* Detroit: US Army Corps of Engineers, Detroit District, 1991.

Woodworth, Craig A. "The Schoellkopf Disaster Aftermath in the Niagara River Gorge." *IEEE Energy and Power Magazine* 10, 6 (November–December 2012): 80–96.

Worster, Donald. *Rivers of Empire: Water, Aridity, and the Growth of the American West.* New York: Pantheon, 1985.

Wrigley, E.A. *The Path to Sustained Growth: England's Transition from an Organic Economy to an Industrial Revolution.* Cambridge: Cambridge University Press, 2016.

Wyckoff, William. *The Developer's Frontier: The Making of the Western New York Landscape.* New Haven, CT: Yale University Press, 1988.

Wyer, Samuel S. *Niagara Falls: Its Power Possibilities and Preservation.* Washington, DC: Smithsonian Institution, 1925.

Zavitz, Sherman. *Niagara Falls: Historical Notes.* St. Catharines, ON: Looking Back Press, 2008.

Zeller, Thomas. "Aiming for Control, Haunted by Its Failure: Towards an Envirotechnical Understanding of Infrastructures." *Global Environment* 10 (2017): 202–28.

Index

Notes: "(f)" following a page number indicates a figure; "(t)" following a page number indicates a table

Claire Elizabeth Campbell, *Shaped by the West Wind: Nature and History in Georgian Bay*
Tina Loo, *States of Nature: Conserving Canada's Wildlife in the Twentieth Century*
Jamie Benidickson, *The Culture of Flushing: A Social and Legal History of Sewage*
John Sandlos, *Hunters at the Margin: Native People and Wildlife Conservation in the Northwest Territories*
William J. Turkel, *The Archive of Place: Unearthing the Pasts of the Chilcotin Plateau*
Greg Gillespie, *Hunting for Empire: Narratives of Sport in Rupert's Land, 1840–70*
James Murton, *Creating a Modern Countryside: Liberalism and Land Resettlement in British Columbia*
Stephen J. Pyne, *Awful Splendour: A Fire History of Canada*
Sharon Wall, *The Nurture of Nature: Childhood, Antimodernism, and Ontario Summer Camps, 1920–55*
Hans M. Carlson, *Home Is the Hunter: The James Bay Cree and Their Land*
Joy Parr, *Sensing Changes: Technologies, Environments, and the Everyday, 1953–2003*
Liza Piper, *The Industrial Transformation of Subarctic Canada*
Jamie Linton, *What Is Water? The History of a Modern Abstraction*
Dean Bavington, *Managed Annihilation: An Unnatural History of the Newfoundland Cod Collapse*
J. Keri Cronin, *Manufacturing National Park Nature: Photography, Ecology, and the Wilderness Industry of Jasper*
Shannon Stunden Bower, *Wet Prairie: People, Land, and Water in Agricultural Manitoba*
Jocelyn Thorpe, *Temagami's Tangled Wild: Race, Gender, and the Making of Canadian Nature*
Sean Kheraj, *Inventing Stanley Park: An Environmental History*
Darcy Ingram, *Wildlife, Conservation, and Conflict in Quebec, 1840–1914*

Caroline Desbiens, *Power from the North: Territory, Identity, and the Culture of Hydroelectricity in Quebec*

Daniel Macfarlane, *Negotiating a River: Canada, the US, and the Creation of the St. Lawrence Seaway*

Justin Page, *Tracking the Great Bear: How Environmentalists Recreated British Columbia's Coastal Rainforest*

Ryan O'Connor, *The First Green Wave: Pollution Probe and the Origins of Environmental Activism in Ontario*

John Thistle, *Resettling the Range: Animals, Ecologies, and Human Communities in British Columbia*

Jessica van Horssen, *A Town Called Asbestos: Environmental Contamination, Health, and Resilience in a Resource Community*

Nancy B. Bouchier and Ken Cruikshank, *The People and the Bay: A Social and Environmental History of Hamilton Harbour*

Carly A. Dokis, *Where the Rivers Meet: Pipelines, Participatory Resource Management, and Aboriginal-State Relations in the Northwest Territories*

Jonathan Peyton, *Unbuilt Environments: Tracing Postwar Development in Northwest British Columbia*

Mark R. Leeming, *In Defence of Home Places: Environmental Activism in Nova Scotia*

Jim Clifford, *West Ham and the River Lea: A Social and Environmental History of London's Industrialized Marshland, 1839–1914*

Michèle Dagenais, *Montreal, City of Water: An Environmental History*

David Calverley, *Who Controls the Hunt? First Nations, Treaty Rights, and Wildlife Conservation in Ontario, 1783–1939*

Jamie Benidickson, *Levelling the Lake: Transboundary Resource Management in the Lake of the Woods Watershed*